T0323725

Structure and Interpretation of Classical Mechanics

Structure and Interpretation of Classical Mechanics

second edition

Gerald Jay Sussman and Jack Wisdom

The MIT Press
Cambridge, Massachusetts
London, England

This book was set in Computer Modern by the authors with the LaTeX typesetting system.

Library of Congress Cataloging-in-Publication Data

Sussman, Gerald Jay.
  Structure and interpretation of classical mechanics / Gerald Jay Sussman and Jack Wisdom.
  [Second edition].
    pages   cm
  Includes bibliographical references and index.
  ISBN 978-0-262-02896-7 (hardcover : alk. paper),978-0-262-55345-2(pb)
  1. Mechanics. I. Wisdom, Jack. II. Title.
  QC125.2.S895   2015

531dc23   2014033586

148702397

This book is dedicated,
in respect and admiration,
to

The Principle of Least Action

"The author has spared himself no pains in his endeavour to present the main ideas in the simplest and most intelligible form, and on the whole, in the sequence and connection in which they actually originated. In the interest of clearness, it appeared to me inevitable that I should repeat myself frequently, without paying the slightest attention to the elegance of the presentation. I adhered scrupulously to the precept of that brilliant theoretical physicist L. Boltzmann, according to whom matters of elegance ought be left to the tailor and to the cobbler."

Albert Einstein, in *Relativity, the Special and General Theory*, (1961), p. v

Contents

Preface

There has been a remarkable revival of interest in classical mechanics in recent years. We now know that there is much more to classical mechanics than previously suspected. The behavior of classical systems is surprisingly rich; derivation of the equations of motion, the focus of traditional presentations of mechanics, is just the beginning. Classical systems display a complicated array of phenomena such as nonlinear resonances, chaotic behavior, and transitions to chaos.

Traditional treatments of mechanics concentrate most of their effort on the extremely small class of symbolically tractable dynamical systems. We concentrate on developing general methods for studying the behavior of systems, whether or not they have a symbolic solution. Typical systems exhibit behavior that is qualitatively different from the solvable systems and surprisingly complicated. We focus on the phenomena of motion, and we make extensive use of computer simulation to explore this motion.

Even when a system is not symbolically tractable, the tools of modern dynamics allow one to extract a qualitative understanding. Rather than concentrating on symbolic descriptions, we concentrate on geometric features of the set of possible trajectories. Such tools provide a basis for the systematic analysis of numerical or experimental data.

Classical mechanics is deceptively simple. It is surprisingly easy to get the right answer with fallacious reasoning or without real understanding. Traditional mathematical notation contributes to this problem. Symbols have ambiguous meanings that depend on context, and often even change within a given context.[1] For example, a fundamental result of mechanics is the Lagrange equations. In traditional notation the Lagrange equations are written

$$\frac{d}{dt}\frac{\partial L}{\partial \dot{q}^i} - \frac{\partial L}{\partial q^i} = 0.$$

The Lagrangian L must be interpreted as a function of the position and velocity components q^i and $\dot{q}^i$, so that the partial derivatives make sense, but then in order for the time derivative d/dt to make sense solution paths must have been inserted into the partial derivatives of the Lagrangian to make functions of time. The traditional use of ambiguous notation is convenient in simple situations, but in more complicated situations it can be a serious handicap to clear reasoning. In order that the reasoning be clear and unambiguous, we have adopted a more precise mathematical notation. Our notation is functional and follows that of modern mathematical presentations.[2] An introduction to our functional notation is in an appendix.

Computation also enters into the presentation of the mathematical ideas underlying mechanics. We require that our mathe-

---

[1]In his book on mathematical pedagogy [17], Hans Freudenthal argues that the reliance on ambiguous, unstated notational conventions in such expressions as $f(x)$ and $df(x)/dx$ makes mathematics, and especially introductory calculus, extremely confusing for beginning students; and he enjoins mathematics educators to use more formal modern notation.

[2]In his beautiful book *Calculus on Manifolds* [40], Michael Spivak uses functional notation. On p. 44 he discusses some of the problems with classical notation. We excerpt a particularly juicy passage:

> The mere statement of [the chain rule] in classical notation requires the introduction of irrelevant letters. The usual evaluation for $D_1(f \circ (g, h))$ runs as follows:
>
> If $f(u, v)$ is a function and $u = g(x, y)$ and $v = h(x, y)$ then
>
> $$\frac{\partial f(g(x,y), h(x,y))}{\partial x} = \frac{\partial f(u,v)}{\partial u}\frac{\partial u}{\partial x} + \frac{\partial f(u,v)}{\partial v}\frac{\partial v}{\partial x}$$
>
> [The symbol $\partial u/\partial x$ means $\partial/\partial x\ g(x,y)$, and $\partial/\partial u\ f(u,v)$ means $D_1 f(u,v) = D_1 f(g(x,y), h(x,y))$.] This equation is often written simply
>
> $$\frac{\partial f}{\partial x} = \frac{\partial f}{\partial u}\frac{\partial u}{\partial x} + \frac{\partial f}{\partial v}\frac{\partial v}{\partial x}.$$
>
> Note that f means something different on the two sides of the equation!

matical notations be explicit and precise enough that they can be interpreted automatically, as by a computer. As a consequence of this requirement the formulas and equations that appear in the text stand on their own. They have clear meaning, independent of the informal context. For example, we write Lagrange's equations in functional notation as follows:[3]

$$D(\partial_2 L \circ \Gamma[q]) - \partial_1 L \circ \Gamma[q] = 0.$$

The Lagrangian L is a real-valued function of time t, coordinates x, and velocities v; the value is $L(t, x, v)$. Partial derivatives are indicated as derivatives of functions with respect to particular argument positions; $\partial_2 L$ indicates the function obtained by taking the partial derivative of the Lagrangian function L with respect to the velocity argument position. The traditional partial derivative notation, which employs a derivative with respect to a "variable," depends on context and can lead to ambiguity.[4] The partial derivatives of the Lagrangian are then explicitly evaluated along a path function q. The time derivative is taken and the Lagrange equations formed. Each step is explicit; there are no implicit substitutions.

Computational algorithms are used to communicate precisely some of the methods used in the analysis of dynamical phenomena. Expressing the methods of variational mechanics in a computer language forces them to be unambiguous and computationally effective. Computation requires us to be precise about the representation of mechanical and geometric notions as computational objects and permits us to represent explicitly the algorithms for manipulating these objects. Also, once formalized as a procedure, a mathematical idea becomes a tool that can be used directly to compute results.

Active exploration on the part of the student is an essential part of the learning experience. Our focus is on understanding the motion of systems; to learn about motion the student must actively explore the motion of systems through simulation and

---

[3] This is presented here without explanation, to give the flavor of the notation. The text gives a full explanation.

[4] "It is necessary to use the apparatus of partial derivatives, in which even the notation is ambiguous." V.I. Arnold, *Mathematical Methods of Classical Mechanics* [5], Section 47, p. 258. See also the footnote on that page.

experiment. The exercises and projects are an integral part of the presentation.

That the mathematics is precise enough to be interpreted automatically allows active exploration to be extended to it. The requirement that the computer be able to interpret any expression provides strict and immediate feedback as to whether the expression is correctly formulated. Experience demonstrates that interaction with the computer in this way uncovers and corrects many deficiencies in understanding.

In this book we express computational methods in Scheme, a dialect of the Lisp family of programming languages that we also use in our introductory computer science subject at MIT. There are many good expositions of Scheme. We provide a short introduction to Scheme in an appendix.

Even in the introductory computer science class we never formally teach the language, because we do not have to. We just use it, and students pick it up in a few days. This is one great advantage of Lisp-like languages: They have very few ways of forming compound expressions, and almost no syntactic structure. All of the formal properties can be covered in an hour, like the rules of chess. After a short time we forget about the syntactic details of the language (because there are none) and get on with the real issues—figuring out what we want to compute.

The advantage of Scheme over other languages for the exposition of classical mechanics is that the manipulation of procedures that implement mathematical functions is easier and more natural in Scheme than in other computer languages. Indeed, many theorems of mechanics are directly representable as Scheme programs.

The version of Scheme that we use in this book is MIT/GNU Scheme, augmented with a large library of software called Scmutils that extends the Scheme operators to be generic over a variety of mathematical objects, including symbolic expressions. The Scmutils library also provides support for the numerical methods we use in this book, such as quadrature, integration of systems of differential equations, and multivariate minimization.

The Scheme system, augmented with the Scmutils library, is free software. We provide this system, complete with documentation and source code, in a form that can be used with the GNU/Linux operating system, on the Internet at mitpress.mit.edu/classical_mech.

This book presents classical mechanics from an unusual perspective. It focuses on understanding motion rather than deriving equations of motion. It weaves recent discoveries in nonlinear dynamics throughout the presentation, rather than presenting them as an afterthought. It uses functional mathematical notation that allows precise understanding of fundamental properties of classical mechanics. It uses computation to constrain notation, to capture and formalize methods, for simulation, and for symbolic analysis.

This book is the result of teaching classical mechanics at MIT. The contents of our class began with ideas from a class on nonlinear dynamics and solar system dynamics by Wisdom and ideas about how computation can be used to formulate methodology developed in an introductory computer science class by Abelson and Sussman. When we started we expected that using this approach to formulate mechanics would be easy. We quickly learned that many things we thought we understood we did not in fact understand. Our requirement that our mathematical notations be explicit and precise enough that they can be interpreted automatically, as by a computer, is very effective in uncovering puns and flaws in reasoning. The resulting struggle to make the mathematics precise, yet clear and computationally effective, lasted far longer than we anticipated. We learned a great deal about both mechanics and computation by this process. We hope others, especially our competitors, will adopt these methods, which enhance understanding while slowing research.

Second Edition

We have taught classical mechanics using this text every year at MIT since the first edition was published. We have learned a great deal about what difficulties students encountered with the material. We have found that some of our explanations needed improvement. This edition is the result of our new understanding.

Our software support has improved substantially over the years, and we have exploited it to provide algebraic proofs of more generality than could be supplied in the first edition. This advantage permeates most of the new edition.

In the first chapter we now go more directly to the coordinate representation of the action, without compromising the importance of the coordinate independence of the action. We also added a simple derivation of the Euler–Lagrange equations from the Principle of Stationary Action, supplementing the more formal derivation of the first edition.

In the chapter on rigid-body motion we now provide an algebraic derivation of the existence of the angular-velocity vector. Our new derivation is in harmony with the development of generalized coordinates for a rigid body as parameters of the transformation from a reference orientation to the actual orientation. We also provide a new section on quaternions as a way of avoiding singularities in the analysis of the motion of rigid bodies.

A canonical transformation is a transformation of phase-space coordinates and an associated transformation of the Hamiltonian that maintains a one-to-one correspondence between trajectories. We allow time-dependent systems and transformations, complicating the treatment of canonical transformations. The chapter on canonical transformations has been extensively revised to clarify the relationship of canonical transformations to symplectic transformations. We split off the treatment of canonical transformations that arise from evolution, including Lie transforms, into a new chapter.

We fixed myriad minor mistakes throughout. We hope that we have not introduced more than we have removed.

Acknowledgments

We would like to thank the many people who have helped us to develop this book and the curriculum it is designed to support. We have had substantial help from the wonderful students who studied with us in our classical mechanics class. They have forced us to be clear; they have found bugs that we had to fix in the software, in the presentation, and in our thinking.

We have had considerable technical help in the development and presentation of the subject matter from Harold Abelson. Abelson is one of the developers of the Scmutils software system. He put mighty effort into some sections of the code. We also consulted him when we were desperately trying to understand the logic of mechanics. He often could propose a direction to lead out of an intellectual maze.

Matthew Halfant started us on the development of the Scmutils system. He encouraged us to get into scientific computation, using Scheme and functional style as an active way to explain the ideas, without the distractions of imperative languages such as C. In the 1980s he wrote some of the early Scheme procedures for numerical computation that we still use.

Dan Zuras helped us with the invention of the unique organization of the Scmutils system. It is because of his insight that the system is organized around a generic extension of the chain rule for taking derivatives. He also helped in the heavy lifting that was required to make a really good polynomial GCD algorithm, based on ideas we learned from Richard Zippel.

This book, and a great deal of other work of our laboratory, could not have been done without the outstanding work of Chris Hanson. Chris developed and maintained the Scheme system underlying this work. More recently, Taylor Campbell and others have continued the development of MIT/GNU Scheme. In addition, Chris took us through a pass of reorganization of the Scmutils system that forced the clarification of many of the ideas of types and of generic operations that make our system as good as it is.

Guillermo Juan Rozas, co-developer of the Scheme system, made major contributions to the Scheme compiler, and implemented a number of other arcane mechanisms that make our system efficient enough to support our work.

Besides contributing to some of the methods for the solution of linear equations in the Scmutils system, Jacob Katzenelson provided valuable feedback that improved the presentation of the material.

Julie Sussman, PPA, provided careful reading and serious criticism that forced us to reorganize and rewrite major parts of the text. Julie worked with first-edition coauthor Meinhard (Hardy) Mayer to create the index. She also developed and maintained Gerald Jay Sussman over these many years.

Cecile Wisdom, saint, is a constant reminder, by her faith and example, of what is really important. This project would not have been possible without the loving support and unfailing encouragement she has given Jack Wisdom. Their children, William, Edward, Thomas, John, and Elizabeth Wisdom, daily enrich his life with theirs.

Many have contributed to our understanding of dynamics over the years. Boris Chirikov, Michel Hénon, Peter Goldreich, and Stan Peale have had particular influence. We also acknowledge the influence of the late Res Jost.

Numerous others have contributed to this work, either in the development of the software or in the development of the content, including Bill Siebert, Panayotis Skordos, Kleanthes Koniaris, Kevin Lin, James McBride, Rebecca Frankel, Thomas F. Knight, Pawan Kumar, Elizabeth Bradley, Alice Seckel, Jihad Touma, and Kenneth Yip. We have had extremely useful feedback from and discussions with Piet Hut, Jon Doyle, David Finkelstein, Peter Fisher, Guy Lewis Steele Jr., and Robert Hermann.

We want to thank the generations of students who have taken our classes and worked through our problems. They have provided exceptional feedback and encouragement. Our students Will Farr, Mark Tobenkin, Keith Winstein, Alexey Radul, Micah Brodsky, Damon Vander Lind, Peter Iannucci, William Throwe, and Leo Stein were especially helpful.

We thank the MIT Computer Science and Artificial Intelligence Laboratory for its hospitality and logistical support. We acknowledge the Panasonic Corporation for support of Gerald Jay Sussman through an endowed chair. We thank Breene M. Kerr for support of Jack Wisdom through an endowed chair. We thank the MIT Mathematics and EECS departments for sabbatical support for Meinhard Mayer, who collaborated with us on the first edition. We are sad to report that Hardy is no longer with us. We sorely miss him.

Structure and Interpretation
of Classical Mechanics

1
Lagrangian Mechanics

> The purpose of mechanics is to describe how
> bodies change their position in space with "time."
> I should load my conscience with grave sins against
> the sacred spirit of lucidity were I to formulate the
> aims of mechanics in this way, without serious
> reflection and detailed explanations. Let us
> proceed to disclose these sins.
>
> Albert Einstein, *Relativity, the Special and
> General Theory* [16], p. 9

The subject of this book is motion and the mathematical tools used to describe it.

Centuries of careful observations of the motions of the planets revealed regularities in those motions, allowing accurate predictions of phenomena such as eclipses and conjunctions. The effort to formulate these regularities and ultimately to understand them led to the development of mathematics and to the discovery that mathematics could be effectively used to describe aspects of the physical world. That mathematics can be used to describe natural phenomena is a remarkable fact.

A pin thrown by a juggler takes a rather predictable path and rotates in a rather predictable way. In fact, the skill of juggling depends crucially on this predictability. It is also a remarkable discovery that the same mathematical tools used to describe the motions of the planets can be used to describe the motion of the juggling pin.

Classical mechanics describes the motion of a system of particles, subject to forces describing their interactions. Complex physical objects, such as juggling pins, can be modeled as myriad particles with fixed spatial relationships maintained by stiff forces of interaction.

There are many conceivable ways a system could move that never occur. We can imagine that the juggling pin might pause in midair or go fourteen times around the head of the juggler before being caught, but these motions do not happen. How can we distinguish motions of a system that can actually occur from other conceivable motions? Perhaps we can invent some mathematical function that allows us to distinguish realizable motions from among all conceivable motions.

The motion of a system can be described by giving the position of every piece of the system at each moment. Such a description of the motion of the system is called a *configuration path*; the configuration path specifies the configuration as a function of time. The juggling pin rotates as it flies through the air; the configuration of the juggling pin is specified by giving the position and orientation of the pin. The motion of the juggling pin is specified by giving the position and orientation of the pin as a function of time.

The path-distinguishing function that we seek takes a configuration path as an input and produces some output. We want this function to have some characteristic behavior when its input is a realizable path. For example, the output could be a number, and we could try to arrange that this number be zero only on realizable paths. Newton's equations of motion are of this form; at each moment Newton's differential equations must be satisfied.

However, there is an alternate strategy that provides more insight and power: we could look for a path-distinguishing function that has a minimum on the realizable paths—on nearby unrealizable paths the value of the function is higher than it is on the realizable path. This is the *variational strategy*: for each physical system we invent a path-distinguishing function that distinguishes realizable motions of the system by having a stationary point for each realizable path.[1] For a great variety of systems realizable motions of the system can be formulated in terms of a variational principle.[2]

---

[1]A *stationary point* of a function is a point where the function's value does not vary as the input is varied. Local maxima or minima are stationary points.

[2]The variational formulation successfully describes all of the Newtonian mechanics of particles and rigid bodies. The variational formulation has also been usefully applied in the description of many other systems such as classical electrodynamics, the dynamics of inviscid fluids, and the design of mechanisms such as four-bar linkages. In addition, modern formulations of quantum mechanics and quantum field theory build on many of the same concepts. However, it appears that not all dynamical systems have a variational formulation. For example, there is no simple prescription to apply the variational apparatus to systems with dissipation, though in special cases variational methods can still be used.

Mechanics, as invented by Newton and others of his era, describes the motion of a system in terms of the positions, velocities, and accelerations of each of the particles in the system. In contrast to the Newtonian formulation of mechanics, the variational formulation of mechanics describes the motion of a system in terms of aggregate quantities that are associated with the motion of the system as a whole.

In the Newtonian formulation the forces can often be written as derivatives of the potential energy of the system. The motion of the system is determined by considering how the individual component particles respond to these forces. The Newtonian formulation of the equations of motion is intrinsically a particle-by-particle description.

In the variational formulation the equations of motion are formulated in terms of the difference of the kinetic energy and the potential energy. The potential energy is a number that is characteristic of the arrangement of the particles in the system; the kinetic energy is a number that is determined by the velocities of the particles in the system. Neither the potential energy nor the kinetic energy depends on how those positions and velocities are specified. The difference is characteristic of the system as a whole and does not depend on the details of how the system is specified. So we are free to choose ways of describing the system that are easy to work with; we are liberated from the particle-by-particle description inherent in the Newtonian formulation.

The variational formulation has numerous advantages over the Newtonian formulation. The equations of motion for those parameters that describe the state of the system are derived in the same way regardless of the choice of those parameters: the method of formulation does not depend on the choice of coordinate system. If there are positional constraints among the particles of a system the Newtonian formulation requires that we consider the forces maintaining these constraints, whereas in the variational formulation the constraints can be built into the coordinates. The variational formulation reveals the association of conservation laws with symmetries. The variational formulation provides a framework for placing any particular motion of a system in the context of all possible motions of the system. We pursue the variational formulation because of these advantages.

1.1   Configuration Spaces

Let us consider mechanical systems that can be thought of as composed of constituent point particles, with mass and position, but with no internal structure.[3] Extended bodies may be thought of as composed of a large number of these constituent particles with specific spatial relationships among them. Extended bodies maintain their shape because of spatial constraints among the constituent particles. Specifying the position of all the constituent particles of a system specifies the *configuration* of the system. The existence of constraints among parts of the system, such as those that determine the shape of an extended body, means that the constituent particles cannot assume all possible positions. The set of all configurations of the system that can be assumed is called the *configuration space* of the system. The *dimension* of the configuration space is the smallest number of parameters that have to be given to completely specify a configuration. The dimension of the configuration space is also called the number of *degrees of freedom* of the system.[4]

For a single unconstrained particle it takes three parameters to specify the configuration; a point particle has a three-dimensional configuration space. If we are dealing with a system with more than one point particle, the configuration space is more complicated. If there are k separate particles we need $3k$ parameters to describe the possible configurations. If there are constraints among the parts of a system the configuration is restricted to a lower-dimensional space. For example, a system consisting of two point particles constrained to move in three dimensions so that the distance between the particles remains fixed has a five-dimensional configuration space: thus with three numbers we can fix the posi-

---

[3]We often refer to a point particle with mass but no internal structure as a *point mass*.

[4]Strictly speaking, the dimension of the configuration space and the number of degrees of freedom are not the same. The number of degrees of freedom is the dimension of the space of configurations that are "locally accessible." For systems with integrable constraints the two are the same. For systems with non-integrable constraints the configuration dimension can be larger than the number of degrees of freedom. For further explanation see the discussion of systems with non-integrable constraints in section 1.10.3. Apart from that discussion, all of the systems we consider have integrable constraints (they are "holonomic"). This is why we have chosen to blur the distinction between the number of degrees of freedom and the dimension of the configuration space.

tion of one particle, and with two others we can give the position of the other particle relative to the first.

Consider a juggling pin. The configuration of the pin is specified if we give the positions of the atoms making up the pin. However, there exist more economical descriptions of the configuration. In the idealization that the juggling pin is truly rigid, the distances among all the atoms of the pin remain constant. So we can specify the configuration of the pin by giving the position of a single atom and the orientation of the pin. Using the constraints, the positions of all the other constituents of the pin can be determined from this information. The dimension of the configuration space of the juggling pin is six: the minimum number of parameters that specify the position in space is three, and the minimum number of parameters that specify an orientation is also three.

As a system evolves with time, the constituent particles move subject to the constraints. The motion of each constituent particle is specified by describing the changing configuration. Thus, the motion of the system may be described as evolving along a path in configuration space. The configuration path may be specified by a function, the configuration-path function, which gives the configuration of the system at any time.

Exercise 1.1: Degrees of freedom

For each of the mechanical systems described below, give the number of degrees of freedom of the configuration space.

a. Three juggling pins.

b. A spherical pendulum, consisting of a point mass (the pendulum bob) hanging from a rigid massless rod attached to a fixed support point. The pendulum bob may move in any direction subject to the constraint imposed by the rigid rod. The point mass is subject to the uniform force of gravity.

c. A spherical double pendulum, consisting of one point mass hanging from a rigid massless rod attached to a second point mass hanging from a second massless rod attached to a fixed support point. The point masses are subject to the uniform force of gravity.

d. A point mass sliding without friction on a rigid curved wire.

e. A top consisting of a rigid axisymmetric body with one point on the symmetry axis of the body attached to a fixed support, subject to a uniform gravitational force.

f. The same as **e**, but not axisymmetric.

1.2   Generalized Coordinates

In order to be able to talk about specific configurations we need
to have a set of parameters that label the configurations. The
parameters used to specify the configuration of the system are
called the *generalized coordinates*. Consider an unconstrained free
particle. The configuration of the particle is specified by giving
its position. This requires three parameters. The unconstrained
particle has three degrees of freedom. One way to specify the position of a particle is to specify its rectangular coordinates relative
to some chosen coordinate axes. The rectangular components of
the position are generalized coordinates for an unconstrained particle. Or consider an ideal planar double pendulum: a point mass
constrained to be a given distance from a fixed point by a rigid
rod, with a second mass constrained to be at a given distance
from the first mass by another rigid rod, all confined to a vertical
plane. The configuration is specified if the orientation of the two
rods is given. This requires at least two parameters; the planar
double pendulum has two degrees of freedom. One way to specify
the orientation of each rod is to specify the angle it makes with a
vertical plumb line. These two angles are generalized coordinates
for the planar double pendulum.

The number of coordinates need not be the same as the dimension of the configuration space, though there must be at least that
many. We may choose to work with more parameters than necessary, but then the parameters will be subject to constraints that
restrict the system to possible configurations, that is, to elements
of the configuration space.

For the planar double pendulum described above, the two angle
coordinates are enough to specify the configuration. We could also
take as generalized coordinates the rectangular coordinates of each
of the masses in the plane, relative to some chosen coordinate axes.
These are also fine coordinates, but we would have to explicitly
keep in mind the constraints that limit the possible configurations
to the actual geometry of the system. Sets of coordinates with
the same dimension as the configuration space are easier to work
with because we do not have to deal with explicit constraints
among the coordinates. So for the time being we will consider
only formulations where the number of configuration coordinates

is equal to the number of degrees of freedom; later we will learn how to handle systems with redundant coordinates and explicit constraints.

In general, the configurations form a space M of some dimension n. The n-dimensional configuration space can be parameterized by choosing a coordinate function χ that maps elements of the configuration space to n-tuples of real numbers.[5] If there is more than one dimension, the function χ is a tuple of n independent coordinate functions[6] χ^i, $i = 0, \ldots, n - 1$, where each χ^i is a real-valued function defined on some region of the configuration space.[7] For a given configuration m in the configuration space M the values $\chi^i(m)$ of the coordinate functions are the generalized coordinates of the configuration. These generalized coordinates permit us to identify points of the n-dimensional configuration space with n-tuples of real numbers.[8] For any given configuration space, there are a great variety of ways to choose generalized coordinates. Even for a single point moving without constraints, we can choose rectangular coordinates, polar coordinates, or any other coordinate system that strikes our fancy.

The motion of the system can be described by a configuration path γ mapping time to configuration-space points. Corresponding to the configuration path is a *coordinate path* $q = \chi \circ \gamma$ mapping time to tuples of generalized coordinates.[9] If there is more than

---

[5] A tuple is an ordered list of elements. An element may itself be a tuple.

[6] A tuple of functions that all have the same domain is itself a function on that domain: Given a point in the domain, the value of the tuple of functions is a tuple of the values of the component functions at that point.

[7] The use of superscripts to index the coordinate components is traditional, even though there is potential confusion with exponents. We use zero-based indexing.

[8] More precisely, the generalized coordinates identify open subsets of the configuration space with open subsets of $\mathbf{R}^n$. It may require more than one set of generalized coordinates to cover the entire configuration space. For example, if the configuration space is a two-dimensional sphere, we could have one set of coordinates that maps (a little more than) the northern hemisphere to a disk, and another set that maps (a little more than) the southern hemisphere to a disk, with a strip near the equator common to both coordinate systems. A space that can be locally parameterized by smooth coordinate functions is called a *differentiable manifold*.

[9] Here $\circ$ denotes composition of functions: $(f \circ g)(t) = f(g(t))$.

one degree of freedom the coordinate path is a structured object: q is a tuple of component coordinate path functions $q^i = \chi^i \circ \gamma$. At each instant of time t, the values $q(t) = (q^0(t), \ldots, q^{n-1}(t))$ are the generalized coordinates of a configuration.

The derivative Dq of the coordinate path q is a function[10] that gives the rate of change of the configuration coordinates at a given time: $Dq(t) = (Dq^0(t), \ldots, Dq^{n-1}(t))$. The rate of change of a generalized coordinate is called a *generalized velocity*.

Exercise 1.2: Generalized coordinates

For each of the systems in exercise 1.1, specify a system of generalized coordinates that can be used to describe the behavior of the system.

1.3   The Principle of Stationary Action

Let us suppose that for each physical system there is a path-distinguishing function that is stationary on realizable paths. We will try to deduce some of its properties.

Experience of motion

Our ordinary experience suggests that physical motion can be described by configuration paths that are continuous and smooth.[11] We do not see the juggling pin jump from one place to another. Nor do we see the juggling pin suddenly change the way it is moving.

Our ordinary experience suggests that the motion of physical systems does not depend upon the entire history of the system. If we enter the room after the juggling pin has been thrown into the air we cannot tell when it left the juggler's hand. The juggler could have thrown the pin from a variety of places at a variety of times with the same apparent result as we walk through the

---

[10]The derivative of a function f is a function, denoted Df. Our notational convention is that D is a high-precedence operator. Thus D operates on the adjacent function before any other application occurs: $Df(x)$ is the same as $(Df)(x)$.

[11]Experience with systems on an atomic scale suggests that at this scale systems do not travel along well-defined configuration paths. To describe the evolution of systems on the atomic scale we employ quantum mechanics. Here, we restrict attention to systems for which the motion is well described by a smooth configuration path.

door.[12] So the motion of the pin does not depend on the details of the history.

Our ordinary experience suggests that the motion of physical systems is deterministic. In fact, a small number of parameters summarize the important aspects of the history of the system and determine its future evolution. For example, at any moment the position, velocity, orientation, and rate of change of the orientation of the juggling pin are enough to completely determine the future motion of the pin.

Realizable paths

From our experience of motion we develop certain expectations about realizable configuration paths. If a path is realizable, then any segment of the path is a realizable path segment. Conversely, a path is realizable if every segment of the path is a realizable path segment. The realizability of a path segment depends on all points of the path in the segment. The realizability of a path segment depends on every point of the path segment in the same way; no part of the path is special. The realizability of a path segment depends only on points of the path within the segment; the realizability of a path segment is a local property.

So the path-distinguishing function aggregates some local property of the system measured at each moment along the path segment. Each moment along the path must be treated in the same way. The contributions from each moment along the path segment must be combined in a way that maintains the independence of the contributions from disjoint subsegments. One method of combination that satisfies these requirements is to add up the contributions, making the path-distinguishing function an integral over the path segment of some local property of the path.[13]

So we will try to arrange that the path-distinguishing function, constructed as an integral of a local property along the path, assumes a stationary value for any realizable path. Such a path-distinguishing function is traditionally called an *action* for the system. We use the word "action" to be consistent with common

---

[12]Extrapolation of the orbit of the Moon backward in time cannot determine the point at which it was placed on this trajectory. To determine the origin of the Moon we must supplement dynamical evidence with other physical evidence such as chemical compositions.

[13]We suspect that this argument can be promoted to a precise constraint on the possible ways of making this path-distinguishing function.

usage. Perhaps it would be clearer to continue to call it "path-distinguishing function," but then it would be more difficult for others to know what we were talking about.[14]

In order to pursue the agenda of variational mechanics, we must invent action functions that are stationary on the realizable trajectories of the systems we are studying. We will consider actions that are integrals of some local property of the configuration path at each moment. Let $q = \chi \circ \gamma$ be a coordinate path in the configuration space; $q(t)$ are the coordinates of the configuration at time t. Then the action of a segment of the path in the time interval from t_1 to t_2 is[15]

$$S[q](t_1, t_2) = \int_{t_1}^{t_2} F[q]. \tag{1.1}$$

where $F[q]$ is a function of time that measures some local property of the path. It may depend upon the value of the function q at that time and the value of any derivatives of q at that time.[16]

The configuration path can be locally described at a moment in terms of the coordinates, the rate of change of the coordinates, and all the higher derivatives of the coordinates at the given moment. Given this information the path can be reconstructed in some interval containing that moment.[17] Local properties of paths can depend on no more than the local description of the path.

---

[14]Historically, Huygens was the first to use the term "action" in mechanics, referring to "the effect of a motion." This is an idea that came from the Greeks. In his manuscript "Dynamica" (1690) Leibniz enunciated a "Least Action Principle" using the "harmless action," which was the product of mass, velocity, and the distance of the motion. Leibniz also spoke of a "violent action" in the case where things collided.

[15]A definite integral of a real-valued function f of a real argument is written $\int_a^b f$. This can also be written $\int_a^b f(x)dx$. The first notation emphasizes that a function is being integrated.

[16]Traditionally, square brackets are put around functional arguments. In this case, the square brackets remind us that the value of S may depend on the function q in complicated ways, such as through its derivatives.

[17]In the case of a real-valued function, the value of the function and its derivatives at some point can be used to construct a power series. For sufficiently nice functions (real analytic), the power series constructed in this way converges in some interval containing the point. Not all functions can be locally represented in this way. For example, the function $f(x) = \exp(-1/x^2)$, with $f(0) = 0$, is zero and has all derivatives zero at $x = 0$, but this infinite number of derivatives is insufficient to determine the function value at any other point.

The function F measures some local property of the coordinate path q. We can decompose $F[q]$ into two parts: a part that measures some property of a local description and a part that extracts a local description of the path from the path function. The function that measures the local property of the system depends on the particular physical system; the method of construction of a local description of a path from a path is the same for any system. We can write $F[q]$ as a composition of these two functions:[18]

$$F[q] = L \circ \Gamma[q]. \tag{1.2}$$

The function Γ takes the coordinate path and produces a function of time whose value is an ordered tuple containing the time, the coordinates at that time, the rate of change of the coordinates at that time, and the values of higher derivatives of the coordinates evaluated at that time. For the path q and time t:

$$\Gamma[q](t) = (t, q(t), Dq(t), \ldots). \tag{1.3}$$

We refer to this tuple, which includes as many derivatives as are needed, as the *local tuple*. The function $\Gamma[q]$ depends only on the coordinate path q and its derivatives; the function $\Gamma[q]$ does not depend on χ or the fact that q is made by composing χ with γ.

The function L depends on the specific details of the physical system being investigated, but does not depend on any particular configuration path. The function L computes a real-valued local property of the path. We will find that L needs only a finite number of components of the local tuple to compute this property: The path can be locally reconstructed from the full local description; that L depends on a finite number of components of the local tuple guarantees that it measures a local property.[19]

The advantage of this decomposition is that the local description of the path is computed by a uniform process from the configuration path, independent of the system being considered. All of the system-specific information is captured in the function L.

---

[18]In our notation the application of a path-dependent function to its path is of higher precedence than the composition, so $L \circ \Gamma[q] = L \circ (\Gamma[q])$.

[19]We will later discover that an initial segment of the local tuple is sufficient to determine the future evolution of the system. That a configuration and a finite number of derivatives determine the future means that there is a way of determining all of the rest of the derivatives of the path from the initial segment.

The function L is called a *Lagrangian*[20] for the system, and the resulting action,

$$S[q](t_1, t_2) = \int_{t_1}^{t_2} L \circ \Gamma[q], \qquad (1.4)$$

is called the *Lagrangian action*. For Lagrangians that depend only on time, positions, and velocities the action can also be written

$$S[q](t_1, t_2) = \int_{t_1}^{t_2} L\left(t, q(t), Dq(t)\right) dt. \qquad (1.5)$$

Lagrangians can be found for a great variety of systems. We will see that for many systems the Lagrangian can be taken to be the difference between kinetic and potential energy. Such Lagrangians depend only on the time, the configuration, and the rate of change of the configuration. We will focus on this class of systems, but will also consider more general systems from time to time.

A realizable path of the system is to be distinguished from others by having stationary action with respect to some set of nearby unrealizable paths. Now some paths near realizable paths will also be realizable: for any motion of the juggling pin there is another that is slightly different. So when addressing the question of whether the action is stationary with respect to variations of the path we must somehow restrict the set of paths we are considering to contain only one realizable path. It will turn out that for Lagrangians that depend only on the configuration and rate of change of configuration it is enough to restrict the set of paths to those that have the same configuration at the endpoints of the path segment.

The *principle of stationary action* asserts that for each dynamical system we can cook up a Lagrangian such that a realizable path connecting the configurations at two times t_1 and t_2 is dis-

---

[20]The classical Lagrangian plays a fundamental role in the path-integral formulation of quantum mechanics (due to Dirac and Feynman), where the complex exponential of the classical action yields the relative probability amplitude for a path. The Lagrangian is the starting point for the Hamiltonian formulation of mechanics (discussed in chapter 3), which is also essential in the Schrödinger and Heisenberg formulations of quantum mechanics and in the Boltzmann–Gibbs approach to statistical mechanics.

tinguished from all conceivable paths by the fact that the action $S[q](t_1, t_2)$ is stationary with respect to variations of the path.[21] For Lagrangians that depend only on the configuration and rate of change of configuration, the variations are restricted to those that preserve the configurations at t_1 and t_2.[22]

Exercise 1.3: Fermat optics

Fermat observed that the laws of reflection and refraction could be accounted for by the following facts: Light travels in a straight line in any particular medium with a velocity that depends upon the medium. The path taken by a ray from a source to a destination through any sequence of media is a path of least total time, compared to neighboring paths. Show that these facts imply the laws of reflection and refraction.[23]

---

[21] The principle becomes the "principle of least action" if the path is sufficiently short. In the more general case the action is stationary. The term "principle of least action" is also commonly used to refer to a result, due to Maupertuis, Euler, and Lagrange, which says that free particles move along paths for which the integral of the kinetic energy is minimized among all paths with the given endpoints. Correspondingly, the term "action" is sometimes used to refer specifically to the integral of the kinetic energy. (Actually, Euler and Lagrange used the *vis viva*, or twice the kinetic energy.)

[22] Other ways of stating the principle of stationary action make it sound teleological and mysterious. For instance, one could imagine that the system considers all possible paths from its initial configuration to its final configuration and then chooses the one with the smallest action. Indeed, the underlying vision of a purposeful, economical, and rational universe played no small part in the philosophical considerations that accompanied the initial development of mechanics. The earliest action principle that remains part of modern physics is Fermat's principle, which states that the path traveled by a light ray between two points is the path that takes the least amount of time. Fermat formulated this principle around 1660 and used it to derive the laws of reflection and refraction. Motivated by this, the French mathematician and astronomer Pierre-Louis Moreau de Maupertuis enunciated the principle of least action as a grand unifying principle in physics. In his *Essai de cosmologie* (1750) Maupertuis appealed to this principle of "economy in nature" as evidence of the existence of God, asserting that it demonstrated "God's intention to regulate physical phenomena by a general principle of the highest perfection." For a historical perspective on Maupertuis's, Euler's, and Lagrange's roles in the formulation of the principle of least action, see [28].

[23] For reflection the angle of incidence is equal to the angle of reflection. Refraction is described by Snell's law: when light passes from one medium to another, the ratio of the sines of the angles made to the normal to the interface is the inverse of the ratio of the refractive indices of the media. The refractive index is the ratio of the speed of light in the vacuum to the speed of light in the medium.

1.4   Computing Actions

To illustrate the above ideas, and to introduce their formulation as computer programs, we consider the simplest mechanical system— a free particle moving in three dimensions. Euler and Lagrange discovered that for a free particle the time integral of the kinetic energy over the particle's actual path is smaller than the same integral along any alternative path between the same points: a free particle moves according to the principle of stationary action, provided we take the Lagrangian to be the kinetic energy. The kinetic energy for a particle of mass m and velocity $\vec{v}$ is $\frac{1}{2}mv^2$, where v is the magnitude of $\vec{v}$. In this case we can choose the generalized coordinates to be the ordinary rectangular coordinates.

Following Euler and Lagrange, the Lagrangian for the free particle is[24]

$$L(t, x, v) = \tfrac{1}{2}m(v \cdot v), \tag{1.6}$$

where the formal parameter x names a tuple of components of the position with respect to a given rectangular coordinate system, and the formal parameter v names a tuple of velocity components.[25]

We can express this formula as a procedure:

```
(define ((L-free-particle mass) local)
  (let ((v (velocity local)))
    (* 1/2 mass (dot-product v v))))
```

The definition indicates that `L-free-particle` is a procedure that takes mass as an argument and returns a procedure that takes a

---

[24]Here we are making a function definition. A definition specifies the value of the function for arbitrarily chosen formal parameters. One may change the name of a formal parameter, so long as the new name does not conflict with any other symbol in the definition. For example, the following definition specifies exactly the same free-particle Lagrangian:

$L(a, b, c) = \tfrac{1}{2}m(c \cdot c).$

[25]The Lagrangian is formally a function of the local tuple, but any particular Lagrangian depends only on a finite initial segment of the local tuple. We define functions of local tuples by explicitly declaring names for the elements of the initial segment of the local tuple that includes the elements upon which the function depends.

local tuple `local`, extracts the generalized velocity with the procedure `velocity`, and uses the velocity to compute the value of the Lagrangian.[26]

Suppose we let q denote a coordinate path function that maps time to position components:[27]

$$q(t) = (x(t), y(t), z(t)) \,. \tag{1.7}$$

We can make this definition[28]

```
(define q
  (up (literal-function 'x)
      (literal-function 'y)
      (literal-function 'z)))
```

where `literal-function` makes a procedure that represents a function of one argument that has no known properties other than the given symbolic name. The symbol `q` now names a procedure of one real argument (time) that produces a tuple of three components representing the coordinates at that time. For example, we can evaluate this procedure for a symbolic time `t` as follows:

```
(q 't)
(up (x t) (y t) (z t))
```

---

[26] We represent the local tuple as a composite data structure, the components of which are the time, the generalized coordinates, the generalized velocities, and possibly higher derivatives. We do not want to be bothered by the details of packing and unpacking the components into these structures, so we provide utilities for doing this.

[27] Be careful. The x in the definition of q is not the same as the x that was used as a formal parameter in the definition of the free-particle Lagrangian above. There are only so many letters in the alphabet, so we are forced to reuse them. We will be careful to indicate where symbols are given new meanings.

[28] A tuple of coordinate or velocity components is made with the procedure `up`. Component `i` of the tuple `q` is (`ref q i`). All indexing is zero based. The word `up` is to remind us that in mathematical notation these components are indexed by superscripts. There are also `down` tuples of components that are indexed by subscripts. See the appendix on notation.

The constructor `up` is also used to package the time, the coordinates, and the velocities into a data structure representing a local tuple. The selectors `time`, `coordinate`, and `velocity` extract the appropriate pieces from the local structure. The procedure `time` is the same as the procedure (`component 0`), and similarly `coordinate` is (`component 1`) and `velocity` is (`component 2`).

The derivative of the coordinate path Dq is the function that maps time to velocity components:

$$Dq(t) = (Dx(t), Dy(t), Dz(t)).$$

We can make and use the derivative of a function.[29] For example, we can write:

```
((D q) 't)
(up ((D x) t) ((D y) t) ((D z) t))
```

The function Γ takes a coordinate path and returns a function of time that gives the local tuple $(t, q(t), Dq(t), \ldots)$. We implement this Γ with the procedure `Gamma`.[30] Here is what `Gamma` does:

```
((Gamma q) 't)
(up t
    (up (x t) (y t) (z t))
    (up ((D x) t) ((D y) t) ((D z) t)))
```

So the composition $L \circ \Gamma$ is a function of time that returns the value of the Lagrangian for this point on the path:[31]

```
((compose (L-free-particle 'm) (Gamma q)) 't)
(+ (* 1/2 m (expt ((D x) t) 2))
   (* 1/2 m (expt ((D y) t) 2))
   (* 1/2 m (expt ((D z) t) 2)))
```

The procedure `show-expression` simplifies the expression and uses TeX to display the result in traditional infix form. We use this method of display to make the boxed expressions in this book.

---

[29] Derivatives of functions yield functions. For example, `((D cube) 2)` => `12` and `((D cube) 'a)` => `(* 3 (expt a 2))`.

[30] Although Γ produces an arbitrarily long local tuple, our procedure `Gamma` produces by default only the first three elements. If a longer local tuple is needed, `Gamma` can be given the length of the required tuple as an extra argument.

[31] In our system, arithmetic operators are generic over symbolic expressions as well as numeric values; arithmetic procedures can work uniformly with numbers or expressions. For example, given the procedure `(define (cube x) (* x x x))` we can obtain its value for a number `(cube 2)` => `8` or for a literal symbol `(cube 'a)` => `(* a a a)`.

The procedure `show-expression` also produces the prefix form, but we usually do not show this.[32]

```
(show-expression
  ((compose (L-free-particle 'm) (Gamma q)) 't))
```

$$\frac{1}{2}m\left(Dx\left(t\right)\right)^2 + \frac{1}{2}m\left(Dy\left(t\right)\right)^2 + \frac{1}{2}m\left(Dz\left(t\right)\right)^2$$

According to equation (1.4) we can compute the Lagrangian action from time t_1 to time t_2 as:

```
(define (Lagrangian-action L q t1 t2)
  (definite-integral (compose L (Gamma q)) t1 t2))
```

`Lagrangian-action` takes as arguments a procedure `L` that computes the Lagrangian, a procedure `q` that computes a coordinate path, and starting and ending times `t1` and `t2`. The `definite-integral` used here takes as arguments a function and two limits `t1` and `t2`, and computes the definite integral of the function over the interval from `t1` to `t2`.[33] Notice that the definition of `Lagrangian-action` does not depend on any particular set of coordinates or even the dimension of the configuration space. The method of computing the action from the coordinate representation of a Lagrangian and a coordinate path does not depend on the coordinate system.

We can now compute the action for the free particle along a path. For example, consider a particle moving at uniform speed along a straight line $t \mapsto (4t + 7, 3t + 5, 2t + 1)$.[34] We represent the path as a procedure

---

[32]For very complicated expressions the prefix notation of Scheme is often better, but simplification is almost always useful. We can separate the functions of simplification and infix display. We will see examples of this later.

[33]Scmutils includes a variety of numerical integration procedures. The examples in this section were computed by rational-function extrapolation of Euler–MacLaurin formulas with a relative error tolerance of 10^{-10}.

[34]For a real physical situation we would have to specify units for these quantities, but in this illustration we leave them unspecified.

```
(define (test-path t)
  (up (+ (* 4 t) 7)
      (+ (* 3 t) 5)
      (+ (* 2 t) 1)))
```

For a particle of mass 3, we obtain the action between $t = 0$ and $t = 10$ as[35]

```
(Lagrangian-action (L-free-particle 3.0) test-path 0.0 10.0)
435.
```

Exercise 1.4: Lagrangian actions

For a free particle an appropriate Lagrangian is[36]

$$L(t, x, v) = \tfrac{1}{2}mv^2. \tag{1.8}$$

Suppose that x is the constant-velocity straight-line path of a free particle, such that $x_a = x(t_a)$ and $x_b = x(t_b)$. Show that the action on the solution path is

$$\frac{m}{2} \frac{(x_b - x_a)^2}{t_b - t_a}. \tag{1.9}$$

Paths of minimum action

We already know that the actual path of a free particle is uniform motion in a straight line. According to Euler and Lagrange, the action is smaller along a straight-line test path than along nearby paths. Let q be a straight-line test path with action $S[q](t_1, t_2)$. Let $q + \epsilon\eta$ be a nearby path, obtained from q by adding a path variation η scaled by the real parameter ϵ.[37] The action on the varied path is $S[q + \epsilon\eta](t_1, t_2)$. Euler and Lagrange found that $S[q+\epsilon\eta](t_1, t_2) > S[q](t_1, t_2)$ for any η that is zero at the endpoints and for any small nonzero ϵ.

---

[35]Here we use numerals with decimal points to specify the parameters. This forces the representations to be floating point, which is efficient for numerical calculation. If symbolic algebra is to be done it is essential that the numbers be exact integers or rational fractions, so that expressions can be reliably reduced to lowest terms. Such numbers are specified without a decimal point.

[36]The squared magnitude of the velocity is $\vec{v} \cdot \vec{v}$, the vector dot product of the velocity with itself, so we write simply $v^2 = v \cdot v$.

[37]Note that we are doing arithmetic on functions. We extend the arithmetic operations so that the combination of two functions of the same type (same domains and ranges) is the function on the same domain that combines the values of the argument functions in the range. For example, if f and g are functions of t, then fg is the function $t \mapsto f(t)g(t)$. A constant multiple of a function is the function whose value is the constant times the value of the function for each argument: cf is the function $t \mapsto cf(t)$.

Let's check this numerically by varying the test path, adding some amount of a test function that is zero at the endpoints $t = t_1$ and $t = t_2$. To make a function η that is zero at the endpoints, given a sufficiently well-behaved function ν, we can use $\eta(t) = (t - t_1)(t - t_2)\nu(t)$. This can be implemented:

```
(define ((make-eta nu t1 t2) t)
  (* (- t t1) (- t t2) (nu t)))
```

We can use this to compute the action for a free particle over a path varied from the given path, as a function of ϵ:[38]

```
(define ((varied-free-particle-action mass q nu t1 t2) eps)
  (let ((eta (make-eta nu t1 t2)))
    (Lagrangian-action (L-free-particle mass)
                       (+ q (* eps eta))
                       t1
                       t2)))
```

The action for the varied path, with $\nu(t) = (\sin t, \cos t, t^2)$ and $\epsilon = 0.001$, is, as expected, larger than for the test path:

```
((varied-free-particle-action 3.0 test-path
                             (up sin cos square)
                             0.0 10.0)
 0.001)
436.29121428571153
```

We can numerically compute the value of ϵ for which the action is minimized. We search between, say, -2 and 1:[39]

```
(minimize
 (varied-free-particle-action 3.0 test-path
                             (up sin cos square)
                             0.0 10.0)
 -2.0 1.0)
(-1.5987211554602254e-14 435.0000000000237 5)
```

---

[38] Note that we are adding procedures. Paralleling our extension of arithmetic operations to functions, arithmetic operations are extended to compatible procedures.

[39] The arguments to `minimize` are a procedure implementing the univariate function in question, and the lower and upper bounds of the region to be searched. Scmutils includes a choice of methods for numerical minimization; the one used here is Brent's algorithm, with an error tolerance of 10^{-5}. The value returned by `minimize` is a list of three numbers: the first is the argument at which the minimum occurred, the second is the minimum obtained, and the third is the number of iterations of the minimization algorithm required to obtain the minimum.

We find exactly what is expected—that the best value for ϵ is zero,[40] and the minimum value of the action is the action along the straight path.

Finding trajectories that minimize the action

We have used the variational principle to determine if a given trajectory is realizable. We can also use the variational principle to find trajectories. Given a set of trajectories that are specified by a finite number of parameters, we can search the parameter space looking for the trajectory in the set that best approximates the real trajectory by finding one that minimizes the action. By choosing a good set of approximating functions we can get arbitrarily close to the real trajectory.[41]

One way to make a parametric path that has fixed endpoints is to use a polynomial that goes through the endpoints as well as a number of intermediate points. Variation of the positions of the intermediate points varies the path; the parameters of the varied path are the coordinates of the intermediate positions. The procedure make-path constructs such a path using a Lagrange interpolation polynomial. The procedure make-path is called with five arguments: (make-path t0 q0 t1 q1 qs), where q0 and q1 are the endpoints, t0 and t1 are the corresponding times, and qs is a list of intermediate points.[42]

---

[40]Yes, *-1.5987211554602254e-14* is zero for the tolerance required of the minimizer. And *435.0000000000237* is arguably the same as *435* obtained before.

[41]There are lots of good ways to make such a parametric set of approximating trajectories. One could use splines or higher-order interpolating polynomials; one could use Chebyshev polynomials; one could use Fourier components. The choice depends upon the kinds of trajectories one wants to approximate.

[42]Here is one way to implement make-path:

```
(define (make-path t0 q0 t1 q1 qs)
  (let ((n (length qs)))
    (let ((ts (linear-interpolants t0 t1 n)))
      (Lagrange-interpolation-function
        (append (list q0) qs (list q1))
        (append (list t0) ts (list t1)))))))
```

The procedure linear-interpolants produces a list of elements that linearly interpolate the first two arguments. We use this procedure here to specify ts, the n evenly spaced intermediate times between t0 and t1 at which the path will be specified. The parameters being adjusted, qs, are the positions at these intermediate times. The procedure Lagrange-interpolation-function takes a list of values and a list of times and produces a procedure that computes the Lagrange interpolation polynomial that passes through these points.

Having specified a parametric path, we can construct a parametric action that is just the action computed along the parametric path:

```
(define ((parametric-path-action Lagrangian t0 q0 t1 q1) qs)
  (let ((path (make-path t0 q0 t1 q1 qs)))
    (Lagrangian-action Lagrangian path t0 t1)))
```

We can find approximate solution paths by finding parameters that minimize the action. We do this minimization with a canned multidimensional minimization procedure:[43]

```
(define (find-path Lagrangian t0 q0 t1 q1 n)
  (let ((initial-qs (linear-interpolants q0 q1 n)))
    (let ((minimizing-qs
           (multidimensional-minimize
            (parametric-path-action Lagrangian t0 q0 t1 q1)
            initial-qs)))
      (make-path t0 q0 t1 q1 minimizing-qs))))
```

The procedure multidimensional-minimize takes a procedure (in this case the value of the call to parametric-path-action) that computes the function to be minimized (in this case the action) and an initial guess for the parameters. Here we choose the initial guess to be equally spaced points on a straight line between the two endpoints, computed with linear-interpolants.

To illustrate the use of this strategy, we will find trajectories of the harmonic oscillator, with Lagrangian[44]

$$L(t, q, v) = \tfrac{1}{2}mv^2 - \tfrac{1}{2}kq^2, \tag{1.10}$$

for mass m and spring constant k. This Lagrangian is implemented by[45]

---

[43]The minimizer used here is the Nelder–Mead downhill simplex method. As usual with numerical procedures, the interface to the nelder-mead procedure is complex, with lots of optional parameters to let the user control errors effectively. For this presentation we have specialized nelder-mead by wrapping it in the more palatable multidimensional-minimize. Unfortunately, you will have to learn to live with complicated numerical procedures someday.

[44]Don't worry. We know that you don't yet know why this is the right Lagrangian. We will get to this in section 1.6.

[45]The square of a structure of components is defined to be the sum of the squares of the individual components.

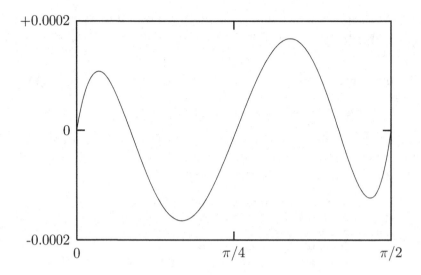

Figure 1.1 The difference between the polynomial approximation with minimum action and the actual trajectory taken by the harmonic oscillator. The abscissa is the time and the ordinate is the error.

```
(define ((L-harmonic m k) local)
  (let ((q (coordinate local))
        (v (velocity local)))
    (- (* 1/2 m (square v)) (* 1/2 k (square q)))))
```

We can find an approximate path taken by the harmonic oscillator for $m = 1$ and $k = 1$ between $q(0) = 1$ and $q(\pi/2) = 0$ as follows:[46]

```
(define q
  (find-path (L-harmonic 1.0 1.0) 0.0 1.0 :pi/2 0.0 3))
```

We know that the trajectories of this harmonic oscillator, for $m = 1$ and $k = 1$, are

$$q(t) = A\cos(t + \varphi) \tag{1.11}$$

where the amplitude A and the phase φ are determined by the initial conditions. For the chosen endpoint conditions the solution is $q(t) = \cos(t)$. The approximate path should be an approximation to cosine over the range from 0 to $\pi/2$. Figure 1.1 shows the error in the polynomial approximation produced by this process.

---

[46]By convention, named constants have names that begin with a colon. The constants named :pi and :-pi are what we would expect from their names.

The maximum error in the approximation with three intermediate points is less than 1.7×10^{-4}. We find, as expected, that the error in the approximation decreases as the number of intermediate points is increased. For four intermediate points it is about a factor of 15 better.

Exercise 1.5: Solution process

We can watch the progress of the minimization by modifying the procedure `parametric-path-action` to plot the path each time the action is computed. Try this:

```
(define win2 (frame 0.0 :pi/2 0.0 1.2))

(define ((parametric-path-action Lagrangian t0 q0 t1 q1)
         intermediate-qs)
  (let ((path (make-path t0 q0 t1 q1 intermediate-qs)))
    ;; display path
    (graphics-clear win2)
    (plot-function win2 path t0 t1 (/ (- t1 t0) 100))
    ;; compute action
    (Lagrangian-action Lagrangian path t0 t1)))

(find-path (L-harmonic 1.0 1.0) 0.0 1.0 :pi/2 0.0 2)
```

Exercise 1.6: Minimizing action

Suppose we try to obtain a path by minimizing an action for an impossible problem. For example, suppose we have a free particle and we impose endpoint conditions on the velocities as well as the positions that are inconsistent with the particle being free. Does the formalism protect itself from such an unpleasant attack? You may find it illuminating to program it and see what happens.

1.5   The Euler–Lagrange Equations

The principle of stationary action characterizes the realizable paths of systems in configuration space as those for which the action has a stationary value. In elementary calculus, we learn that the critical points of a function are the points where the derivative vanishes. In an analogous way, the paths along which the action is stationary are solutions of a system of differential equations. This system, called the *Euler–Lagrange equations* or just the *Lagrange equations*, is the link that permits us to use the principle of stationary action to compute the motions of mechanical systems, and to relate the variational and Newtonian formulations of mechanics.

Lagrange equations

We will find that if L is a Lagrangian for a system that depends on time, coordinates, and velocities, and if q is a coordinate path for which the action $S[q](t_1, t_2)$ is stationary (with respect to any variation in the path that keeps the endpoints of the path fixed), then

$$D(\partial_2 L \circ \Gamma[q]) - \partial_1 L \circ \Gamma[q] = 0. \tag{1.12}$$

Here L is a real-valued function of a local tuple; $\partial_1 L$ and $\partial_2 L$ denote the partial derivatives of L with respect to its generalized position argument and generalized velocity argument respectively.[47] The function $\partial_2 L$ maps a local tuple to a structure whose components are the derivatives of L with respect to each component of the generalized velocity. The function $\Gamma[q]$ maps time to the local tuple: $\Gamma[q](t) = (t, q(t), Dq(t), \ldots)$. Thus the compositions $\partial_1 L \circ \Gamma[q]$ and $\partial_2 L \circ \Gamma[q]$ are functions of one argument, time. The Lagrange equations assert that the derivative of $\partial_2 L \circ \Gamma[q]$ is equal to $\partial_1 L \circ \Gamma[q]$, at any time. Given a Lagrangian, the Lagrange equations form a system of ordinary differential equations that must be satisfied by realizable paths.

Lagrange's equations are traditionally written as a separate equation for each component of q:

$$\frac{d}{dt}\frac{\partial L}{\partial \dot{q}^i} - \frac{\partial L}{\partial q^i} = 0 \qquad i = 0, \ldots, n-1 \,.$$

In this way of writing Lagrange's equations the notation does not distinguish between L, which is a real-valued function of three variables $(t, q, \dot{q})$, and $L \circ \Gamma[q]$, which is a real-valued function of one real variable t. If we do not realize this notational pun, the equations don't make sense as written—$\partial L/\partial \dot{q}$ is a function of three variables, so we must regard the arguments $q, \dot{q}$ as functions of t before taking d/dt of the expression. Similarly, $\partial L/\partial q$ is a function of three variables, which we must view as a function of t before setting it equal to $d/dt(\partial L/\partial \dot{q})$.

---

[47]The derivative or partial derivative of a function that takes structured arguments is a new function that takes the same number and type of arguments. The range of this new function is itself a structure with the same number of components as the argument with respect to which the function is differentiated. See the appendix on notation for more.

A correct use of the traditional notation is more explicit:

$$\frac{d}{dt}\left(\frac{\partial L(t,w,\dot{w})}{\partial \dot{w}^i}\bigg|_{\substack{w = q(t) \\ \dot{w} = \frac{dq(t)}{dt}}}\right) - \frac{\partial L(t,w,\dot{w})}{\partial w^i}\bigg|_{\substack{w = q(t) \\ \dot{w} = \frac{dq(t)}{dt}}} = 0,$$

where $i = 0, \ldots, n - 1$. In these equations we see that the partial derivatives of the Lagrangian function are taken, then the path and its derivative are substituted for the position and velocity arguments of the Lagrangian, resulting in an expression in terms of the time.

1.5.1   Derivation of the Lagrange Equations

We will show that the principle of stationary action implies that realizable paths satisfy the Euler–Lagrange equations.

A Direct Derivation

Let q be a realizable coordinate path from $(t_1, q(t_1))$ to $(t_2, q(t_2))$. Consider nearby paths $q + \epsilon\eta$ where $\eta(t_1) = \eta(t_2) = 0$. Let

$$g(\epsilon) = S[q + \epsilon\eta](t_1, t_2)$$
$$= \int_{t_1}^{t_2} L(t, q(t) + \epsilon\eta(t), Dq(t) + \epsilon D\eta(t))dt. \tag{1.13}$$

Expanding as a power series in ϵ

$$g(\epsilon) = g(0) + \epsilon Dg(0) + \cdots \tag{1.14}$$

and using the chain rule we get

$$Dg(0) = \int_{t_1}^{t_2} (\partial_1 L(t, q(t), Dq(t))\eta(t)) \, dt$$
$$+ \int_{t_1}^{t_2} (\partial_2 L(t, q(t), Dq(t))D\eta(t)) \, dt. \tag{1.15}$$

Integrating the second term by parts we obtain

$$Dg(0) = \int_{t_1}^{t_2} (\partial_1 L(t, q(t), Dq(t))\eta(t)) \, dt$$
$$+ \partial_2 L(t, q(t), Dq(t))\eta(t)\big|_{t_1}^{t_2}$$
$$- \int_{t_1}^{t_2} \frac{d}{dt} (\partial_2 L(t, q(t), Dq(t))) \, \eta(t)dt. \tag{1.16}$$

The increment ΔS in the action due to the variation in the path is, to first order in ϵ, $\epsilon Dg(0)$. Because η is zero at the endpoints the integrated term is zero. Collecting together the other two terms, and reverting to functional notation, we find the increment to be

$$\Delta S = \epsilon \int_{t_1}^{t_2} \{\partial_1 L \circ \Gamma[q] - D(\partial_2 L \circ \Gamma[q])\}\, \eta. \qquad (1.17)$$

If ΔS is zero the action is stationary. We retain enough freedom in the choice of the variation that the factor in the integrand multiplying η is forced to be zero at each point along the path. We argue by contradiction: Suppose this factor were nonzero at some particular time. Then it would have to be nonzero in at least one of its components. But if we choose our η to be a bump that is nonzero only in that component in a neighborhood of that time, and zero everywhere else, then the integral will be nonzero. So we may conclude that the factor in curly brackets is identically zero and thus obtain Lagrange's equations:[48]

$$D(\partial_2 L \circ \Gamma[q]) - \partial_1 L \circ \Gamma[q] = 0. \qquad (1.18)$$

The Variation Operator

First we will develop tools for investigating how path-dependent functions vary as the paths are varied. We will then apply these tools to the action, to derive the Lagrange equations.

Suppose that we have a function $f[q]$ that depends on a path q. How does the function vary as the path is varied? Let q be a coordinate path and $q + \epsilon\eta$ be a varied path, where the function η is a path-like function that can be added to the path q, and the factor ϵ is a scale factor. We define the *variation* $\delta_\eta f[q]$ of the function f on the path q by[49]

$$\delta_\eta f[q] = \lim_{\epsilon \to 0} \left(\frac{f[q + \epsilon\eta] - f[q]}{\epsilon} \right). \qquad (1.19)$$

The variation of f is a linear approximation to the change in the function f for small variations in the path. The variation of f depends on η.

---

[48]To make this argument more precise requires careful analysis.

[49]The variation operator δ_η is like the derivative operator in that it acts on the immediately following function: $\delta_\eta f[q] = (\delta_\eta f)[q]$.

A simple example is the variation of the identity path function: $I[q] = q$. Applying the definition, we find

$$\delta_\eta I[q] = \lim_{\epsilon \to 0} \left(\frac{(q + \epsilon\eta) - q}{\epsilon} \right) = \eta. \qquad (1.20)$$

It is traditional to write $\delta_\eta I[q]$ simply as δq. Another example is the variation of the path function that returns the derivative of the path. We have[50]

$$\delta_\eta g[q] = \lim_{\epsilon \to 0} \left(\frac{D(q + \epsilon\eta) - Dq}{\epsilon} \right) = D\eta \quad \text{with} \quad g[q] = Dq. \qquad (1.21)$$

It is traditional to write $\delta_\eta g[q]$ as δDq.

The variation may be represented in terms of a derivative. Let $g(\epsilon) = f[q + \epsilon\eta]$; then

$$\delta_\eta f[q] = \lim_{\epsilon \to 0} \left(\frac{g(\epsilon) - g(0)}{\epsilon} \right) = Dg(0). \qquad (1.22)$$

Variations have the following derivative-like properties. For path-dependent functions f and g and constant c:

$$\delta_\eta (f\, g)[q] = \delta_\eta f[q]\, g[q] + f[q]\, \delta_\eta g[q] \qquad (1.23)$$
$$\delta_\eta (f + g)[q] = \delta_\eta f[q] + \delta_\eta g[q] \qquad (1.24)$$
$$\delta_\eta (cf)[q] = c\, \delta_\eta f[q]. \qquad (1.25)$$

Let F be a path-independent function and g be a path-dependent function; then

$$\delta_\eta h[q] = (DF \circ g[q])\, \delta_\eta g[q] \quad \text{with} \quad h[q] = F \circ g[q]. \qquad (1.26)$$

The operators D (differentiation) and δ (variation) commute in the following sense:

$$D\delta_\eta f[q] = \delta_\eta g[q] \quad \text{with} \quad g[q] = D(f[q]). \qquad (1.27)$$

Variations also commute with integration in a similar sense.

If a path-dependent function f is stationary for a particular path q with respect to small changes in that path, then it must be

---

[50]We separate out the definition of g: We cannot substitute Dq for $g[q]$ in $\delta_\eta g[q]$ because δ_η applies to g not $g[q]$.

stationary for a subset of those variations that results from adding small multiples of a particular function η to q. So the statement $\delta_\eta f[q] = 0$ for arbitrary η implies the function f is stationary for small variations of the path around q.

Exercise 1.7: Properties of δ

Show that δ has the properties 1.23–1.27.

Exercise 1.8: Implementation of δ

a. Suppose we have a procedure f that implements a path-dependent function: for path q and time t it has the value ((f q) t). The procedure delta computes the variation $(\delta_\eta f)[q](t)$ as the value of the expression ((((delta eta) f) q) t). Complete the definition of delta:

```
(define (((delta eta) f) q)
   ...
   )
```

b. Use your delta procedure to verify the properties of δ listed in exercise 1.7 for simple functions such as implemented by the procedure f:[51]

```
(define (f q)
  (compose
    (literal-function 'F
                      (-> (UP Real (UP* Real) (UP* Real)) Real))
    (Gamma q)))
```

This implements an n-degree-of-freedom path-dependent function that depends on the local tuple of the path at each moment. You can define a literal two-dimensional path by

```
(define q (literal-function 'q (-> Real (UP Real Real))))
```

You should compute both sides of the equalities and subtract the results. The answer should be zero.

A Derivation with the Variation Operator

The action is the integral of the Lagrangian along a path:

$$S[q](t_1, t_2) = \int_{t_1}^{t_2} L \circ \Gamma[q]. \tag{1.28}$$

---

[51]The type of a literal function is described by a function signature. The default function signature is (-> Real Real) indicating a real-valued function of a real argument. In this case F is declared as a function that is shaped like a Lagrangian, with an unspecified number of degrees of freedom. For more information about function signatures see the appendix on notation.

For a realizable path q the variation of the action with respect to any variation η that preserves the endpoints, $\eta(t_1) = \eta(t_2) = 0$, is zero:

$$\delta_\eta S[q](t_1, t_2) = 0. \tag{1.29}$$

Variation commutes with integration, so the variation of the action is

$$\delta_\eta S[q](t_1, t_2) = \int_{t_1}^{t_2} \delta_\eta h[q] \quad \text{where} \quad h[q] = L \circ \Gamma[q]. \tag{1.30}$$

Using the fact that

$$\delta_\eta \Gamma[q](t) = (0, \eta(t), D\eta(t)), \tag{1.31}$$

which follows from equations (1.20) and (1.21), and using the chain rule for variations (1.26), we get[52]

$$\delta_\eta S[q](t_1, t_2) = \int_{t_1}^{t_2} (DL \circ \Gamma[q]) \delta_\eta \Gamma[q]$$

$$= \int_{t_1}^{t_2} ((\partial_1 L \circ \Gamma[q])\eta + (\partial_2 L \circ \Gamma[q])D\eta) . \tag{1.32}$$

Integrating the last term of equation (1.32) by parts gives

$$\delta_\eta S[q](t_1, t_2) = (\partial_2 L \circ \Gamma[q])\eta|_{t_1}^{t_2}$$

$$+ \int_{t_1}^{t_2} \{(\partial_1 L \circ \Gamma[q]) - D(\partial_2 L \circ \Gamma[q])\} \eta. \tag{1.33}$$

For our variation η we have $\eta(t_1) = \eta(t_2) = 0$, so the first term vanishes.

Thus the variation of the action is zero if and only if

$$0 = \int_{t_1}^{t_2} \{(\partial_1 L \circ \Gamma[q]) - D(\partial_2 L \circ \Gamma[q])\} \eta. \tag{1.34}$$

---

[52] A function of multiple arguments is considered a function of an up tuple of its arguments. Thus, the derivative of a function of multiple arguments is a down tuple of the partial derivatives of that function with respect to each of the arguments. So in the case of a Lagrangian L,

$$DL(t, q, v) = [\partial_0 L(t, q, v), \partial_1 L(t, q, v), \partial_2 L(t, q, v)] .$$

The variation of the action is zero because, by assumption, q is a realizable path. Thus (1.34) must be true for *any* function η that is zero at the endpoints. Since η is arbitrary, except for being zero at the endpoints, the bracketed factor of the integrand is zero. So

$$D\left(\partial_2 L \circ \Gamma[q]\right) - \left(\partial_1 L \circ \Gamma[q]\right) = 0. \tag{1.35}$$

This is just what we set out to obtain, the Lagrange equations.

A path satisfying Lagrange's equations is one for which the action is stationary, and the fact that the action is stationary depends only on the values of L at each point of the path (and at each point on nearby paths), not on the coordinate system we use to compute these values. So if the system's path satisfies Lagrange's equations in some particular coordinate system, it must satisfy Lagrange's equations in *any* coordinate system. Thus the equations of variational mechanics are derived the same way in any configuration space and any coordinate system.

Harmonic oscillator

For an example, consider the harmonic oscillator. A Lagrangian is

$$L(t, x, v) = \tfrac{1}{2}mv^2 - \tfrac{1}{2}kx^2. \tag{1.36}$$

Then

$$\partial_1 L(t, x, v) = -kx \quad \text{and} \quad \partial_2 L(t, x, v) = mv. \tag{1.37}$$

The Lagrangian is applied to a tuple of the time, a coordinate, and a velocity. The symbols t, x, and v are arbitrary; they are used to specify formal parameters of the Lagrangian.

Now suppose we have a configuration path y, which gives the coordinate of the oscillator $y(t)$ for each time t. The initial segment of the corresponding local tuple at time t is

$$\Gamma[y](t) = (t, y(t), Dy(t)). \tag{1.38}$$

So

$$(\partial_1 L \circ \Gamma[y])(t) = -ky(t) \quad \text{and} \quad (\partial_2 L \circ \Gamma[y])(t) = mDy(t), \tag{1.39}$$

and

$$D(\partial_2 L \circ \Gamma[y])(t) = mD^2 y(t), \tag{1.40}$$

so the Lagrange equation is

$$mD^2y(t) + ky(t) = 0, \tag{1.41}$$

which is the equation of motion of the harmonic oscillator.

Orbital motion
As another example, consider the two-dimensional motion of a particle of mass m orbiting a fixed center of attraction, with gravitational potential energy $-\mu/r$, where r is the distance to the center of attraction. This is called the *Kepler problem*.

A Lagrangian for this problem is[53]

$$L(t; \xi, \eta; v_\xi, v_\eta) = \frac{1}{2}m(v_\xi^2 + v_\eta^2) + \frac{\mu}{\sqrt{\xi^2 + \eta^2}}, \tag{1.42}$$

where ξ and η are formal parameters for rectangular coordinates of the particle, and v_ξ and v_η are formal parameters for corresponding rectangular velocity components. Then

$$\partial_1 L(t; \xi, \eta; v_\xi, v_\eta) = [\partial_{1,0}L(t; \xi, \eta; v_\xi, v_\eta), \partial_{1,1}L(t; \xi, \eta; v_\xi, v_\eta)]$$
$$= \left[\frac{-\mu\xi}{(\xi^2 + \eta^2)^{3/2}}, \frac{-\mu\eta}{(\xi^2 + \eta^2)^{3/2}} \right]. \tag{1.43}$$

Similarly,

$$\partial_2 L(t; \xi, \eta; v_\xi, v_\eta) = [mv_\xi, mv_\eta]. \tag{1.44}$$

Now suppose we have a configuration path $q = (x, y)$, so that the coordinate tuple at time t is $q(t) = (x(t), y(t))$. The initial segment of the local tuple at time t is

$$\Gamma[q](t) = (t; x(t), y(t); Dx(t), Dy(t)). \tag{1.45}$$

So

$$(\partial_1 L \circ \Gamma[q])(t) = \left[\frac{-\mu x(t)}{((x(t))^2 + (y(t))^2)^{3/2}}, \frac{-\mu y(t)}{((x(t))^2 + (y(t))^2)^{3/2}} \right]$$
$$(\partial_2 L \circ \Gamma[q])(t) = [mDx(t), mDy(t)] \tag{1.46}$$

and

$$D(\partial_2 L \circ \Gamma[q])(t) = [mD^2x(t), mD^2y(t)]. \tag{1.47}$$

---

[53] When we write a definition that names the components of the local tuple, we indicate that these are grouped into time, position, and velocity components by separating the groups with semicolons.

The component Lagrange equations at time t are

$$mD^2x(t) + \frac{\mu x(t)}{((x(t))^2 + (y(t))^2)^{3/2}} = 0$$

$$mD^2y(t) + \frac{\mu y(t)}{((x(t))^2 + (y(t))^2)^{3/2}} = 0. \qquad (1.48)$$

Exercise 1.9: Lagrange's equations

Derive the Lagrange equations for the following systems, showing all of the intermediate steps as in the harmonic oscillator and orbital motion examples.

a. An ideal planar pendulum consists of a bob of mass m connected to a pivot by a massless rod of length l subject to uniform gravitational acceleration g. A Lagrangian is $L(t, \theta, \dot{\theta}) = \frac{1}{2}ml^2\dot{\theta}^2 + mgl\cos\theta$. The formal parameters of L are t, θ, and $\dot{\theta}$; θ measures the angle of the pendulum rod to a plumb line and $\dot{\theta}$ is the angular velocity of the rod.[54]

b. A particle of mass m moves in a two-dimensional potential $V(x, y) = (x^2 + y^2)/2 + x^2y - y^3/3$, where x and y are rectangular coordinates of the particle. A Lagrangian is $L(t; x, y; v_x, v_y) = \frac{1}{2}m(v_x^2 + v_y^2) - V(x, y)$.

c. A Lagrangian for a particle of mass m constrained to move on a sphere of radius R is $L(t; \theta, \varphi; \alpha, \beta) = \frac{1}{2}mR^2(\alpha^2 + (\beta\sin\theta)^2)$. The angle θ is the colatitude of the particle and φ is the longitude; the rate of change of the colatitude is α and the rate of change of the longitude is β.

Exercise 1.10: Higher-derivative Lagrangians

Derive Lagrange's equations for Lagrangians that depend on accelerations. In particular, show that the Lagrange equations for Lagrangians of the form $L(t, q, \dot{q}, \ddot{q})$ with $\ddot{q}$ terms are[55]

$$D^2(\partial_3 L \circ \Gamma[q]) - D(\partial_2 L \circ \Gamma[q]) + \partial_1 L \circ \Gamma[q] = 0. \qquad (1.49)$$

In general, these equations, first derived by Poisson, will involve the fourth derivative of q. Note that the derivation is completely analogous

---

[54]The symbol $\dot{\theta}$ is just a mnemonic symbol; the dot over the θ does not indicate differentiation. To define L we could have just as well have written: $L(a, b, c) = \frac{1}{2}ml^2c^2 + mgl\cos b$. However, we use a dotted symbol to remind us that the argument matching a formal parameter, such as $\dot{\theta}$, is a rate of change of an angle, such as θ.

[55]In traditional notation these equations read

$$\frac{d^2}{dt^2}\frac{\partial L}{\partial \ddot{q}} - \frac{d}{dt}\frac{\partial L}{\partial \dot{q}} + \frac{\partial L}{\partial q} = 0.$$

to the derivation of the Lagrange equations without accelerations; it is just longer. What restrictions must we place on the variations so that the critical path satisfies a differential equation?

1.5.2 Computing Lagrange's Equations

The procedure for computing Lagrange's equations mirrors the functional expression (1.12), where the procedure `Gamma` implements Γ:[56]

```
(define ((Lagrange-equations Lagrangian) q)
  (- (D (compose ((partial 2) Lagrangian) (Gamma q)))
     (compose ((partial 1) Lagrangian) (Gamma q))))
```

The argument of `Lagrange-equations` is a procedure that computes a Lagrangian. The `Lagrange-equations` procedure returns a procedure that when applied to a path q returns a procedure of one argument (time) that computes the left-hand side of the Lagrange equations (1.12). These residual values are zero if q is a path for which the Lagrangian action is stationary.

Observe that the `Lagrange-equations` procedure, like the Lagrange equations themselves, is valid for *any* generalized coordinate system. When we write programs to investigate particular systems, the procedures that implement the Lagrangian function and the path q will reflect the actual coordinates chosen to represent the system, but we use the same `Lagrange-equations` procedure in each case. This abstraction reflects the important fact that the method of derivation of Lagrange's equations from a Lagrangian is always the same; it is independent of the number of degrees of freedom, the topology of the configuration space, and the coordinate system used to describe points in the configuration space.

The free particle
Consider again the case of a free particle. The Lagrangian is implemented by the procedure `L-free-particle`. Rather than numerically integrating and minimizing the action, as we did in section 1.4, we can check Lagrange's equations for an arbitrary straight-line path $t \mapsto (at + a_0, bt + b_0, ct + c_0)$:

---

[56]The `Lagrange-equations` procedure uses the operations (`partial 1`) and (`partial 2`), which implement the partial derivative operators with respect to the second and third argument positions (those with indices 1 and 2).

```
(define (test-path t)
  (up (+ (* 'a t) 'a0)
      (+ (* 'b t) 'b0)
      (+ (* 'c t) 'c0)))

(((Lagrange-equations (L-free-particle 'm))
  test-path)
 't)
(down 0 0 0)
```

That the residuals are zero indicates that the test path satisfies the Lagrange equations.[57]

We can also apply the `Lagrange-equations` procedure to an arbitrary function:[58]

```
(show-expression
 (((Lagrange-equations (L-free-particle 'm))
   (literal-function 'x))
  't))
(* (((expt D 2) x) t) m)
```

$$mD^2x\,(t)$$

The result is an expression containing the arbitrary time t and mass m, so it is zero precisely when $D^2x = 0$, which is the expected equation for a free particle.

The harmonic oscillator

Consider the harmonic oscillator again, with Lagrangian (1.10). We know that the motion of a harmonic oscillator is a sinusoid with a given amplitude, frequency, and phase:

$$x(t) = a\cos(\omega t + \varphi). \tag{1.50}$$

---

[57]There is a Lagrange equation for every degree of freedom. The residuals of all the equations are zero if the path is realizable. The residuals are arranged in a **down** tuple because they result from derivatives of the Lagrangian with respect to argument slots that take up tuples. See the appendix on notation.

[58]Observe that the second derivative is indicated as the square of the derivative operator (`expt D 2`). Arithmetic operations in Scmutils extend over operators as well as functions.

Suppose we have forgotten how the constants in the solution relate to the mass m and spring constant k of the oscillator. Let's plug in the proposed solution and look at the residual:

```
(define (proposed-solution t)
  (* 'A (cos (+ (* 'omega t) 'phi))))

(show-expression
 (((Lagrange-equations (L-harmonic 'm 'k))
   proposed-solution)
  't))
```

$$\cos\left(\omega t + \varphi\right) A \left(k - m\omega^2\right)$$

The residual here shows that for nonzero amplitude, the only solutions allowed are ones where $(k - m\omega^2) = 0$ or $\omega = \sqrt{k/m}$.

Exercise 1.11: Kepler's third law

A Lagrangian suitable for studying the relative motion of two particles, of masses m_1 and m_2, with potential energy V, is:

```
(define ((L-central-polar m V) local)
  (let ((q (coordinate local))
        (qdot (velocity local)))
    (let ((r (ref q 0))       (phi (ref q 1))
          (rdot (ref qdot 0)) (phidot (ref qdot 1)))
      (- (* 1/2 m
            (+ (square rdot) (square (* r phidot))) )
         (V r)))))
```

The argument m is the *reduced mass* of the system

$$m = \frac{m_1 m_2}{m_1 + m_2}. \tag{1.51}$$

For gravity, the potential energy function is

```
(define ((gravitational-energy G m1 m2) r)
  (- (/ (* G m1 m2) r)))
```

where r is the distance between the two particles.

Consider the simple situation of the particles in circular orbits around their common center of mass. Construct a circular orbit and plug it into the Lagrange equations. Show that the residual gives Kepler's law:

$$n^2 a^3 = G(m_1 + m_2) \tag{1.52}$$

where n is the angular frequency of the orbit and a is the distance between the particles.

Exercise 1.12: Lagrange's Equations

Compute Lagrange's equations for the Lagrangians in exercise 1.9 using the `Lagrange-equations` procedure. Additionally, use the computer to perform each of the steps in the `Lagrange-equations` procedure and show the intermediate results. Relate these steps to the ones you showed in the hand derivation of exercise 1.9.

Exercise 1.13: Higher-derivative Lagrangians

a. Write a procedure to compute the Lagrange equations for Lagrangians that depend upon acceleration, as in exercise 1.10. Note that `Gamma` can take an optional argument giving the length of the initial segment of the local tuple needed. The default length is 3, giving components of the local tuple up to and including the velocities.

b. Use your procedure to compute the Lagrange equations for the Lagrangian

$$L(t, x, v, a) = -\tfrac{1}{2}mxa - \tfrac{1}{2}kx^2.$$

Do you recognize the resulting equation of motion?

c. For more fun, write the general Lagrange equation procedure that takes a Lagrangian that depends on any number of derivatives, and the number of derivatives, to produce the required equations of motion.

1.6 How to Find Lagrangians

Lagrange's equations are a system of second-order differential equations. In order to use them to compute the evolution of a mechanical system, we must find a suitable Lagrangian for the system. There is no general way to construct a Lagrangian for every system, but there is an important class of systems for which we can identify Lagrangians in a straightforward way in terms of kinetic and potential energy. The key idea is to construct a Lagrangian L such that Lagrange's equations are Newton's equations $\vec{F} = m\vec{a}$.

Suppose our system consists of N particles indexed by α, with mass m_α and vector position $\vec{x}_\alpha(t)$. Suppose further that the forces acting on the particles can be written in terms of a gradient of a potential energy $\mathcal{V}$ that is a function of the positions of the particles and possibly time, but does not depend on the velocities. In other words, the force on particle α is $\vec{F}_\alpha = -\vec{\nabla}_{\vec{x}_\alpha} \mathcal{V}$, where $\vec{\nabla}_{\vec{x}_\alpha} \mathcal{V}$ is the gradient of $\mathcal{V}$ with respect to the position of the particle with index α. We can write Newton's equations as

$$D(m_\alpha\, D\vec{x}_\alpha)(t) + \vec{\nabla}_{\vec{x}_\alpha} \mathcal{V}(t, \vec{x}_0(t), \dots, \vec{x}_{N-1}(t)) = 0. \tag{1.53}$$

Vectors can be represented as tuples of components of the vectors on a rectangular basis. So $\vec{x}_1(t)$ is represented as the tuple $\mathbf{x}_1(t)$. Let V be the potential energy function expressed in terms of components:

$$V(t; \mathbf{x}_0(t), \dots, \mathbf{x}_{N-1}(t)) = \mathcal{V}(t, \vec{x}_0(t), \dots, \vec{x}_{N-1}(t)). \tag{1.54}$$

Newton's equations are

$$D(m_\alpha\, D\mathbf{x}_\alpha)(t) + \partial_{1,\alpha} V(t; \mathbf{x}_0(t), \dots, \mathbf{x}_\alpha(t), \dots, \mathbf{x}_{N-1}(t)) = 0, \tag{1.55}$$

where $\partial_{1,\alpha} V$ is the partial derivative of V with respect to the $\mathbf{x}_\alpha(t)$ argument slot.

To form the Lagrange equations we collect all the position components of all the particles into one tuple $x(t)$, so $x(t) = (\mathbf{x}_0(t), \dots, \mathbf{x}_{N-1}(t))$. The Lagrange equations for the coordinate path x are

$$D\left(\partial_2 L \circ \Gamma[x]\right) - \partial_1 L \circ \Gamma[x] = 0. \tag{1.56}$$

Observe that Newton's equations (1.55) are just the components of the Lagrange equations (1.56) if we choose L to have the properties

$$(\partial_2 L \circ \Gamma[x])(t) = [m_0 D\mathbf{x}_0(t), \dots, m_{N-1} D\mathbf{x}_{N-1}(t)]$$
$$(\partial_1 L \circ \Gamma[x])(t) = [-\partial_{1,0} V(t, x(t)), \dots, -\partial_{1,N-1} V(t, x(t))]; \tag{1.57}$$

here $V(t, x(t)) = V(t; \mathbf{x}_0(t), \dots, \mathbf{x}_{N-1}(t))$ and $\partial_{1,\alpha} V(t, x(t))$ is the tuple of the components of the derivative of V with respect to the coordinates of the particle with index α, evaluated at time t and coordinates $x(t)$. These conditions are satisfied if for every $\mathbf{x}_\alpha$ and $\mathbf{v}_\alpha$

$$\partial_2 L(t; \mathbf{x}_0, \ldots, \mathbf{x}_{N-1}; \mathbf{v}_0, \ldots, \mathbf{v}_{N-1})$$
$$= [m_0 \mathbf{v}_0, \ldots, m_{N-1} \mathbf{v}_{N-1}] \tag{1.58}$$

and

$$\partial_1 L(t; \mathbf{x}_0, \ldots, \mathbf{x}_{N-1}; \mathbf{v}_0, \ldots, \mathbf{v}_{N-1})$$
$$= [-\partial_{1,0} V(t, x), \ldots, -\partial_{1,N-1} V(t, x)], \tag{1.59}$$

where $x = (\mathbf{x}_0, \ldots, \mathbf{x}_{N-1})$. One choice for L that has the required properties (1.58–1.59) is

$$L(t, x, v) = \frac{1}{2} \sum_\alpha m_\alpha v_\alpha^2 - V(t, x), \tag{1.60}$$

where v_α^2 is the sum of the squares of the components of $\mathbf{v}_\alpha$.[59]

The first term is the kinetic energy, conventionally denoted T. So this choice for the Lagrangian is $L(t, x, v) = T(t, x, v) - V(t, x)$, the difference of the kinetic and potential energy. We will often extend the arguments of the potential energy function to include the velocities so that we can write $L = T - V$.[60]

Hamilton's principle

Given a system of point particles for which we can identify the force as the (negative) derivative of a potential energy V that is independent of velocity, we have shown that the system evolves along a path that satisfies Lagrange's equations with $L = T - V$. Having identified a Lagrangian for this class of systems, we can restate the principle of stationary action in terms of energies. This statement is known as *Hamilton's principle*: A point-particle system for which the force is derived from a velocity-independent potential energy evolves along a path q for which the action

$$S[q](t_1, t_2) = \int_{t_1}^{t_2} L \circ \Gamma[q]$$

---

[59] Remember that x and v are just formal parameters of the Lagrangian. This x is not the path x used earlier in the derivation, though it could be the value of that path at a particular time.

[60] We can always give a function extra arguments that are not used so that it can be algebraically combined with other functions of the same shape.

is stationary with respect to variations of the path q that leave the endpoints fixed, where $L = T - V$ is the difference between kinetic and potential energy.[61]

It might seem that we have reduced Lagrange's equations to nothing more than $\vec{F} = m\vec{a}$, and indeed, the principle is motivated by comparing the two equations for this special class of systems. However, the Lagrangian formulation of the equations of motion has an important advantage over $\vec{F} = m\vec{a}$. Our derivation used the rectangular components $\mathbf{x}_\alpha$ of the positions of the constituent particles for the generalized coordinates, but if the system's path satisfies Lagrange's equations in some particular coordinate system, it must satisfy the equations in *any* coordinate system. Thus we see that $L = T - V$ is suitable as a Lagrangian with *any* set of generalized coordinates. The equations of variational mechanics are derived the same way in any configuration space and any coordinate system. In contrast, the Newtonian formulation is based on elementary geometry: In order for $D^2\vec{x}(t)$ to be meaningful as an acceleration, $\vec{x}(t)$ must be a vector in physical space. Lagrange's equations have no such restriction on the meaning of the coordinate q. The generalized coordinates can be any parameters that conveniently describe the configurations of the system.

---

[61] William Rowan Hamilton formulated the fundamental variational principle for time-independent systems in 1834–1835. Jacobi gave this principle the name "Hamilton's principle." For systems subject to generic, nonstationary constraints Hamilton's principle was investigated in 1848 by Ostrogradsky, and in the Russian literature Hamilton's principle is often called the Hamilton–Ostrogradsky principle.

Hamilton (1805–1865) was a brilliant mathematician. His early work on geometric optics (based on Fermat's principle) was so impressive that he was elected to the post of Professor of Astronomy at Trinity College and Royal Astronomer of Ireland while he was still an undergraduate. He produced two monumental works of mathematics. His discovery of quaternions revitalized abstract algebra and sparked the development of vector techniques in physics. His 1835 memoir "On a General Method in Dynamics" put variational mechanics on a firm footing, finally giving substance to Maupertuis's vaguely stated Principle of Least Action of 100 years before. Hamilton also wrote poetry and carried on an extensive correspondence with Wordsworth, who advised him to put his energy into writing mathematics rather than poetry.

In addition to the formulation of the fundamental variational principle, Hamilton also emphasized the analogy between geometric optics and mechanics, and stressed the importance of the momentum variables (which were earlier introduced by Lagrange and Cauchy), leading to the "canonical" form of mechanics discussed in chapter 3.

Constant acceleration

Consider a particle of mass m in a uniform gravitational field with acceleration g. The potential energy is mgh where h is the height of the particle. The kinetic energy is just $\frac{1}{2}mv^2$. A Lagrangian for the system is the difference of the kinetic and potential energies. In rectangular coordinates, with y measuring the vertical position and x measuring the horizontal position, the Lagrangian is $L(t; x, y; v_x, v_y) = \frac{1}{2}m\left(v_x^2 + v_y^2\right) - mgy$. We have[62]

```
(define ((L-uniform-acceleration m g) local)
  (let ((q (coordinate local))
        (v (velocity local)))
    (let ((y (ref q 1)))
      (- (* 1/2 m (square v)) (* m g y)))))

(show-expression
 (((Lagrange-equations
    (L-uniform-acceleration 'm 'g))
   (up (literal-function 'x)
       (literal-function 'y)))
  't))
```

$$\begin{bmatrix} mD^2x\,(t) \\[2mm] gm + mD^2y\,(t) \end{bmatrix}$$

This equation describes unaccelerated motion in the horizontal direction $(mD^2x(t) = 0)$ and constant acceleration in the vertical direction $(mD^2y(t) = -gm)$.

Central force field

Consider planar motion of a particle of mass m in a central force field, with an arbitrary potential energy $U(r)$ depending only upon the distance r to the center of attraction. We will derive the Lagrange equations for this system in both rectangular coordinates and polar coordinates.

In rectangular coordinates (x, y), with origin at the center of attraction, the potential energy is $V(t; x, y) = U(\sqrt{x^2 + y^2})$ and

---

[62]When applied to a tuple, `square` means the sum of the squares of the components of the tuple.

the kinetic energy is $T(t; x, y; v_x, v_y) = \frac{1}{2}m(v_x^2 + v_y^2)$. A Lagrangian for the system is $L = T - V$:

$$L(t; x, y; v_x, v_y) = \frac{1}{2}m(v_x^2 + v_y^2) - U(\sqrt{x^2 + y^2}). \tag{1.61}$$

As a procedure:

```
(define ((L-central-rectangular m U) local)
   (let ((q (coordinate local))
         (v (velocity local)))
    (- (* 1/2 m (square v))
       (U (sqrt (square q))))))
```

The Lagrange equations are

```
(show-expression
 (((Lagrange-equations
    (L-central-rectangular 'm (literal-function 'U)))
   (up (literal-function 'x)
       (literal-function 'y)))
  't))
```

$$\begin{bmatrix} mD^2x\,(t) + \dfrac{DU\left(\sqrt{(y\,(t))^2 + (x\,(t))^2}\right) x\,(t)}{\sqrt{(y\,(t))^2 + (x\,(t))^2}} \\[2em] mD^2y\,(t) + \dfrac{DU\left(\sqrt{(x\,(t))^2 + (y\,(t))^2}\right) y\,(t)}{\sqrt{(x\,(t))^2 + (y\,(t))^2}} \end{bmatrix}$$

We can rewrite these Lagrange equations as:

$$mD^2x(t) = -\frac{x(t)}{r(t)}DU(r(t)) \tag{1.62}$$

$$mD^2y(t) = -\frac{y(t)}{r(t)}DU(r(t)), \tag{1.63}$$

where $r(t) = \sqrt{(x(t))^2 + (y(t))^2}$. We can interpret these as follows. The particle is subject to a radially directed force with magnitude $-DU(r)$. Newton's equations equate the force with the product of the mass and the acceleration. The two Lagrange equations are just the rectangular components of Newton's equations.

We can describe the same system in polar coordinates. The relationship between rectangular coordinates (x, y) and polar coordinates (r, φ) is

$$x = r \cos \varphi$$
$$y = r \sin \varphi. \tag{1.64}$$

The relationship of the generalized velocities is derived from the coordinate transformation. Consider a configuration path that is represented in both rectangular and polar coordinates. Let $\widetilde{x}$ and $\widetilde{y}$ be components of the rectangular coordinate path, and let $\widetilde{r}$ and $\widetilde{\varphi}$ be components of the corresponding polar coordinate path. The rectangular components at time t are $(\widetilde{x}(t), \widetilde{y}(t))$ and the polar coordinates at time t are $(\widetilde{r}(t), \widetilde{\varphi}(t))$. They are related by (1.64):

$$\widetilde{x}(t) = \widetilde{r}(t) \cos \widetilde{\varphi}(t)$$
$$\widetilde{y}(t) = \widetilde{r}(t) \sin \widetilde{\varphi}(t). \tag{1.65}$$

The rectangular velocity at time t is $(D\widetilde{x}(t), D\widetilde{y}(t))$. Differentiating (1.65) gives the relationship among the velocities

$$D\widetilde{x}(t) = D\widetilde{r}(t) \cos \widetilde{\varphi}(t) - \widetilde{r}(t) D\widetilde{\varphi}(t) \sin \widetilde{\varphi}(t)$$
$$D\widetilde{y}(t) = D\widetilde{r}(t) \sin \widetilde{\varphi}(t) + \widetilde{r}(t) D\widetilde{\varphi}(t) \cos \widetilde{\varphi}(t). \tag{1.66}$$

These relations are valid for any configuration path at any moment, so we can abstract them to relations among coordinate representations of an arbitrary velocity. Let v_x and v_y be the rectangular components of the velocity and $\dot{r}$ and $\dot{\varphi}$ be the rate of change of r and φ. Then

$$v_x = \dot{r} \cos \varphi - r\dot{\varphi} \sin \varphi$$
$$v_y = \dot{r} \sin \varphi + r\dot{\varphi} \cos \varphi. \tag{1.67}$$

The kinetic energy is $\frac{1}{2}m(v_x^2 + v_y^2)$:

$$T(t; r, \varphi; \dot{r}, \dot{\varphi}) = \tfrac{1}{2}m(\dot{r}^2 + r^2\dot{\varphi}^2), \tag{1.68}$$

and the Lagrangian is

$$L(t; r, \varphi; \dot{r}, \dot{\varphi}) = \tfrac{1}{2}m(\dot{r}^2 + r^2\dot{\varphi}^2) - U(r). \tag{1.69}$$

We express this Lagrangian as follows:

```
(define ((L-central-polar m U) local)
  (let ((q (coordinate local))
        (qdot (velocity local)))
    (let ((r (ref q 0)) (phi (ref q 1))
          (rdot (ref qdot 0)) (phidot (ref qdot 1)))
      (- (* 1/2 m
            (+ (square rdot)
               (square (* r phidot))) )
         (U r)))))
```

Lagrange's equations are

```
(show-expression
 (((Lagrange-equations
    (L-central-polar 'm (literal-function 'U)))
   (up (literal-function 'r)
       (literal-function 'phi)))
  't))
```

$$\left[\begin{array}{c} mD^2r\left(t\right) - mr\left(t\right)\left(D\varphi\left(t\right)\right)^2 + DU\left(r\left(t\right)\right) \\ 2mDr\left(t\right)r\left(t\right)D\varphi\left(t\right) + mD^2\varphi\left(t\right)\left(r\left(t\right)\right)^2 \end{array}\right]$$

We can interpret the first equation as saying that the product of the mass and the radial acceleration is the sum of the force due to the potential and the centrifugal force. The second equation can be interpreted as saying that the derivative of the angular momentum $mr^2D\varphi$ is zero, so angular momentum is conserved.

Note that we used the same `Lagrange-equations` procedure for the derivation in both coordinate systems. Coordinate representations of the Lagrangian are different for different coordinate systems, and the Lagrange equations in different coordinate systems look different. Yet the same method is used to derive the Lagrange equations in any coordinate system.

Exercise 1.14: Coordinate-independence of Lagrange equations

Check that the Lagrange equations for central force motion in polar coordinates and in rectangular coordinates are equivalent. Determine the relationship among the second derivatives by substituting paths into the transformation equations and computing derivatives, then substitute these relations into the equations of motion.

1.6.1   Coordinate Transformations

The motion of a system is independent of the coordinates we use to describe it. This coordinate-free nature of the motion is apparent in the action principle. The action depends only on the value of the Lagrangian along the path and not on the particular coordinates used in the representation of the Lagrangian. We can use this property to find a Lagrangian in one coordinate system in terms of a Lagrangian in another coordinate system.

Suppose we have a mechanical system whose motion is described by a Lagrangian L that depends on time, coordinates, and velocities. And suppose we have a coordinate transformation F such that $x = F(t, x')$. The Lagrangian L is expressed in terms of the unprimed coordinates. We want to find a Lagrangian L' expressed in the primed coordinates that describes the same system. One way to do this is to require that the value of the Lagrangian along any configuration path be independent of the coordinate system. If q is a path in the unprimed coordinates and q' is the corresponding path in primed coordinates, then the Lagrangians must satisfy:

$$L' \circ \Gamma[q'] = L \circ \Gamma[q]. \tag{1.70}$$

We have seen that the transformation from rectangular to polar coordinates implies that the generalized velocities transform in a certain way. The velocity transformation can be deduced from the requirement that a path in polar coordinates and a corresponding path in rectangular coordinates are consistent with the coordinate transformation. In general, the requirement that paths in two different coordinate systems be consistent with the coordinate transformation can be used to deduce how all of the components of the local tuple transform. Given a coordinate transformation F, let C be the corresponding function that maps local tuples in the primed coordinate system to corresponding local tuples in the unprimed coordinate system:

$$C \circ \Gamma[q'] = \Gamma[q]. \tag{1.71}$$

We will deduce the general form of C below.

Given such a local-tuple transformation C, a Lagrangian L' that satisfies equation (1.70) is

$$L' = L \circ C. \tag{1.72}$$

We can see this by substituting for L' in equation (1.70):

$$L' \circ \Gamma[q'] = L \circ C \circ \Gamma[q'] = L \circ \Gamma[q]. \tag{1.73}$$

To find the local-tuple transformation C given a coordinate transformation F, we deduce how each component of the local tuple transforms. The coordinate transformation specifies how the coordinate component of the local tuple transforms

$$x = F(t, x'). \tag{1.74}$$

The generalized-velocity component of the local-tuple transformation can be deduced as follows. Let q and q' be the same configuration path expressed in the two coordinate systems. Substituting these paths into the coordinate transformation and computing the derivative, we find

$$Dq(t) = \partial_0 F(t, q'(t)) + \partial_1 F(t, q'(t)) Dq'(t). \tag{1.75}$$

Through any point there is always a path of any given velocity, so we may generalize and conclude that along corresponding coordinate paths the generalized velocities satisfy

$$v = \partial_0 F(t, x') + \partial_1 F(t, x') v'. \tag{1.76}$$

If needed, rules for higher-derivative components of the local tuple can be determined in a similar fashion. The local-tuple transformation that takes a local tuple in the primed system to a local tuple in the unprimed system is constructed from the component transformations:

$$(t,\ x,\ v,\ \ldots) = C(t,\ x',\ v',\ \ldots)$$
$$= (t,\ F(t, x'),\ \partial_0 F(t, x') + \partial_1 F(t, x') v',\ \ldots) . \tag{1.77}$$

So if we take the Lagrangian L' to be

$$L' = L \circ C, \tag{1.78}$$

then the action has a value that is independent of the coordinate
system used to compute it. The configuration path of stationary
action does not depend on which coordinate system is used to
describe the path. The Lagrange equations derived from these
Lagrangians will in general look very different from one another,
but they must be equivalent.

Exercise 1.15: Equivalence

Show by direct calculation that the Lagrange equations for L' are satisfied if the Lagrange equations for L are satisfied.

Given a coordinate transformation F, we can use (1.77) to find
the function C that transforms local tuples. The procedure F->C
implements this:[63]

```
(define ((F->C F) local)
  (up (time local)
      (F local)
      (+ (((partial 0) F) local)
         (* (((partial 1) F) local)
            (velocity local)))))
```

As an illustration, consider the transformation from polar to
rectangular coordinates, $x = r\cos\varphi$ and $y = r\sin\varphi$, with the
following implementation:

```
(define (p->r local)
  (let ((polar-tuple (coordinate local)))
    (let ((r (ref polar-tuple 0))
          (phi (ref polar-tuple 1)))
      (let ((x (* r (cos phi)))
            (y (* r (sin phi))))
        (up x y)))))
```

In terms of the polar coordinates and the rates of change of the polar coordinates, the rates of change of the rectangular components
are[64]

---

[63] As described in footnote 26 above, the procedure **up** constructs a local tuple
from an initial segment of time, coordinates, and velocities.

[64] We hope you appreciate the TeX magic here. Symbols with carets, under-
lines, the names of Greek letters, and those terminating in the characters "dot"
are converted by **show-expression** to the corresponding TeX expression.

```
(show-expression
 (velocity
  ((F->C p->r)
   (up 't (up 'r 'phi) (up 'rdot 'phidot)))))
```

$$\begin{pmatrix} -\dot{\varphi}r\sin{(\varphi)} + \dot{r}\cos{(\varphi)} \\ \dot{\varphi}r\cos{(\varphi)} + \dot{r}\sin{(\varphi)} \end{pmatrix}$$

We can use F->C to find the Lagrangian for central force motion in polar coordinates from the Lagrangian in rectangular components, using equation (1.72):

```
(define (L-central-polar m U)
  (compose (L-central-rectangular m U) (F->C p->r)))
```

```
(show-expression
  ((L-central-polar 'm (literal-function 'U))
   (up 't (up 'r 'phi) (up 'rdot 'phidot))))
```

$$\frac{1}{2}m\dot{\varphi}^2r^2 + \frac{1}{2}m\dot{r}^2 - U(r)$$

The result is the same as Lagrangian (1.69).

Exercise 1.16: Central force motion

Find Lagrangians for central force motion in three dimensions in rectangular coordinates and in spherical coordinates. First, find the Lagrangians analytically, then check the results with the computer by generalizing the programs that we have presented.

Coriolis and centrifugal forces

The equations of motion of a free particle in a rotating coordinate system have additional terms. Consider a free particle moving in two dimensions. A Lagrangian is:

```
(define ((L-free-rectangular m) local)
  (let ((vx (ref (velocities local) 0))
        (vy (ref (velocities local) 1)))
    (* 1/2 m (+ (square vx) (square vy)))))
```

The rotation will be easy to describe in polar coordinates, so we transform to polar coordinates:

```
(define (L-free-polar m)
  (compose (L-free-rectangular m) (F->C p->r)))
```

Now we can make a simple time-dependent transformation to rotating coordinates, with rate of rotation Omega:

```
(define ((F Omega) local)
  (let ((t (time local))
        (r (ref (coordinates local) 0))
        (theta (ref (coordinates local) 1)))
    (up r (+ theta (* Omega t)))))
```

```
(define (L-rotating-polar m Omega)
  (compose (L-free-polar m) (F->C (F Omega))))
```

Now let's transform back to rectangular coordinates:

```
(define (L-rotating-rectangular m Omega)
  (compose (L-rotating-polar m Omega) (F->C r->p)))
```

The new Lagrangian, in the rotating rectangular coordinate system is:

```
((L-rotating-rectangular 'm 'Omega)
 (up 't (up 'x_r 'y_r) (up 'xdot_r 'ydot_r)))
(+ (* 1/2 (expt Omega 2) m (expt x_r 2))
   (* 1/2 (expt Omega 2) m (expt y_r 2))
   (* -1 Omega m xdot_r y_r)
   (* Omega m ydot_r x_r)
   (* 1/2 m (expt xdot_r 2))
   (* 1/2 m (expt ydot_r 2)))
```

Although the transformation of coordinates is time dependent the resulting Lagrangian is independent of time.

The Lagrange equations for the free particle in the rotating coordinate system have force terms involving the angular velocity Ω:

```
(((Lagrange-equations (L-rotating-rectangular 'm 'Omega))
  (up (literal-function 'x_r) (literal-function 'y_r)))
 't)
(down
 (+ (* -1 (expt Omega 2) m (x_r t))
    (* -2 Omega m ((D y_r) t))
    (* m (((expt D 2) x_r) t)))
 (+ (* -1 (expt Omega 2) m (y_r t))
    (* 2 Omega m ((D x_r) t))
    (* m (((expt D 2) y_r) t))))
```

The terms that are proportional to Ω^2 are called *centrifugal force* terms, and the ones that are proportional to Ω are called *Coriolis* force terms. Note that the centrifugal force terms are radial, pointing away from the center of rotation. These additional force terms are derived from the corresponding terms in the Lagrangian. The terms in the Lagrangian that are proportional to Ω^2 can be thought of as the negation of a *centrifugal potential energy.*

Because this is a free particle the velocity in the original unrotated coordinates is constant. In rotating coordinates, the Coriolis terms describe an acceleration that is perpendicular to the velocity, causing the trajectory to curve.

1.6.2   Systems with Rigid Constraints

We have found that $L = T - V$ is a suitable Lagrangian for a system of point particles subject to forces derived from a potential. Extended bodies can sometimes be conveniently idealized as a system of point particles connected by rigid constraints. We will find that $L = T - V$, expressed in irredundant coordinates, is also a suitable Lagrangian for modeling systems of point particles with rigid constraints. We will first illustrate the method and then provide a justification.

Lagrangians for rigidly constrained systems
The system is presumed to be made of N point masses, indexed by α, in ordinary three-dimensional space. The first step is to choose a convenient set of irredundant generalized coordinates q and redescribe the system in terms of these. In terms of the generalized coordinates the rectangular coordinates of particle α are

$$\mathbf{x}_\alpha = f_\alpha(t, q). \tag{1.79}$$

For irredundant coordinates q all the coordinate constraints are built into the functions f_α. We deduce the relationship of the generalized velocities v to the velocities of the constituent particles $\mathbf{v}_\alpha$ by inserting path functions into equation (1.79), differentiating, and abstracting to arbitrary velocities (see section 1.6.1). We find

$$\mathbf{v}_\alpha = \partial_0 f_\alpha(t, q) + \partial_1 f_\alpha(t, q) v. \tag{1.80}$$

We use equations (1.79) and (1.80) to express the kinetic energy in terms of the generalized coordinates and velocities. Let $\tilde{T}$ be the kinetic energy as a function of the rectangular coordinates and velocities:

$$\widetilde{T}(t; \mathbf{x}_0, \ldots, \mathbf{x}_{N-1}; \mathbf{v}_0, \ldots, \mathbf{v}_{N-1}) = \sum_\alpha \frac{1}{2} m_\alpha \mathbf{v}_\alpha^2, \tag{1.81}$$

where $\mathbf{v}_\alpha^2$ is the squared magnitude of $\mathbf{v}_\alpha$. As a function of the generalized coordinate tuple q and the generalized velocity tuple v, the kinetic energy is

$$T(t, q, v) = \widetilde{T}(t, f(t, q), \partial_0 f(t, q) + \partial_1 f(t, q)v)$$
$$= \sum_\alpha \frac{1}{2} m_\alpha (\partial_0 f_\alpha(t, q) + \partial_1 f_\alpha(t, q)v)^2. \tag{1.82}$$

Similarly, we use equation (1.79) to reexpress the potential energy in terms of the generalized coordinates. Let $\widetilde{V}(t, x)$ be the potential energy at time t in the configuration specified by the tuple of rectangular coordinates x. Expressed in generalized coordinates the potential energy is

$$V(t, q, v) = \widetilde{V}(t, f(t, q)). \tag{1.83}$$

We take the Lagrangian to be the difference of the kinetic energy and the potential energy: $L = T - V$.

A pendulum driven at the pivot

Consider a pendulum (see figure 1.2) of length l and mass m, modeled as a point mass, supported by a pivot that is driven in the vertical direction by a given function of time y_s.

The dimension of the configuration space for this system is one; we choose θ, shown in figure 1.2, as the generalized coordinate.

The position of the bob is given, in rectangular coordinates, by

$$x = l \sin \theta \quad \text{and} \quad y = y_s(t) - l \cos \theta. \tag{1.84}$$

The velocities are

$$v_x = l \dot{\theta} \cos \theta \quad \text{and} \quad v_y = D y_s(t) + l \dot{\theta} \sin \theta, \tag{1.85}$$

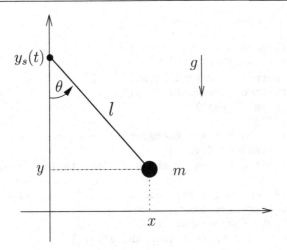

Figure 1.2 The pendulum is driven by vertical motion of the pivot. The pivot slides on the y-axis. Although the bob is drawn as a blob it is modeled as a point mass. The bob is acted on by the uniform acceleration g of gravity in the negative $\hat{y}$ direction.

obtained by differentiating along a path and abstracting to velocities at the moment.

The kinetic energy is $\widetilde{T}(t; x, y; v_x, v_y) = \frac{1}{2}m(v_x^2 + v_y^2)$. Expressed in generalized coordinates the kinetic energy is

$$T(t, \theta, \dot{\theta}) = \frac{1}{2}m\left(l^2\dot{\theta}^2 + (Dy_s(t))^2 + 2lDy_s(t)\dot{\theta}\sin\theta\right). \qquad (1.86)$$

The potential energy is $\widetilde{V}(t; x, y) = mgy$. Expressed in generalized coordinates the potential energy is

$$V(t, \theta, \dot{\theta}) = gm\left(y_s(t) - l\cos\theta\right). \qquad (1.87)$$

A Lagrangian is $L = T - V$:

$$\begin{aligned} L(t, \theta, \dot{\theta}) =& \frac{1}{2}m\left(l^2\dot{\theta}^2 + (Dy_s(t))^2 + 2lDy_s(t)\dot{\theta}\sin\theta\right) \\ & - gm\left(y_s(t) - l\cos\theta\right). \end{aligned} \qquad (1.88)$$

The Lagrangian is expressed as

```
(define ((T-pend m l g ys) local)
  (let ((t (time local))
        (theta (coordinate local))
        (thetadot (velocity local)))
    (let ((vys (D ys)))
      (* 1/2 m
         (+ (square (* l thetadot))
            (square (vys t))
            (* 2 l (vys t) thetadot (sin theta)))))))

(define ((V-pend m l g ys) local)
  (let ((t (time local))
        (theta (coordinate local)))
    (* m g (- (ys t) (* l (cos theta))))))

(define L-pend (- T-pend V-pend))
```

Lagrange's equation for this system is

```
(show-expression
 (((Lagrange-equations
    (L-pend 'm 'l 'g (literal-function 'y_s)))
   (literal-function 'theta))
  't))
```

$$D^2\theta(t)\,l^2 m + D^2 y_s(t)\sin(\theta(t))\,lm + \sin(\theta(t))\,glm$$

Exercise 1.17: Bead on a helical wire

A bead of mass m is constrained to move on a frictionless helical wire. The helix is oriented so that its axis is horizontal. The diameter of the helix is d and its pitch (turns per unit length) is h. The system is in a uniform gravitational field with vertical acceleration g. Formulate a Lagrangian that describes the system and find the Lagrange equations of motion.

Exercise 1.18: Bead on a triaxial surface

A bead of mass m moves without friction on a triaxial ellipsoidal surface. In rectangular coordinates the surface satisfies

$$\frac{x^2}{a^2} + \frac{y^2}{b^2} + \frac{z^2}{c^2} = 1 \tag{1.89}$$

for some constants a, b, and c. Identify suitable generalized coordinates, formulate a Lagrangian, and find Lagrange's equations.

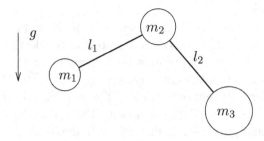

Figure 1.3   A two-bar linkage is modeled by three point masses connected by rigid massless struts. This linkage is subject to a uniform vertical gravitational acceleration.

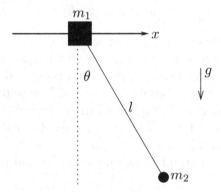

Figure 1.4   This pendulum is pivoted on a point particle of mass m_1 that is allowed to slide on a horizontal rail. The pendulum bob is a point particle of mass m_2 that is acted on by the vertical force of gravity.

Exercise 1.19: Two-bar linkage

The two-bar linkage shown in figure 1.3 is constrained to move in the plane. It is composed of three small massive bodies interconnected by two massless rigid rods in a uniform gravitational field with vertical acceleration g. The rods are pinned to the central body by a hinge that allows the linkage to fold. The system is arranged so that the hinge is completely free: the members can go through all configurations without collision. Formulate a Lagrangian that describes the system and find the Lagrange equations of motion. Use the computer to do this, because the equations are rather big.

Exercise 1.20: Sliding pendulum

Consider a pendulum of length l attached to a support that is free to move horizontally, as shown in figure 1.4. Let the mass of the support be m_1 and the mass of the pendulum bob be m_2. Formulate a Lagrangian and derive Lagrange's equations for this system.

Why it works

In this section we show that $L = T - V$ is in fact a suitable Lagrangian for rigidly constrained systems. We do this by requiring that the Lagrange equations be equivalent to the Newtonian vectorial dynamics with vector constraint forces.[65]

We consider a system of particles. The particle with index α has mass m_α and position $\vec{x}_\alpha(t)$ at time t. There may be a very large number of these particles, or just a few. Some of the positions may also be specified functions of time, such as the position of the pivot of a driven pendulum. There are rigid position constraints among some of the particles. We assume that all of these constraints are of the form

$$(\vec{x}_\alpha(t) - \vec{x}_\beta(t)) \cdot (\vec{x}_\alpha(t) - \vec{x}_\beta(t)) = l_{\alpha\beta}^2 \, ; \tag{1.90}$$

that is, the distance between particles α and β is $l_{\alpha\beta}$.

The Newtonian equation of motion for particle α says that the mass times the acceleration of particle α is equal to the sum of the potential forces and the constraint forces. The potential forces are derived as the negative gradient of the potential energy, and may depend on the positions of the other particles and the time. The constraint forces $\vec{F}_{\alpha\beta}$ are the vector constraint forces associated with the rigid constraint between particle α and particle β. So

$$D(m_\alpha \, D\vec{x}_\alpha)(t)$$
$$= -\vec{\nabla}_{\vec{x}_\alpha} \mathcal{V}(t, \vec{x}_0(t), \dots, \vec{x}_{N-1}(t)) + \sum_{\{\beta|\beta\leftrightarrow\alpha\}} \vec{F}_{\alpha\beta}(t), \tag{1.91}$$

where in the summation β ranges over only those particle indices for which there are rigid constraints with the particle indexed by α; we use the notation $\beta \leftrightarrow \alpha$ for the relation that there is a rigid constraint between the indicated particles.

---

[65] We will simply accept the Newtonian procedure for systems with rigid constraints and find Lagrangians that are equivalent. Of course, actual bodies are never truly rigid, so we may wonder what detailed approximations have to be made to treat them as if they were truly rigid. For instance, a more satisfying approach would be to replace the rigid distance constraints by very stiff springs. We could then immediately write the Lagrangian as $L = T - V$, and we should be able to *derive* the Newtonian procedure for systems with rigid constraints as an approximation. However, this is too complicated to do at this stage, so we accept the Newtonian idealization.

The force of constraint is directed along the line between the particles, so we may write

$$\vec{F}_{\alpha\beta}(t) = F_{\alpha\beta}(t)\frac{\vec{x}_\beta(t) - \vec{x}_\alpha(t)}{l_{\alpha\beta}} \tag{1.92}$$

where $F_{\alpha\beta}(t)$ is the scalar magnitude of the tension in the constraint at time t. Note that $\vec{F}_{\alpha\beta} = -\vec{F}_{\beta\alpha}$. In general, the scalar constraint forces change as the system evolves.

Formally, we can reproduce Newton's equations with the Lagrangian[66]

$$L(t; x, F; \dot{x}, \dot{F}) = \sum_\alpha \tfrac{1}{2}m_\alpha \dot{\mathbf{x}}_\alpha^2 - V(t, x)$$

$$- \sum_{\{\alpha,\beta|\alpha<\beta,\alpha\leftrightarrow\beta\}} \frac{F_{\alpha\beta}}{2l_{\alpha\beta}}\left[(\mathbf{x}_\beta - \mathbf{x}_\alpha)^2 - l_{\alpha\beta}^2\right] \tag{1.93}$$

where the constraint forces are being treated as additional generalized coordinates. Here x is a structure composed of all the rectangular components $\mathbf{x}_\alpha$ of all the $\vec{x}_\alpha$, $\dot{x}$ is a structure composed of all the rectangular components $\dot{\mathbf{x}}_\alpha$ of all the velocity vectors $\vec{v}_\alpha$, and F is a structure composed of all the $F_{\alpha\beta}$. The velocity of F does not appear in the Lagrangian, and F itself appears only linearly. So the Lagrange equations associated with F are

$$(\mathbf{x}_\beta(t) - \mathbf{x}_\alpha(t))^2 - l_{\alpha\beta}^2 = 0 \tag{1.94}$$

but this is just a restatement of the constraints. The Lagrange equations for the coordinates of the particles are Newton's equations (1.91)

$$D(mD\mathbf{x}_\alpha)(t) = -\partial_{1,\alpha}V(t, x(t))$$

$$+ \sum_{\{\beta|\alpha\leftrightarrow\beta\}} F_{\alpha\beta}(t)\frac{\mathbf{x}_\beta(t) - \mathbf{x}_\alpha(t)}{l_{\alpha\beta}}. \tag{1.95}$$

---

[66]This Lagrangian is purely formal and does not represent a model of the constraint forces. In particular, note that the constraint terms do not add up to a potential energy with a minimum when the constraints are exactly satisfied. Rather, the constraint terms in the Lagrangian are zero when the constraint is satisfied, and can be either positive or negative depending on whether the distance between the particles is larger or smaller than the constraint distance.

Now that we have a suitable Lagrangian, we can use the fact that Lagrangians can be reexpressed in any generalized coordinates to find a simpler Lagrangian. The strategy is to choose a new set of coordinates for which many of the coordinates are constants and the remaining coordinates are irredundant.

Let q be a tuple of generalized coordinates that specify the degrees of freedom of the system without redundancy. Let c be a tuple of other generalized coordinates that specify the distances between particles for which constraints are specified. The c coordinates will have constant values. The combination of q and c replaces the redundant rectangular coordinates x.[67] In addition, we still have the F coordinates, which are the scalar constraint forces. Our new coordinates are the components of q, c, and F.

There exist functions f_α that give the rectangular coordinates of the constituent particles in terms of q and c:

$$\mathbf{x}_\alpha = f_\alpha(t, q, c). \tag{1.96}$$

To reexpress the Lagrangian in terms of q, c, and F, we need to find $\mathbf{v}_\alpha$ in terms of the generalized velocities $\dot{q}$ and $\dot{c}$: we do this by differentiating f_α along a path and abstracting to arbitrary velocities (see section 1.6.1):

$$\mathbf{v}_\alpha = \partial_0 f_\alpha(t, q, c) + \partial_1 f_\alpha(t, q, c)\, \dot{q} + \partial_2 f_\alpha(t, q, c)\, \dot{c}. \tag{1.97}$$

Substituting these into Lagrangian (1.93), and using

$$c_{\alpha\beta}^2 = (\mathbf{x}_\beta - \mathbf{x}_\alpha)^2, \tag{1.98}$$

we find

$$
\begin{aligned}
L'(t; q, c, F; \dot{q}, \dot{c}, \dot{F}) \\
= \sum_\alpha \tfrac{1}{2} m_\alpha \left(\partial_0 f_\alpha(t, q, c) + \partial_1 f_\alpha(t, q, c)\, \dot{q} + \partial_2 f_\alpha(t, q, c)\, \dot{c} \right)^2 \\
- V(t, f(t, q, c)) - \sum_{\{\alpha, \beta \mid \alpha < \beta, \alpha \leftrightarrow \beta\}} \frac{F_{\alpha\beta}}{2 l_{\alpha\beta}} \left[c_{\alpha\beta}^2 - l_{\alpha\beta}^2 \right]. \quad (1.99)
\end{aligned}
$$

---

[67]Typically the number of components of x is equal to the sum of the number of components of q and c; adding a strut removes a degree of freedom and adds a distance constraint. However, there are singular cases in which the addition of single strut can remove more than a single degree of freedom. We do not consider the singular cases here.

The Lagrange equations are derived by the usual procedure. Rather than write out all the gory details, let's think about how it will go.

The Lagrange equations associated with F just restate the constraints:

$$0 = c_{\alpha\beta}^2(t) - l_{\alpha\beta}^2 \tag{1.100}$$

and consequently we know that along a solution path, $c(t) = l$ and $Dc(t) = D^2c(t) = 0$. We can use this result to simplify the Lagrange equations associated with q and c.

The Lagrange equations associated with q are the same as if they were derived from the Lagrangian[68]

$$
\begin{aligned}
L''(t, q, \dot{q}) = & \sum_\alpha \tfrac{1}{2} m_\alpha \left(\partial_0 f_\alpha(t, q, l) + \partial_1 f_\alpha(t, q, l)\, \dot{q} \right)^2 \\
& - V(t, f(t, q, l)),
\end{aligned} \tag{1.101}
$$

but this is exactly $T - V$ where T and V are computed from the generalized coordinates q, with fixed constraints. Notice that the constraint forces do not appear in the Lagrange equations for q because in the Lagrange equations they are multiplied by a term that is identically zero on the solution paths. So the Lagrange equations for $T - V$ with irredundant generalized coordinates q and fixed constraints are equivalent to Newton's equations with vector constraint forces.

The Lagrange equations for c can be used to find the constraint forces. The Lagrange equations are a big mess so we will not show them explicitly, but in general they are equations in D^2c, Dc, and c that will depend upon q, Dq, and F. The dependence on F is linear, so we can solve for F in terms of the solution path q and Dq, with $c = l$ and $Dc = D^2c = 0$.

If we are not interested in the constraint forces, we can abandon the full Lagrangian (1.99) in favor of Lagrangian (1.101), which is

---

[68]Consider a function g of, say, three arguments, and let g_0 be a function of two arguments satisfying $g_0(x, y) = g(x, y, 0)$. Then $(\partial_0 g_0)(x, y) = (\partial_0 g)(x, y, 0)$. The substitution of a value in an argument commutes with the taking of the partial derivative with respect to a different argument. In deriving the Lagrange equations for q we can set $c = l$ and $\dot{c} = 0$ in the Lagrangian, but we cannot do this in deriving the Lagrange equations associated with c, because we have to take derivatives with respect to those arguments.

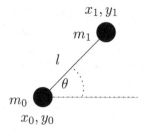

Figure 1.5 A rigid rod of length l constrains two massive particles in a plane.

equivalent as far as the evolution of the generalized coordinates q is concerned.

The same derivation goes through even if the lengths $l_{\alpha\beta}$ specified in the interparticle distance constraints are a function of time. It can also be generalized to allow distance constraints to time-dependent positions, by making some of the positions of particles $\vec{x}_\beta$ be specified functions of time.

Exercise 1.21: A dumbbell

In this exercise we will recapitulate the derivation of the Lagrangian for constrained systems for a particular simple system.

Consider two massive particles in the plane constrained by a massless rigid rod to remain a distance l apart, as in figure 1.5. There are apparently four degrees of freedom for two massive particles in the plane, but the rigid rod reduces this number to three.

We can uniquely specify the configuration with the redundant coordinates of the particles, say $x_0(t), y_0(t)$ and $x_1(t), y_1(t)$. The constraint $(x_1(t) - x_0(t))^2 + (y_1(t) - y_0(t))^2 = l^2$ eliminates one degree of freedom.

a. Write Newton's equations for the balance of forces for the four rectangular coordinates of the two particles, given that the scalar tension in the rod is F.

b. Write the formal Lagrangian

$$L(t; x_0, y_0, x_1, y_1, F; \dot{x}_0, \dot{y}_0, \dot{x}_1, \dot{y}_1, \dot{F})$$

such that Lagrange's equations will yield the Newton's equations you derived in part **a**.

c. Make a change of coordinates to a coordinate system with center of mass coordinates x_{CM}, y_{CM}, angle θ, distance between the particles c, and tension force F. Write the Lagrangian in these coordinates, and write the Lagrange equations.

d. You may deduce from one of these equations that $c(t) = l$. From this fact we get that $Dc = 0$ and $D^2c = 0$. Substitute these into the Lagrange equations you just computed to get the equation of motion for x_{CM}, y_{CM}, θ.

e. Make a Lagrangian $(= T - V)$ for the system described with the irredundant generalized coordinates x_{CM}, y_{CM}, θ and compute the Lagrange equations from this Lagrangian. They should be the same equations as you derived for the same coordinates in part **d**.

Exercise 1.22: Driven pendulum

Show that the Lagrangian (1.93) can be used to describe the driven pendulum (section 1.6.2), where the position of the pivot is a specified function of time: Derive the equations of motion using the Newtonian constraint force prescription, and show that they are the same as the Lagrange equations. Be sure to examine the equations for the constraint forces as well as the position of the pendulum bob.

Exercise 1.23: Fill in the details

Show that the Lagrange equations for Lagrangian (1.101) are the same as the Lagrange equations for Lagrangian (1.99) with the substitution $c(t) = l$, $Dc(t) = D^2c(t) = 0$.

Exercise 1.24: Constraint forces

Find the tension in an undriven planar pendulum.

1.6.3   Constraints as Coordinate Transformations

The derivation of a Lagrangian for a constrained system involves steps that are analogous to those in the derivation of a coordinate transformation.

We can make a Lagrangian for the unconstrained system of particles in rectangular coordinates. In general there will be more coordinates than real degrees of freedom; the constraints will eliminate the redundancy. We then choose a convenient set of irredundant generalized coordinates that incorporate the constraints to describe our system. We express the redundant rectangular coordinates and velocities in terms of the irredundant generalized coordinates and generalized velocities, and we use these transformations to reexpress the Lagrangian in the generalized coordinates.

To carry out a coordinate transformation we specify how the configuration of a system expressed in one set of generalized coordinates can be reexpressed in terms of another set of generalized coordinates. We then determine the transformation of generalized velocities implied by the transformation of generalized coordinates. A Lagrangian that is expressed in terms of one of the sets of generalized coordinates can then be reexpressed in terms of the other set of generalized coordinates.

These are really two applications of the same process, so we can make Lagrangians for constrained systems by composing a Lagrangian for unconstrained particles with a coordinate transformation that incorporates the constraint. Our deduction that $L = T - V$ is a suitable Lagrangian for a constrained systems was in fact based on a coordinate transformation from a set of coordinates subject to constraints to a set of irredundant coordinates plus constraint coordinates that are constant.

Let $\mathbf{x}_\alpha$ be the tuple of rectangular components of the constituent particle with index α, and let $\mathbf{v}_\alpha$ be its velocity. The Lagrangian

$$L_f(t; \mathbf{x}_0, \ldots, \mathbf{x}_{N-1}; \mathbf{v}_0, \ldots, \mathbf{v}_{N-1})$$
$$= \sum_\alpha \tfrac{1}{2} m_\alpha \mathbf{v}_\alpha^2 - V(t; \mathbf{x}_0, \ldots, \mathbf{x}_{N-1}; \mathbf{v}_0, \ldots, \mathbf{v}_{N-1}) \qquad (1.102)$$

is the difference of kinetic and potential energies of the constituent particles. This is a suitable Lagrangian for a set of unconstrained free particles with potential energy V.

Let q be a tuple of irredundant generalized coordinates and v be the corresponding generalized velocity tuple. The coordinates q are related to $\mathbf{x}_\alpha$, the coordinates of the constituent particles, by $\mathbf{x}_\alpha = f_\alpha(t, q)$, as before. The constraints among the constituent particles are taken into account in the definition of the f_α. Here we view this as a coordinate transformation. What is unusual about this as a coordinate transformation is that the dimension of x is not the same as the dimension of q. From this coordinate transformation we can find the local-tuple transformation function (see section 1.6.1)

$$(t; \mathbf{x}_0, \ldots, \mathbf{x}_{N-1}; \mathbf{v}_0, \ldots, \mathbf{v}_{N-1}) = C(t, q, v). \qquad (1.103)$$

A Lagrangian for the constrained system can be obtained from the Lagrangian for the unconstrained system by composing it with the local-tuple transformation function from constrained coordinates to unconstrained coordinates:

$$L = L_f \circ C. \tag{1.104}$$

The constraints enter only in the transformation.

To illustrate this we will find a Lagrangian for the driven pendulum introduced in section 1.6.2. As we saw on page 40, the $T - V$ Lagrangian for a free particle of mass m in a vertical plane subject to a gravitational potential with acceleration g is

$$L_f(t; x, y; v_x, v_y) = \tfrac{1}{2}m(v_x^2 + v_y^2) - mgy, \tag{1.105}$$

where y measures the height of the point mass. A program that computes this Lagrangian is

```
(define ((L-uniform-acceleration m g) local)
  (let ((q (coordinate local))
        (v (velocity local)))
    (let ((y (ref q 1)))
      (- (* 1/2 m (square v)) (* m g y)))))
```

The coordinate transformation from generalized coordinate θ to rectangular coordinates is $x = l\sin\theta$, $y = y_s(t) - l\cos\theta$, where l is the length of the pendulum and y_s gives the height of the support as a function of time. It is interesting that the drive enters only through the specification of the constraints. A program implementing this coordinate transformation is

```
(define ((dp-coordinates l y_s) local)
  (let ((t (time local))
        (theta (coordinate local)))
    (let ((x (* l (sin theta)))
          (y (- (y_s t) (* l (cos theta)))))
      (up x y))))
```

Using F->C we can deduce the local-tuple transformation and define the Lagrangian for the driven pendulum by composition:

```
(define (L-pend m l g y_s)
  (compose (L-uniform-acceleration m g)
           (F->C (dp-coordinates l y_s))))
```

The Lagrangian is

```
(show-expression
 ((L-pend 'm 'l 'g (literal-function 'y_s))
  (up 't 'theta 'thetadot)))
```

$$glm\cos\left(\theta\right)-gmy_s\left(t\right)+\frac{1}{2}l^2m\dot\theta^2+lm\dot\theta Dy_s\left(t\right)\sin\left(\theta\right)+\frac{1}{2}m\left(Dy_s\left(t\right)\right)^2$$

This is the same as the Lagrangian of equation (1.88) on page 51.

We have found a very interesting decomposition of the Lagrangian for constrained systems. One part consists of the difference of the kinetic and potential energy of the constituents. The other part describes the constraints that are specific to the configuration of a particular system.

Exercise 1.25: Foucault pendulum Lagrangian

A Foucault pendulum is a long-period pendulum of length l and mass m that is suspended at a height l above the surface of the Earth (radius R) at colatitude ϕ. If the pendulum is released, at rest, with non-zero displacement from the local vertical, it will oscillate in an apparent plane. However, the apparent plane of oscillation precesses as the Earth rotates. The Earth rotates with angular speed Ω.

One way to specify the position of the bob is to erect a Foucault pendulum at the North Pole and rotate it to a point on the surface of the Earth at the appropriate colatitude and a fixed longitude. Because the Earth is rotating this is a time-varying transformation. There are two parts of this transformation.

First, we relate the generalized coordinates θ and λ to the coordinates of the pendulum bob with the pendulum at the North pole. Let θ be the angle of the bob relative to the line through the center of the Earth and let λ be the precession angle. The rectangular coordinates of the bob for a pendulum at the North Pole are:

$x_0 = l\sin\theta\cos\lambda$

$y_0 = l\sin\theta\sin\lambda$

$z_0 = (R + l) - l\cos\theta.$

Next, we rotate the pendulum to its actual location at colatitude ϕ. We can choose the longitude to be zero, so the angular position of the support of the bob is rotated by Ωt. The transformation of coordinates is:[69]

$$(x, y, z) = R_z(\Omega t)R_y(\phi)(x_0, y_0, z_0).$$

---

[69] $R_z(\alpha)$ yields a function that rotates its argument about the $\hat{z}$ axis by the angle α, and R_y is similar.

This second transformation can be implemented with the following code:

```
((compose (Rz (* 'Omega 't)) (Ry 'phi))
 (up 'x_0 'y_0 'z_0))
(up
 (+ (* x_0 (cos phi) (cos (* Omega t)))
    (* z_0 (sin phi) (cos (* Omega t)))
    (* -1 y_0 (sin (* Omega t))))
 (+ (* x_0 (cos phi) (sin (* Omega t)))
    (* z_0 (sin phi) (sin (* Omega t)))
    (* y_0 (cos (* Omega t))))
 (+ (* -1 x_0 (sin phi)) (* z_0 (cos phi))))
```

Construct a coordinate transformation F from these parts that you can use with F->C to compose with the free Lagrangian for a particle in a gravitational potential to make a Lagrangian for the Foucault pendulum. The Newtonian potential energy is $-GMm/r$, where r is the distance of the bob from the center of the Earth, and M is the mass of the Earth.

1.6.4   The Lagrangian Is Not Unique

Lagrangians are not in a one-to-one relationship with physical systems—many Lagrangians can be used to describe the same physical system. In this section we will demonstrate this by showing that the addition to the Lagrangian of a "total time derivative" of a function of the coordinates and time does not change the paths of stationary action or the equations of motion deduced from the action principle.

Total time derivatives

Let's first explain what we mean by a "total time derivative." Let F be a function of time and coordinates. The function F on the path at time t is

$$(F \circ \Gamma[q])(t) = F(t, q(t)). \tag{1.106}$$

So, by the chain rule

$$D(F \circ \Gamma[q])(t) = \partial_0 F(t, q(t)) + \partial_1 F(t, q(t)) Dq(t). \tag{1.107}$$

More formally, the time derivative of F along a path q is

$$D(F \circ \Gamma[q]) = (DF \circ \Gamma[q]) \, D\Gamma[q]. \tag{1.108}$$

Because F depends only on time and coordinates, we have

$$DF \circ \Gamma[q] = [\partial_0 F \circ \Gamma[q], \partial_1 F \circ \Gamma[q]] . \tag{1.109}$$

So we need only the first two components of $D\Gamma[q]$,

$$(D\Gamma[q])(t) = \left(1, Dq(t), D^2 q(t), \ldots \right),\qquad(1.110)$$

to form the product

$$\begin{aligned}D(F \circ \Gamma[q]) &= \partial_0 F \circ \Gamma[q] + (\partial_1 F \circ \Gamma[q]) Dq \\ &= (\partial_0 F + (\partial_1 F)\dot{Q}) \circ \Gamma[q],\end{aligned}\qquad(1.111)$$

where $\dot{Q} = I_2$ is a selector function:[70] $c = \dot{Q}(a, b, c)$, so $Dq = \dot{Q} \circ \Gamma[q]$.

The function

$$D_t F = \partial_0 F + (\partial_1 F)\dot{Q}\qquad(1.112)$$

is called the *total time derivative* of F; it is a function of three arguments: the time, the generalized coordinates, and the generalized velocities.

In general, the total time derivative of a local-tuple function F is that function $D_t F$ that when composed with a local-tuple path is the time derivative of the composition of the function F with the same local-tuple path:

$$D_t F \circ \Gamma[q] = D(F \circ \Gamma[q]).\qquad(1.113)$$

The total time derivative $D_t F$ is explicitly given by

$$\begin{aligned}D_t F(t, q, v, a, \ldots) = {}& \partial_0 F(t, q, v, a, \ldots) \\ &+ \partial_1 F(t, q, v, a, \ldots)\, v \\ &+ \partial_2 F(t, q, v, a, \ldots)\, a + \cdots,\end{aligned}\qquad(1.114)$$

where we take as many terms as needed to exhaust the arguments of F.

Exercise 1.26: Properties of D_t

The total time derivative $D_t F$ is not the derivative of the function F. Nevertheless, the total time derivative shares many properties with the derivative. Demonstrate that D_t has the following properties for local-tuple functions F and G, number c, and a function H with domain containing the range of G.

---

[70]Components of a tuple structure, such as the value of $\Gamma[q](t)$, can be selected with selector functions: I_i gets the element with index i from the tuple.

a. $D_t(F + G) = D_t F + D_t G$

b. $D_t(cF) = cD_t F$

c. $D_t(FG) = F D_t G + (D_t F)G$

d. $D_t(H \circ G) = (DH \circ G)D_t G$

Adding total time derivatives to Lagrangians

Consider two Lagrangians L and L' that differ by the addition of a total time derivative of a function F that depends only on the time and the coordinates

$$L' = L + D_t F. \tag{1.115}$$

The corresponding action integral is

$$
\begin{aligned}
S'[q](t_1, t_2) &= \int_{t_1}^{t_2} L' \circ \Gamma[q] \\
&= \int_{t_1}^{t_2} (L + D_t F) \circ \Gamma[q] \\
&= \int_{t_1}^{t_2} L \circ \Gamma[q] + \int_{t_1}^{t_2} D(F \circ \Gamma[q]) \\
&= S[q](t_1, t_2) + (F \circ \Gamma[q])|_{t_1}^{t_2}. \tag{1.116}
\end{aligned}
$$

The variational principle states that the action integral along a realizable trajectory is stationary with respect to variations of the trajectory that leave the configuration at the endpoints fixed. The action integrals $S[q](t_1, t_2)$ and $S'[q](t_1, t_2)$ differ by a term

$$(F \circ \Gamma[q])|_{t_1}^{t_2} = F(t_2, q(t_2)) - F(t_1, q(t_1)) \tag{1.117}$$

that depends only on the coordinates and time at the endpoints and these are not allowed to vary. Thus, if $S[q](t_1, t_2)$ is stationary for a path, then $S'[q](t_1, t_2)$ will also be stationary. So either Lagrangian can be used to distinguish the realizable paths.

The addition of a total time derivative to a Lagrangian does not affect whether the action is stationary for a given path. So if we have two Lagrangians that differ by a total time derivative, the corresponding Lagrange equations are equivalent in that the same paths satisfy each. Moreover, the additional terms introduced into the action by the total time derivative appear only in the endpoint condition and thus do not affect the Lagrange equations derived from the variation of the action, so the Lagrange equations are the same. The Lagrange equations are not changed by the addition of a total time derivative to a Lagrangian.

Exercise 1.27: Lagrange equations for total time derivatives

Let $F(t, q)$ be a function of t and q only, with total time derivative

$$D_t F = \partial_0 F + (\partial_1 F)\dot{Q}. \tag{1.118}$$

Show explicitly that the Lagrange equations for $D_t F$ are identically zero, and thus that the addition of $D_t F$ to a Lagrangian does not affect the Lagrange equations.

The driven pendulum provides a nice illustration of adding total time derivatives to Lagrangians. The equation of motion for the driven pendulum (see section 1.6.2),

$$ml^2 D^2\theta(t) + ml(g + D^2 y_s(t)) \sin \theta(t) = 0, \tag{1.119}$$

has an interesting and suggestive interpretation: it is the same as the equation of motion of an undriven pendulum, except that the acceleration of gravity g is augmented by the acceleration of the pivot $D^2 y_s$. This intuitive interpretation was not apparent in the Lagrangian derived as the difference of the kinetic and potential energies in section 1.6.2. However, we can write an alternate Lagrangian with the same equation of motion that is as easy to interpret as the equation of motion:

$$L'(t, \theta, \dot{\theta}) = \tfrac{1}{2}ml^2\dot{\theta}^2 + ml(g + D^2 y_s(t)) \cos \theta. \tag{1.120}$$

With this Lagrangian it is apparent that the effect of the accelerating pivot is to modify the acceleration of gravity. Note, however, that it is not the difference of the kinetic and potential energies. Let's compare the two Lagrangians for the driven pendulum. The difference $\Delta L = L - L'$ is

$$\begin{aligned}
\Delta L(t, \theta, \dot{\theta}) = {}& \tfrac{1}{2}m(Dy_s(t))^2 + mlDy_s(t)\dot{\theta} \sin \theta \\
& - gmy_s(t) - mlD^2 y_s(t) \cos \theta.
\end{aligned} \tag{1.121}$$

The two terms in ΔL that depend on neither θ nor $\dot{\theta}$ do not affect the equations of motion. The remaining two terms are the total time derivative of the function $F(t, \theta) = -mlDy_s(t) \cos \theta$, which does not depend on $\dot{\theta}$. The addition of such terms to a Lagrangian does not affect the equations of motion.

Properties of total time derivatives

If the local-tuple function G, with arguments (t, q, v), is the total time derivative of a function F, with arguments (t, q), then G must have certain properties.

From equation (1.112), we see that G must be linear in the generalized velocities

$$G(t, q, v) = G_0(t, q, v) + G_1(t, q, v) \, v \qquad (1.122)$$

where neither G_1 nor G_0 depends on the generalized velocities: $\partial_2 G_1 = \partial_2 G_0 = 0$.

If G is the total time derivative of F then $G_1 = \partial_1 F$ and $G_0 = \partial_0 F$, so

$$\partial_0 G_1 = \partial_0 \partial_1 F$$
$$\partial_1 G_0 = \partial_1 \partial_0 F. \qquad (1.123)$$

The partial derivative with respect to the time argument does not have structure, so $\partial_0 \partial_1 F = \partial_1 \partial_0 F$. So if G is the total time derivative of F then

$$\partial_0 G_1 = \partial_1 G_0. \qquad (1.124)$$

Furthermore, $G_1 = \partial_1 F$, so

$$\partial_1 G_1 = \partial_1 \partial_1 F. \qquad (1.125)$$

If there is more than one degree of freedom these partials are actually structures of partial derivatives with respect to each coordinate. The partial derivatives with respect to two different coordinates must be the same independent of the order of the differentiation. So $\partial_1 G_1$ must be symmetric.

Note that we have not shown that these conditions are sufficient for determining that a function is a total time derivative, only that they are necessary.

Exercise 1.28: Identifying total time derivatives

For each of the following functions, either show that it is not a total time derivative or produce a function from which it can be derived.

a. $G(t, x, v_x) = mv_x$

b. $G(t, x, v_x) = mv_x \cos t$

c. $G(t, x, v_x) = v_x \cos t - x \sin t$

d. $G(t, x, v_x) = v_x \cos t + x \sin t$

e. $G(t; x, y; v_x, v_y) = 2(xv_x + yv_y) \cos t - (x^2 + y^2) \sin t$

f. $G(t; x, y; v_x, v_y) = 2(xv_x + yv_y) \cos t - (x^2 + y^2) \sin t + y^3 v_x + xv_y$

Exercise 1.29: Galilean invariance of kinetic energy

We have taken the kinetic energy of a set of particles indexed by α to be $\sum_\alpha \frac{1}{2} m_\alpha v_\alpha^2$. This form is Galilean invariant.

a. Start with a Lagrangian for free particles, which is only the sum of their kinetic energies:

$$L(t, x, v) = \sum_\alpha \tfrac{1}{2} m_\alpha v_\alpha^2. \tag{1.126}$$

Carry out a coordinate transformation from old to new coordinates that consists of a shift and a uniform translation

$$x_\alpha = x'_\alpha + \Delta x + \Delta vt. \tag{1.127}$$

Derive the Lagrangian in new coordinates.

b. The new Lagrangian can be put in the form $\sum_\alpha \frac{1}{2} m_\alpha (v'_\alpha)^2$ plus some additional terms. Show that the additional terms are a total time derivative.

Thus the kinetic energy can be taken to be $\sum_\alpha \frac{1}{2} m_\alpha v_\alpha^2$ in any uniformly moving coordinate system.

1.7 Evolution of Dynamical State

Lagrange's equations are ordinary differential equations that the path must satisfy. They can be used to test if a proposed path is a realizable path of the system. However, we can also use them to develop a path, starting with initial conditions.

The *state* of a system is defined to be the information that must be specified for the subsequent evolution to be determined. Remember our juggler: he or she must throw the pin in a certain way for it to execute the desired motion. The juggler has control of the initial position and orientation of the pin, and the initial velocity and spin of the pin. Our experience with juggling and similar systems suggests that the initial configuration and the rate of change of the configuration are sufficient to determine the subsequent motion. Other systems may require higher derivatives of the configuration.

For Lagrangians that are written in terms of a set of generalized coordinates and velocities we have shown that Lagrange's equations are second-order ordinary differential equations. If the differential equations can be solved for the highest-order derivatives and if the differential equations satisfy the Lipschitz conditions,[71] then there is a unique solution to the initial-value problem: given values of the solution and the lower derivatives of the solution at a particular moment, there is a unique solution function. Given irredundant coordinates the Lagrange equations satisfy these conditions.[72] Thus a trajectory is determined by the generalized coordinates and the generalized velocities at any time. This is the information required to specify the dynamical state.

A complete local description of a path consists of the path and all of its derivatives at a moment. The complete local description of a path can be reconstructed from an initial segment of the local tuple, given a prescription for computing higher-order derivatives of the path in terms of lower-order derivatives. The state of the system is specified by that initial segment of the local tuple from which the rest of the complete local description can be deduced. The complete local description gives us the path near that moment. Actually, all we need is a rule for computing the next higher derivative; we can get all the rest from this. Assume that the state of a system is given by the tuple (t, q, v). If we are

---

[71] The Lipschitz condition is that the rate of change of the state is bounded by a constant in an open set around each state. See [25] for a good treatment of the Lipschitz condition.

[72] If the coordinates are redundant we cannot, in general, solve for the highest-order derivative. However, since we can transform to irredundant coordinates, since we can solve the initial-value problem in the irredundant coordinates, and since we can construct the redundant coordinates from the irredundant coordinates, we can in general solve the initial-value problem for redundant coordinates. The only hitch is that we cannot specify arbitrary initial conditions: the initial conditions must be consistent with the constraints.

given a prescription for computing the acceleration $a = A(t, q, v)$, then

$$D^2 q = A \circ \Gamma[q],   \tag{1.128}$$

and we have as a consequence

$$D^3 q = D(A \circ \Gamma[q]) = D_t A \circ \Gamma[q],   \tag{1.129}$$

and so on. So the higher-derivative components of the local tuple are given by functions $D_t A$, $D_t^2 A$, $\ldots$. Each of these functions depends on lower-derivative components of the local tuple. All we need to deduce the path from the state is a function that gives the next-higher derivative component of the local description from the state. We use the Lagrange equations to find this function.

First, we expand the Lagrange equations

$$\partial_1 L \circ \Gamma[q] = D(\partial_2 L \circ \Gamma[q])$$

so that the second derivative appears explicitly:

$$\partial_1 L \circ \Gamma[q]$$
$$= \partial_0 \partial_2 L \circ \Gamma[q] + (\partial_1 \partial_2 L \circ \Gamma[q]) Dq + (\partial_2 \partial_2 L \circ \Gamma[q]) D^2 q.$$

Solving this system for $D^2 q$, one obtains the generalized acceleration along a solution path q:

$$D^2 q =$$
$$[\partial_2 \partial_2 L \circ \Gamma[q]]^{-1} [\partial_1 L \circ \Gamma[q] - (\partial_1 \partial_2 L \circ \Gamma[q]) Dq - \partial_0 \partial_2 L \circ \Gamma[q]]$$

where $[\partial_2 \partial_2 L \circ \Gamma]$ is a structure that can be represented by a symmetric square matrix, so we can compute its inverse.[73] The function that gives the acceleration is

$$A = (\partial_2 \partial_2 L)^{-1} \left[\partial_1 L - \partial_0 \partial_2 L - (\partial_1 \partial_2 L) \dot{Q} \right],   \tag{1.130}$$

where $\dot{Q} = I_2$ is the velocity component selector.

---

[73] We may encounter singularities that make this matrix uninvertable, but in real systems these singularities are isolated and can be avoided by changing coordinates.

That initial segment of the local tuple that specifies the state is called the local state tuple, or, more simply, the state tuple.[74]

We can express the function that gives the acceleration as a function of the state tuple as the following procedure. It takes a procedure that computes the Lagrangian, and returns a procedure that takes a state tuple as its argument and returns the acceleration.[75]

```
(define (Lagrangian->acceleration L)
  (let ((P ((partial 2) L)) (F ((partial 1) L)))
    (solve-linear-left
      ((partial 2) P)
      (- F
         (+ ((partial 0) P)
            (* ((partial 1) P) velocity))))))
```

Once we have a way of computing the acceleration from the coordinates and the velocities, we can give a prescription for computing the derivative of the state as a function of the state. For the state $(t, q(t), Dq(t))$ at the moment t the derivative of the state is $(1, Dq(t), D^2q(t)) = (1, Dq(t), A(t, q(t), Dq(t)))$. The procedure Lagrangian->state-derivative takes a Lagrangian and returns a procedure that takes a state and returns the derivative of the state:

```
(define (Lagrangian->state-derivative L)
  (let ((acceleration (Lagrangian->acceleration L)))
    (lambda (state)
      (up 1
          (velocity state)
          (acceleration state)))))
```

We represent a state by an up tuple of the components of that initial segment of the local tuple that determine the state.

---

[74]For Lagrangians that depend on time, coordinates, and velocities the state is specified by time, coordinates, and velocities. However, if a Lagrangian depends on the first four components of the local tuple (time, coordinates, velocities, and accelerations) the state of the system will be specified by the first five components of the local tuple.

[75]The procedure solve-linear-left multiplies its second argument by the inverse of its first argument on the left. So, if $u = Mv$ then $v = M^{-1}u$; (solve-linear-left M u) produces v.

For example, the parametric state derivative for a harmonic oscillator is

```
(define (harmonic-state-derivative m k)
  (Lagrangian->state-derivative (L-harmonic m k)))

((harmonic-state-derivative 'm 'k)
 (up 't (up 'x 'y) (up 'v_x 'v_y)))
(up 1 (up v_x v_y) (up (/ (* -1 k x) m) (/ (* -1 k y) m)))
```

The Lagrange equations are a second-order system of differential equations that constrain realizable paths q. We can use the state derivative to express the Lagrange equations as a first-order system of differential equations that constrain realizable coordinate paths q and velocity paths v:

```
(define ((Lagrange-equations-first-order L) q v)
  (let ((state-path (qv->state-path q v)))
    (- (D state-path)
       (compose (Lagrangian->state-derivative L)
                state-path))))

(define ((qv->state-path q v) t)
  (up t (q t) (v t)))
```

For example, we can find the first-order form of the equations of motion of a two-dimensional harmonic oscillator:

```
(show-expression
 (((Lagrange-equations-first-order (L-harmonic 'm 'k))
   (up (literal-function 'x)
       (literal-function 'y))
   (up (literal-function 'v_x)
       (literal-function 'v_y)))
  't))
```

$$
\begin{pmatrix}
0 \\
\begin{pmatrix} Dx\left(t\right) - v_x\left(t\right) \\ Dy\left(t\right) - v_y\left(t\right) \end{pmatrix} \\
\begin{pmatrix} \dfrac{kx\left(t\right)}{m} + Dv_x\left(t\right) \\ \dfrac{ky\left(t\right)}{m} + Dv_y\left(t\right) \end{pmatrix}
\end{pmatrix}
$$

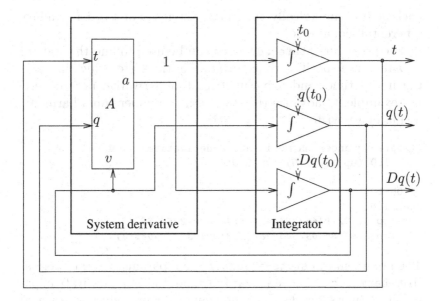

Figure 1.6   The input to the system derivative is the state. The function A gives the acceleration as a function of the components that determine the state. The output of the system derivative is the derivative of the state. The integrator takes the derivative of the state as its input and produces the integrated state, starting at the initial conditions. Notice how the second-order system is put into first-order form by the routing of the $Dq(t)$ components in the system derivative.

The zero in the first element of the structure of the Lagrange equations residuals is just the tautology that time advances uniformly: the time function is just the identity, so its derivative is one and the residual is zero. The equations in the second element constrain the velocity path to be the derivative of the coordinate path. The equations in the third element give the rate of change of the velocity in terms of the applied forces.

Numerical integration

A set of first-order ordinary differential equations that give the state derivative in terms of the state can be integrated to find the state path that emanates from a given initial state. Numerical integrators find approximate solutions of such differential equations by a process illustrated in figure 1.6. The state derivative produced by `Lagrangian->state-derivative` can be used by a package that numerically integrates systems of first-order ordinary differential equations.

The procedure `state-advancer` can be used to find the state of a system at a specified time, given an initial state, which includes the initial time, and a parametric state-derivative procedure.[76] For example, to advance the state of a two-dimensional harmonic oscillator we write[77]

```
((state-advancer harmonic-state-derivative 2.0 1.0)
 (up 1.0 (up 1.0 2.0) (up 3.0 4.0))
 10.0
 1.0e-12)
(up 11.0
    (up 3.7127916645844437 5.420620823651583)
    (up 1.6148030925459782 1.8189103724750855))
```

The arguments to `state-advancer` are a parametric state derivative, `harmonic-state-derivative`, and the state-derivative parameters (mass 2 and spring constant 1). A procedure is returned that takes an initial state, (up 1 (up 1 2) (up 3 4)); a time increment, 10; and a relative error tolerance, 1.0e-12. The output is an approximation to the state at the specified final time.

Consider the driven pendulum described in section 1.6.2 with a periodic drive. We choose $y_s(t) = A \cos \omega t$.

```
(define ((periodic-drive amplitude frequency phase) t)
  (* amplitude (cos (+ (* frequency t) phase))))

(define (L-periodically-driven-pendulum m l g A omega)
  (let ((ys (periodic-drive A omega 0)))
    (L-pend m l g ys)))
```

---

[76] The Scmutils system provides a variety of numerical integration routines that can be accessed through this interface. These include quality-controlled Runge–Kutta and Bulirsch–Stoer. The default integration method is Bulirsch–Stoer.

[77] The procedure `state-advancer` automatically compiles state-derivative procedures the first time they are encountered. The first time a new state derivative is used there is a delay while compilation occurs.

Lagrange's equation for this system is

```
(show-expression
 (((Lagrange-equations
    (L-periodically-driven-pendulum 'm 'l 'g 'A 'omega))
   (literal-function 'theta))
  't))
```

$$D^2\theta\,(t)\,l^2 m - \cos\,(\omega t)\sin\,(\theta\,(t))\,Alm\omega^2 + \sin\,(\theta\,(t))\,glm$$

The parametric state derivative for the periodically driven pendulum is

```
(define (pend-state-derivative m l g A omega)
  (Lagrangian->state-derivative
   (L-periodically-driven-pendulum m l g A omega)))

(show-expression
  ((pend-state-derivative 'm 'l 'g 'A 'omega)
   (up 't 'theta 'thetadot)))
```

$$\begin{pmatrix} 1 \\ \dot{\theta} \\ \dfrac{A\omega^2\cos\,(\omega t)\sin\,(\theta)}{l} - \dfrac{g\sin\,(\theta)}{l} \end{pmatrix}$$

To examine the evolution of the driven pendulum we need a mechanism that evolves a system for some interval while monitoring aspects of the system as it evolves. The procedure evolve provides this service, using state-advancer repeatedly to advance the state to the required moments. The procedure evolve takes a parametric state derivative and its parameters and returns a procedure that evolves the system from a specified initial state to a number of other times, monitoring some aspect of the state at those times. To generate a plot of the angle versus time we make a monitor procedure that generates the plot as the evolution proceeds:[78]

```
(define ((monitor-theta win) state)
  (let ((theta ((principal-value :pi) (coordinate state))))
    (plot-point win (time state) theta)))

(define plot-win (frame 0.0 100.0 :-pi :pi))

((evolve pend-state-derivative
          1.0                    ;m=1kg
          1.0                    ;l=1m
          9.8                    ;g=9.8m/s^2
          0.1                    ;a=1/10 m
          (* 2.0 (sqrt 9.8)) )   ;omega
    (up 0.0                      ;t_0=0
        1.0                      ;theta_0=1 radian
        0.0)                     ;thetadot_0=0 radians/s
    (monitor-theta plot-win)
    0.01                         ;step between plotted points
    100.0                        ;final time
    1.0e-13)                     ;local error tolerance
```

Figure 1.7 shows the angle θ versus time for a couple of orbits for the driven pendulum. The initial conditions for the two runs are the same except that in one the bob is given a tiny velocity equal to 10^{-10}m/s, about one atom width per second. The initial segments of the two orbits are indistinguishable. After about 75 seconds the two orbits diverge and become completely different. This extreme sensitivity to tiny changes in initial conditions is characteristic of what is called *chaotic behavior*. Later, we will investigate this example further, using other tools such as Lyapunov exponents, phase space, and Poincaré sections.

---

[78]The results are plotted in a plot window created by the procedure `frame` with arguments `xmin`, `xmax`, `ymin`, and `ymax` that specify the limits of the plotting area. Points are added to the plot with the procedure `plot-point` that takes a plot window and the abscissa and ordinate of the point to be plotted.

The procedure `principal-value` is used to reduce an angle to a standard interval. The argument to `principal-value` is the point at which the circle is to be cut. Thus (`principal-value :pi`) is a procedure that reduces an angle θ to the interval $-\pi \leq \theta < \pi$.

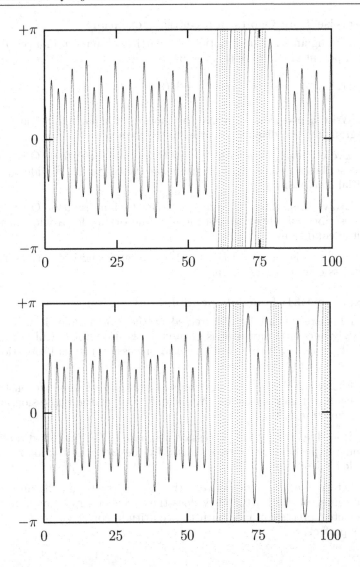

Figure 1.7 Orbits of the driven pendulum. The angle θ is plotted against time. Because angles are periodic, this plot may be thought of as being wound around a cylinder. The upper plot shows the results of a simulation with initial conditions $\theta = 1$ and $\dot{\theta} = 0$. The orbit oscillates for a while, then circulates, then resumes oscillating. In the lower plot we show the result for a slightly different initial angular velocity, $\dot{\theta} = 10^{-10}$. The initial behavior is indistinguishable from the top figure, but the two trajectories become uncorrelated after the transition between oscillation and circulation. This extreme sensitivity to initial conditions is characteristic of systems with chaotic behavior.

Exercise 1.30: Orbits in a central potential

A Lagrangian for planar motion of a particle of mass m in a central field with potential energy $V(r) = -\beta r^\alpha$ is

$$L(t; r, \theta; \dot{r}, \dot{\theta}) = \tfrac{1}{2}m(\dot{r}^2 + (r\dot{\theta})^2) + V(r).$$

a. Write a program to evolve the motion of a particle subject to this Lagrangian and display the orbit in the plane.

b. Evolve this system with $\alpha = +2$ (harmonic oscillator). Observe that it describes an ellipse with its center at the origin, for a wide variety of initial conditions.

c. Evolve this system with $\alpha = -1$ (Newtonian gravity). Observe that it describes an ellipse with a focus at the origin, for a wide variety of initial conditions.

d. Evolve this system with $\alpha = +1/4$. Observe that it describes a trefoil with its center at the origin.

Exercise 1.31: Foucault pendulum evolution

If a Foucault pendulum is erected at the North Pole, it will precess exactly once in a day. If it is erected at the Equator it will not precess at all. It is widely reported that the precession rate is proportional to the cosine of the colatitude.

a. Evolve the Foucault pendulum, using the Lagrangian you constructed in exercise 1.25 on page 62. You should look at the precession angle λ as a function of time.

b. How does the rate of precession compare to the predicted rate? You should expect to see an error caused by the fact that the local vertical, as defined by a plumb bob, is not directed to the center of the Earth.

c. Let $\Delta\phi$ be the angle between the local vertical and the direction to the center of the Earth. How does the precession rate compare to the predicted precession rate with the colatitude corrected to $\phi - \Delta\phi$? Is this perfect?

1.8   Conserved Quantities

A function of the state of the system that is constant along a solution path is called a *conserved quantity* or a *constant of motion.*[79] If C is a conserved quantity, then

$$D(C \circ \Gamma[q]) = D_t C \circ \Gamma[q] = 0 \qquad (1.131)$$

---

[79]In older literature conserved quantities are sometimes called *first integrals*.

for solution paths q. In this section, we will investigate systems with symmetry and find that symmetries are associated with conserved quantities. For instance, linear momentum is conserved in a system with translational symmetry, angular momentum is conserved if there is rotational symmetry, energy is conserved if the system does not depend on the origin of time. We first consider systems for which a coordinate system can be chosen that expresses the symmetry naturally, and later discuss systems for which no coordinate system can be chosen that simultaneously expresses all symmetries.

1.8.1   Conserved Momenta

If a Lagrangian $L(t, q, v)$ does not depend on some particular coordinate q^i, then

$$(\partial_1 L)_i = 0, \tag{1.132}$$

and the corresponding ith component of the Lagrange equations is

$$(D(\partial_2 L \circ \Gamma[q]))_i = 0. \tag{1.133}$$

The derivative of a component is equal to the component of the derivative, so this is the same as

$$D\left((\partial_2 L)_i \circ \Gamma[q]\right) = 0, \tag{1.134}$$

and we can see that

$$\mathcal{P}_i = (\partial_2 L)_i \tag{1.135}$$

is a conserved quantity. The function $\mathcal{P}$ is called the *momentum state function*. The value of the momentum state function is the *generalized momentum*. We refer to the ith component of the generalized momentum as the momentum *conjugate* to the ith coordinate.[80] The momenta depend on the choice of Lagrangian used to describe the system.[81] A generalized coordinate compo-

---

[80]Observe that we indicate a component of the generalized momentum with a subscript, and a component of the generalized coordinates with a superscript. These conventions are consistent with those commonly used in tensor algebra, which is sometimes helpful in working out complex problems.

[81]For example, we may construct equivalent Lagrangians that differ only by a total time derivative. The momentum state functions for these Lagrangians are different.

nent that does not appear explicitly in the Lagrangian is called
a *cyclic coordinate*. The generalized momentum component conjugate to any cyclic coordinate is a constant of the motion. Its
value is constant along realizable paths; it may have different values on different paths. As we will see, momentum is an important
quantity even when it is not conserved.

Given the coordinate path q and the Lagrangian L, the momentum path p is

$$p = \partial_2 L \circ \Gamma[q] = \mathcal{P} \circ \Gamma[q], \tag{1.136}$$

with components

$$p_i = \mathcal{P}_i \circ \Gamma[q]. \tag{1.137}$$

The momentum path is well defined for any path q. If the path is
realizable and the Lagrangian does not depend on q^i, then p_i is a
constant function

$$D p_i = 0. \tag{1.138}$$

The constant value of p_i may be different for different trajectories.

Examples of conserved momenta

The free-particle Lagrangian $L(t, x, v) = \frac{1}{2}mv^2$ is independent
of x. So the momentum state function, $\mathcal{P}(t, q, v) = mv$, is conserved along realizable paths. The momentum path p for the
coordinate path q is $p(t) = \mathcal{P} \circ \Gamma[q](t) = m\, Dq(t)$. For a realizable
path $Dp(t) = 0$. For the free particle the usual linear momentum
is conserved for realizable paths.

For a particle in a central force field (section 1.6), the Lagrangian

$$L(t; r, \varphi; \dot{r}, \dot{\varphi}) = \frac{1}{2}m(\dot{r}^2 + r^2\dot{\varphi}^2) - V(r)$$

depends on r but is independent of φ. The momentum state
function is

$$\mathcal{P}(t; r, \varphi; \dot{r}, \dot{\varphi}) = \left[m\dot{r}, mr^2\dot{\varphi} \right].$$

It has two components. The first component, the "radial momentum," is not conserved. The second component, the "angular
momentum," is conserved along any solution trajectory.

If the central-potential problem had been expressed in rectangular coordinates, then all of the coordinates would have appeared in the Lagrangian. In that case there would not be any obvious conserved quantities. Nevertheless, the motion of the system does not depend on the choice of coordinates, so the angular momentum is still conserved.

We see that there is great advantage in making a judicious choice for the coordinate system. If we can choose the coordinates so that a symmetry of the system is reflected in the Lagrangian by the absence of some coordinate component, then the existence of a corresponding conserved quantity will be evident.[82]

1.8.2   Energy Conservation

Momenta are conserved by the motion if the Lagrangian does not depend on the corresponding coordinate. There is another constant of the motion, the energy, if the Lagrangian $L(t, q, \dot{q})$ does not depend explicitly on the time: $\partial_0 L = 0$.

Consider the time derivative of the Lagrangian along a solution path q:

$$D(L \circ \Gamma[q]) = \partial_0 L \circ \Gamma[q] + (\partial_1 L \circ \Gamma[q]) Dq + (\partial_2 L \circ \Gamma[q]) D^2 q. \quad (1.139)$$

Using Lagrange's equations to rewrite the second term yields

$$D(L \circ \Gamma[q]) = (\partial_0 L) \circ \Gamma[q] + D(\partial_2 L \circ \Gamma[q]) Dq + (\partial_2 L \circ \Gamma[q]) D^2 q. \quad (1.140)$$

Isolating $\partial_0 L$ and combining the other terms we get

$$
\begin{aligned}
(\partial_0 L) \circ \Gamma[q] &= D(L \circ \Gamma[q]) - D((\partial_2 L \circ \Gamma[q]) Dq) \\
&= D(L \circ \Gamma[q]) - D((\partial_2 L \circ \Gamma[q])(\dot{Q} \circ \Gamma[q])) \\
&= D((L - \mathcal{P}\dot{Q}) \circ \Gamma[q]),
\end{aligned}
\quad (1.141)
$$

where, as before, $\dot{Q}$ selects the velocity from the state. So we see that if $\partial_0 L = 0$ then

$$\mathcal{E} = \mathcal{P}\dot{Q} - L \quad (1.142)$$

---

[82]In general, conserved quantities in a physical system are associated with continuous symmetries, whether or not one can find a coordinate system in which the symmetry is apparent. This powerful notion was formalized and a theorem linking conservation laws with symmetries was proved by Noether early in the 20th century. See section 1.8.5 on Noether's theorem.

is conserved along realizable paths. The function $\mathcal{E}$ is called the *energy state function*.[83] The energy state function for a system depends on the choice of Lagrangian used to describe the system.[84] Let $E = \mathcal{E} \circ \Gamma[q]$ denote the energy function on the path q. The energy function has a constant value along any realizable trajectory if the Lagrangian has no explicit time dependence; the energy E may have a different value for different trajectories. A system that has no explicit time dependence is called *autonomous*.

Given a Lagrangian procedure L, we may construct the energy function:

```
(define (Lagrangian->energy L)
  (let ((P ((partial 2) L)))
    (- (* P velocity) L)))
```

Energy in terms of kinetic and potential energies

In some cases the energy can be written as the sum of kinetic and potential energies. Suppose the system is composed of particles with rectangular coordinates $\mathbf{x}_\alpha$, the movement of which may be subject to constraints, and that these rectangular coordinates are some functions of the generalized coordinates q and possibly time t: $\mathbf{x}_\alpha = f_\alpha(t, q)$. We form the Lagrangian as $L = T - V$ and compute the kinetic energy in terms of q by writing the rectangular velocities in terms of the generalized velocities:

$$\mathbf{v}_\alpha = \partial_0 f_\alpha(t, q) + \partial_1 f_\alpha(t, q)v. \tag{1.143}$$

The kinetic energy is

$$T(t, q, v) = \tfrac{1}{2} \sum_\alpha m_\alpha v_\alpha^2, \tag{1.144}$$

where v_α is the magnitude of $\mathbf{v}_\alpha$.

If the f_α functions do not depend explicitly on time ($\partial_0 f_\alpha = 0$), then the rectangular velocities are homogeneous functions of the generalized velocities of degree 1, and T is a homogeneous function of the generalized velocities of degree 2, because it is formed by

---

[83]The sign of the energy state function is a matter of convention.

[84]We may construct equivalent Lagrangians that differ only by a total time derivative. The energy state functions for these Lagrangians are different.

summing the square of homogeneous functions of degree 1. If T is a homogeneous function of degree 2 in the generalized velocities then

$$P\dot{Q} = (\partial_2 T)\dot{Q} = 2T, \qquad (1.145)$$

where the second equality follows from Euler's theorem on homogeneous functions.[85] The energy state function is

$$\mathcal{E} = P\dot{Q} - L = 2T - T + V. \qquad (1.146)$$

So if f_α is independent of time, the energy function can be rewritten

$$\mathcal{E} = 2T - T + V = T + V. \qquad (1.147)$$

Notice that if V depends on time the energy is still the sum of the kinetic energy and potential energy, but the energy is not conserved.

The energy state function is always well defined, whether or not it can be written in the form $T + V$, and whether or not it is conserved along realizable paths.

Exercise 1.32: Time-dependent constraints

An analogous result holds when the f_α depend explicitly on time.

a.   Show that in this case the kinetic energy contains terms that are linear in the generalized velocities.

b.   By adding a total time derivative, show that the Lagrangian can be written in the form $L = A - B$, where A is a homogeneous quadratic form in the generalized velocities and B is independent of velocity.

c.   Show, using Euler's theorem, that the energy function is $\mathcal{E} = A + B$.

An example in which terms that were linear in the velocity were removed from the Lagrangian by adding a total time derivative has already been given: the driven pendulum.

---

[85] A function f is homogenous of degree n if and only if $f(ax) = a^n f(x)$. Euler's theorem says that if f is a homogeneous function of degree n, then $Df(x)x = nf(x)$. The proof is as follows: Let $g_x(a) = f(ax)$. Then $Dg_x(a) = Df(ax)x$. But $g_x(a) = a^n f(x)$ by the definition of homogeneity. Therefore $Dg_x(a) = na^{n-1}f(x)$. Equating these, we find $Df(ax)x = na^{n-1}f(x)$. Specializing to $a = 1$ we obtain $Df(x)x = nf(x)$ as required.

Exercise 1.33: Falling off a log

A particle of mass m slides off a horizontal cylinder of radius R in a uniform gravitational field with acceleration g. If the particle starts close to the top of the cylinder with zero initial speed, with what angular velocity does it leave the cylinder? Use the method of incorporating constraint forces that we introduced in section 1.6.2, together with conservation of energy.

1.8.3   Central Forces in Three Dimensions

One important physical system is the motion of a particle in a central field in three dimensions, with an arbitrary potential energy $V(r)$ depending only on the radius. We will describe this system in spherical coordinates r, θ, and φ, where θ is the colatitude and φ is the longitude. The kinetic energy has three terms:

$$T(t; r, \theta, \varphi; \dot{r}, \dot{\theta}, \dot{\varphi}) = \tfrac{1}{2} m (\dot{r}^2 + r^2 \dot{\theta}^2 + r^2 (\sin \theta)^2 \dot{\varphi}^2).$$

As a procedure:

```
(define ((T3-spherical m) state)
  (let ((q (coordinate state))
        (qdot (velocity state)))
    (let ((r (ref q 0))
          (theta (ref q 1))
          (rdot (ref qdot 0))
          (thetadot (ref qdot 1))
          (phidot (ref qdot 2)))
      (* 1/2 m
         (+ (square rdot)
            (square (* r thetadot))
            (square (* r (sin theta) phidot)))))))
```

A Lagrangian is then formed by subtracting the potential energy:

```
(define (L3-central m Vr)
  (define (Vs state)
    (let ((r (ref (coordinate state) 0)))
      (Vr r)))
  (- (T3-spherical m) Vs))
```

Let's first look at the generalized forces (the derivatives of the Lagrangian with respect to the generalized coordinates). We compute these with a partial derivative with respect to the coordinate argument of the Lagrangian:

```
(show-expression
 (((partial 1) (L3-central 'm (literal-function 'V)))
  (up 't
      (up 'r 'theta 'phi)
      (up 'rdot 'thetadot 'phidot))))
```

$$\begin{bmatrix} m\dot{\varphi}^2 r \left(\sin\left(\theta\right)\right)^2 + mr\dot{\theta}^2 - DV\left(r\right) \\ m\dot{\varphi}^2 r^2 \cos\left(\theta\right)\sin\left(\theta\right) \\ 0 \end{bmatrix}$$

The φ component of the force is zero because φ does not appear
in the Lagrangian (it is a cyclic coordinate). The corresponding
momentum component is conserved. Compute the momenta:

```
(show-expression
 (((partial 2) (L3-central 'm (literal-function 'V)))
  (up 't
      (up 'r 'theta 'phi)
      (up 'rdot 'thetadot 'phidot))))
```

$$\begin{bmatrix} m\dot{r} \\ mr^2\dot{\theta} \\ mr^2\dot{\varphi}\left(\sin\left(\theta\right)\right)^2 \end{bmatrix}$$

The momentum conjugate to φ is conserved. This is the z component of the angular momentum $\vec{r} \times (m\vec{v})$, for vector position
$\vec{r}$ and linear momentum $m\vec{v}$. We can show this by writing the z
component of the angular momentum in spherical coordinates:

```
(define ((ang-mom-z m) rectangular-state)
  (let ((xyz (coordinate rectangular-state))
        (v (velocity rectangular-state)))
    (ref (cross-product xyz (* m v)) 2)))

(define (s->r spherical-state)
  (let ((q (coordinate spherical-state)))
    (let ((r (ref q 0))
          (theta (ref q 1))
          (phi (ref q 2)))
      (let ((x (* r (sin theta) (cos phi)))
            (y (* r (sin theta) (sin phi)))
            (z (* r (cos theta))))
        (up x y z)))))
```

```
(show-expression
  ((compose (ang-mom-z 'm) (F->C s->r))
   (up 't
       (up 'r 'theta 'phi)
       (up 'rdot 'thetadot 'phidot)))) 
```

$$mr^2\dot{\varphi}\left(\sin\left(\theta\right)\right)^2$$

The choice of the z-axis is arbitrary, so the conservation of any component of the angular momentum implies the conservation of all components. Thus the total angular momentum is conserved. We can choose the z-axis so all of the angular momentum is in the z component. Since $\vec{x} \cdot (\vec{x} \times \vec{v}) = \vec{v} \cdot (\vec{x} \times \vec{x}) = 0$, the motion is confined to the plane perpendicular to the angular momentum: $\theta = \pi/2$, and $\dot{\theta} = 0$. Planar motion in a central-force field was discussed in section 1.6.

We can also see that the energy state function computed from the Lagrangian for a central field is in fact $T + V$:

```
(show-expression
  ((Lagrangian->energy (L3-central 'm (literal-function 'V)))
   (up 't
       (up 'r 'theta 'phi)
       (up 'rdot 'thetadot 'phidot)))) 
```

$$\frac{1}{2}m\dot{\varphi}^2 r^2 \left(\sin\left(\theta\right)\right)^2 + \frac{1}{2}mr^2\dot{\theta}^2 + \frac{1}{2}mr^2 + V\left(r\right)$$

The energy is conserved because the Lagrangian has no explicit time dependence.

Exercise 1.34: Driven spherical pendulum

A spherical pendulum is a massive bob, subject to uniform gravity, that may swing in three dimensions, but remains at a given distance from the pivot. Formulate a Lagrangian for a spherical pendulum driven by vertical motion of the pivot. What symmetry(ies) can you find? Find coordinates that express the symmetry(ies). What is conserved? Give analytic expression(s) for the conserved quantity(ies).

1.8.4 The Restricted Three-Body Problem

Consider the situation of two bodies of masses M_0 and M_1 in circular orbit about their common center of mass. What is the

behavior of a third particle, gravitationally attracted to the other
two, that must move in the plane of their circular orbit? Assume
that the third particle has such small mass that we can neglect its
effect on the orbits of the two massive particles.

The third particle, of mass m, moves in a field derived from a
time-varying gravitational potential energy. We have:

```
(define ((L0 m V) local)
  (let ((t (time local))
        (q (coordinates local))
        (v (velocities local)))
    (- (* 1/2 m (square v)) (V t q))))
```

Let a be the constant distance between the two bodies. If we put
the center of mass at the origin of the coordinate system then the
distances of the two particles from the origin are:

$$a_0 = \frac{M_1}{M_0 + M_1} a \quad \text{and} \quad a_1 = \frac{M_0}{M_0 + M_1} a \qquad (1.148)$$

Each massive particle revolves in a circle about their common
center of mass with angular frequency Ω. The radii of the circles
are the distances given above. Kepler's law gives the angular
frequency of the orbit:

$$\Omega^2 a^3 = G(M_0 + M_1) \qquad (1.149)$$

We choose our axes so that at $t = 0$ the body with mass M_1 is on
the positive $\hat{x}$ axis and the body with mass M_0 is on the negative
$\hat{x}$ axis. The gravitational potential energy function is:

```
(define ((V a GM0 GM1 m) t xy)
  (let ((Omega (sqrt (/ (+ GM0 GM1) (expt a 3))))
        (a0 (* (/ GM1 (+ GM0 GM1)) a))
        (a1 (* (/ GM0 (+ GM0 GM1)) a)))
    (let ((x (ref xy 0)) (y (ref xy 1))
          (x0 (* -1 a0 (cos (* Omega t))))
          (y0 (* -1 a0 (sin (* Omega t))))
          (x1 (* +1 a1 (cos (* Omega t))))
          (y1 (* +1 a1 (sin (* Omega t)))))
      (let ((r0
             (sqrt (+ (square (- x x0)) (square (- y y0)))))
            (r1
             (sqrt (+ (square (- x x1)) (square (- y y1))))))
        (- (+ (/ (* GM0 m) r0) (/ (* GM1 m) r1)))))))
```

It is convenient to examine the motion of the third particle in a rotating coordinate system where the massive particles are fixed. We can place the rotating axes so that the two massive particles are on the $\hat{x}'$ axis, and we can choose the rotating and nonrotating axes to be coincident at $t = 0$. We can transform to the rotating rectangular coordinates as we did on page 48. The resulting Lagrangian is the Lagrangian for the free particle with the addition of two gravitational potential energy terms:

$$L_r(t; x_r, y_r; \dot{x}_r, \dot{y}_r)$$
$$= \tfrac{1}{2}m(\dot{x}_r^2 + \dot{y}_r^2) + \tfrac{1}{2}m\Omega^2(x_r^2 + y_r^2) + m\Omega(x_r\dot{y}_r - \dot{x}_r y_r)$$
$$+ \frac{GM_0 m}{r_0} + \frac{GM_1 m}{r_1} \tag{1.150}$$

where now $r_0^2 = (x_r + a_0)^2 + y_r^2$ and $r_1^2 = (x_r - a_1)^2 + y_r^2$. As a program we can write:

```
(define ((LR3B m a GM0 GM1) local)
  (let ((q (coordinates local))
        (qdot (velocities local))
        (Omega (sqrt (/ (+ GM0 GM1) (expt a 3))))
        (a0 (* (/ GM1 (+ GM0 GM1)) a))
        (a1 (* (/ GM0 (+ GM0 GM1)) a)))
    (let ((x (ref q 0))          (y (ref q 1))
          (xdot (ref qdot 0)) (ydot (ref qdot 1)))
      (let ((r0 (sqrt (+ (square (+ x a0)) (square y))))
            (r1 (sqrt (+ (square (- x a1)) (square y)))))
        (+ (* 1/2 m (square qdot))
           (* 1/2 m (square Omega) (square q))
           (* m Omega (- (* x ydot) (* xdot y)))
           (/ (* GM0 m) r0) (/ (* GM1 m) r1))))))
```

Notice that the Lagrangian in rotating coordinates is independent of time. So the energy state function defined by this Lagrangian is a conserved quantity. Let's compute it. It is clearest if we express the result in terms of Ω, a_0, and a_1, so we make those into explicit parameters of the Lagrangian:

```
(define ((LR3B1 m a0 a1 Omega GM0 GM1) local)
  (let ((q (coordinates local))
        (qdot (velocities local)))
    (let ((x (ref q 0))          (y (ref q 1))
          (xdot (ref qdot 0)) (ydot (ref qdot 1)))
      (let ((r0 (sqrt (+ (square (+ x a0)) (square y))))
            (r1 (sqrt (+ (square (- x a1)) (square y)))))
        (+ (* 1/2 m (square qdot))
           (* 1/2 m (square Omega) (square q))
           (* m Omega (- (* x ydot) (* xdot y)))
           (/ (* GM0 m) r0) (/ (* GM1 m) r1))))))
```

And we compute the energy state function (with a bit of hand simplification):

```
((Lagrangian->energy (LR3B1 'm 'a_0 'a_1 'Omega 'GM_0 'GM_1))
 (up 't (up 'x_r 'y_r) (up 'v_r^x 'v_r^y)))
(+ (* 1/2 m (expt v_r^x 2))
   (* 1/2 m (expt v_r^y 2))
   (/ (* -1 GM_0 m)
      (sqrt (+ (expt (+ x_r a_0) 2) (expt y_r 2))))
   (/ (* -1 GM_1 m)
      (sqrt (+ (expt (- x_r a_1) 2) (expt y_r 2))))
   (* -1/2 m (expt Omega 2) (expt x_r 2))
   (* -1/2 m (expt Omega 2) (expt y_r 2)))
```

If we separate this into a velocity-dependent part and a velocity-independent part we get

$$\mathcal{E}(t; x_r, y_r; \dot{x}_r, \dot{y}_r) = \tfrac{1}{2}m\left(\dot{x}_r^2 + \dot{y}_r^2\right) + mU_r(x_r, y_r) \qquad (1.151)$$

where

$$U_r(x_r, y_r) = -\left(\frac{GM_0}{r_0} + \frac{GM_1}{r_1} + \frac{1}{2}\Omega^2(x_r^2 + y_r^2)\right). \qquad (1.152)$$

This constant of motion of the restricted three-body problem is called the *Jacobi constant*.[86] Notice that the energy function is a positive definite quadratic form in the components of the velocity (in rotating coordinates) plus a function that depends only on the rotating coordinates. Note that the energy state function does not have terms that are linear in the velocities $\dot{x}_r$ and $\dot{y}_r$, although such terms appear in the Lagrangian (1.150).

---

[86]Traditionally the Jacobi constant is defined as $C_J = -2\mathcal{E}$.

Exercise 1.35: Restricted equations of motion

Derive the Lagrange equations for the restricted three-body problem, given the Lagrangian (1.150). Identify the Coriolis and centrifugal force terms in your equations of motion.

1.8.5 Noether's Theorem

We have seen that if a system has a symmetry and a coordinate system can be chosen so that the Lagrangian does not depend on the coordinate associated with that symmetry, then there is a conserved quantity associated with the symmetry. However, there are more general symmetries that no coordinate system can fully express. For example, motion in a central potential is spherically symmetric (the dynamical system is invariant under rotations about any axis), but the expression of the Lagrangian for the system in spherical coordinates exhibits symmetry around only one axis. More generally, a Lagrangian has a symmetry if there is a coordinate transformation that leaves the Lagrangian unchanged. A continuous symmetry is a parametric family of symmetries. Noether proved that for any continuous symmetry there is a conserved quantity.

Consider a parametric coordinate transformation $\widetilde{F}$ with parameter s:

$$x = \widetilde{F}(s)(t, x'). \tag{1.153}$$

To this parametric coordinate transformation there corresponds a parametric state transformation $\widetilde{C}$:

$$(t, x, v) = \widetilde{C}(s)(t, x', v'). \tag{1.154}$$

We require that the transformation $\widetilde{F}(0)$ be the identity coordinate transformation $x' = \widetilde{F}(0)(t, x')$, and as a consequence $\widetilde{C}(0)$ is the identity state transformation $(t, x', v') = \widetilde{C}(0)(t, x', v')$. The Lagrangian L has a continuous symmetry corresponding to $\widetilde{F}$ if it is invariant under the transformations

$$\widetilde{L}(s) = L \circ \widetilde{C}(s) = L \tag{1.155}$$

for any s. The Lagrangian L is the same function as the transformed Lagrangian $\widetilde{L}(s)$.

That $\widetilde{L}(s) = L$ for any s implies $D\widetilde{L}(s) = 0$. Explicitly, $\widetilde{L}(s)$ is

$$\widetilde{L}(s)(t, x', v') = L(t, \widetilde{F}(s)(t, x'), D_t(\widetilde{F}(s))(t, x', v')), \qquad (1.156)$$

where we have rewritten the velocity component of $\widetilde{C}(s)$ in terms of the total time derivative. The derivative of $\widetilde{L}$ is zero:

$$\begin{aligned}
0 &= D\widetilde{L}(s)(t, x', v') \\
&= \partial_1 L(t, x, v) \, (D\widetilde{F})(s)(t, x') + \partial_2 L(t, x, v) \, D_t(D\widetilde{F}(s))(t, x'),
\end{aligned}$$
$$(1.157)$$

where we have used the fact that[87]

$$D_t(D\widetilde{F}(s)) = DG(s) \quad \text{with} \quad G(s) = D_t(\widetilde{F}(s)). \qquad (1.158)$$

On a realizable path q we can use the Lagrange equations to rewrite the first term of equation (1.157):

$$\begin{aligned}
0 &= (D_t \partial_2 L \circ \Gamma[q]) \, ((D\widetilde{F})(s) \circ \Gamma[q']) \\
&\quad + (\partial_2 L \circ \Gamma[q]) \, (D_t(D\widetilde{F}(s)) \circ \Gamma[q']).
\end{aligned}$$
$$(1.159)$$

For $s = 0$ the paths q and q' are the same, because $\widetilde{F}(0)$ is the identity, so $\Gamma[q] = \Gamma[q']$ and this equation becomes

$$\begin{aligned}
0 &= ((D_t \partial_2 L) \, ((D\widetilde{F})(0)) + (\partial_2 L) \, (D_t(D\widetilde{F}(0)))) \circ \Gamma[q] \\
&= D_t((\partial_2 L) \, (D\widetilde{F}(0))) \circ \Gamma[q].
\end{aligned}$$
$$(1.160)$$

Thus the state function $\mathcal{I}$,

$$\mathcal{I} = (\partial_2 L)(D\widetilde{F}(0)), \qquad (1.161)$$

is conserved along solution trajectories. This conserved quantity is called *Noether's integral*. It is the product of the momentum and a vector associated with the symmetry.

---

[87] The total time derivative is like a derivative with respect to a real-number argument in that it does not generate structure, so it can commute with derivatives that generate structure. Be careful, though: it may not commute with some derivatives for other reasons. For example, $D_t \partial_1(\widetilde{F}(s))$ is the same as $\partial_1 D_t(\widetilde{F}(s))$, but $D_t \partial_2(\widetilde{F}(s))$ is not the same as $\partial_2 D_t(\widetilde{F}(s))$. The reason is that $\widetilde{F}(s)$ does not depend on the velocity, but $D_t(\widetilde{F}(s))$ does.

Illustration: motion in a central potential

For example, consider the central-potential Lagrangian in rectangular coordinates:

$$L(t; x, y, z; v_x.v_y, v_z)$$
$$= \tfrac{1}{2}m\left(v_x^2 + v_y^2 + v_z^2\right) - U\left(\sqrt{x^2 + y^2 + z^2}\right), \qquad (1.162)$$

and a parametric rotation $R_z(s)$ about the z axis

$$\begin{pmatrix} x \\ y \\ z \end{pmatrix} = R_z(s)\begin{pmatrix} x' \\ y' \\ z' \end{pmatrix} = \begin{pmatrix} x'\cos s - y'\sin s \\ x'\sin s + y'\cos s \\ z' \end{pmatrix}. \qquad (1.163)$$

The rotation is an orthogonal transformation so

$$x^2 + y^2 + z^2 = (x')^2 + (y')^2 + (z')^2. \qquad (1.164)$$

Differentiating along a path, we get

$$(v_x, v_y, v_z) = R_z(s)(v_x', v_y', v_z'), \qquad (1.165)$$

so the velocities also transform by an orthogonal transformation, and $v_x^2 + v_y^2 + v_z^2 = (v_x')^2 + (v_y')^2 + (v_z')^2$. Thus

$$L'(t; x', y', z'; v_x', v_y', v_z')$$
$$= \tfrac{1}{2}m\left((v_x')^2 + (v_y')^2 + (v_z')^2\right)$$
$$- U\left(\sqrt{(x')^2 + (y')^2 + (z')^2}\right), \qquad (1.166)$$

and we see that L' is precisely the same function as L.

The momenta are

$$\partial_2 L(t; x, y, z; v_x, v_y, v_z) = [mv_x, mv_y, mv_z] \qquad (1.167)$$

and

$$D\widetilde{F}(0)(t; x, y, z) = D\widetilde{R}_z(0)(x, y, z) = (y, -x, 0). \qquad (1.168)$$

So the Noether integral is

$$\mathcal{I}(t; x, y, z; v_x, v_y, v_z) = ((\partial_2 L)(D\widetilde{F}(0)))(t; x, y, z; v_x, v_y, v_z)$$
$$= m(yv_x - xv_y), \qquad (1.169)$$

which we recognize as minus the z component of the angular momentum: $\vec{x} \times (m\vec{v})$. Since the Lagrangian is preserved by any continuous rotational symmetry, all components of the vector angular momenta are conserved for the central-potential problem.

The procedure calls `((Rx angle-x) q)`, `((Ry angle-y) q)`, and `((Rz angle-z) q)` rotate the rectangular tuple q about the indicated axis by the indicated angle.[88] We use these to make a parametric coordinate transformation `F-tilde`:

```
(define (F-tilde angle-x angle-y angle-z)
  (compose (Rx angle-x) (Ry angle-y) (Rz angle-z) coordinate))
```

A Lagrangian for motion in a central potential is

```
(define ((L-central-rectangular m U) state)
  (let ((q (coordinate state))
        (v (velocity state)))
    (- (* 1/2 m (square v))
       (U (sqrt (square q))))))
```

The Noether integral is then

```
(define the-Noether-integral
  (let ((L (L-central-rectangular
             'm (literal-function 'U))))
    (* ((partial 2) L) ((D F-tilde) 0 0 0))))
```

```
(the-Noether-integral
 (up 't
     (up 'x 'y 'z)
     (up 'vx 'vy 'vz)))
(down (+ (* m vy z) (* -1 m vz y))
      (+ (* m vz x) (* -1 m vx z))
      (+ (* m vx y) (* -1 m vy x)))
```

We get all three components of the angular momentum.

---

[88]The definition of the procedure `Rx` is

```
(define ((Rx angle) q)
  (let ((ca (cos angle)) (sa (sin angle)))
    (let ((x (ref q 0)) (y (ref q 1)) (z (ref q 2)))
      (up x
          (- (* ca y) (* sa z))
          (+ (* sa y) (* ca z))))))
```

The definitions of `Ry` and `Rz` are similar. See footnote 69.

Exercise 1.36: Noether integral

Consider motion on an ellipsoidal surface. The surface is specified by:

$$x^2/a^2 + y^2/b^2 + z^2/c^2 = 1$$

Formulate a Lagrangian for frictionless motion on this surface. Assume that two of the axes of the ellipsoid are equal: $b = c$.

Using angular coordinates (θ, ϕ), where θ is colatitude from the $\hat{z}$-axis, and ϕ is longitude measured from the $\hat{x}$-axis, formulate a Lagrangian that captures the symmetry of this ellipsoid: rotational symmetry around the $\hat{x}$-axis. Formulate a parametric transformation that represents this symmetry and show that the Lagrangian you formulated is invariant under this transformation. Compute the Noether integral associated with this symmetry.

Note that the choice of coordinates does not build in this symmetry.

1.9   Abstraction of Path Functions

An essential step in the derivation of the local-tuple transformation function C from the coordinate transformation F was the deduction of the relationship between the velocities in the two coordinate systems. We did this by inserting coordinate paths into the coordinate transformation function F, differentiating, and then generalizing the results on the path to arbitrary velocities at a moment. The last step is an example of a more general problem of abstracting a local-tuple function from a path function. Given a function f of a local tuple, a corresponding path-dependent function $\bar{f}[q]$ is $\bar{f}[q] = f \circ \Gamma[q]$. Given $\bar{f}$, how can we reconstitute f? The local-tuple function f depends on only a finite number of components of the local tuple, and $\bar{f}$ depends only on the corresponding local components of the path. So $\bar{f}$ has the same value for all paths that have that number of components of the local tuple in common. Given $\bar{f}$ we can reconstitute f by taking the argument of f, which is a finite initial segment of a local tuple, constructing a path that has this local description, and finding the value of $\bar{f}$ for this path.

Two paths that have the same local description up to the nth derivative are said to *osculate with order n contact*. For example, a path and the truncated power series representation of the path up to order n have order n contact; if fewer than n derivatives are needed by a local-tuple function, the path and the truncated power series representation are equivalent. Let O be a function that generates an osculating path with the given local-tuple components. So $O(t, q, v, \ldots)(t) = q$, $D(O(t, q, v, \ldots))(t) = v$, and in general

$$(t, q, v, \ldots) = \Gamma[O(t, q, v, \ldots)](t). \tag{1.170}$$

The number of components of the local tuple that are required is finite, but unspecified. One way of constructing O is through the truncated power series

$$O(t, q, v, a, \ldots)(t') = q + v(t' - t) + \tfrac{1}{2}a(t' - t)^2 + \cdots, \tag{1.171}$$

where the number of terms is the same as the number of components of the local tuple that are specified.

Given the path function $\bar{f}$ we reconstitute the f function as follows. We take the argument of f and construct an osculating path with this local description. Then the value of f is the value of $\bar{f}$ for this osculating path:

$$\begin{aligned} f(t, q, v, \ldots) &= (f \circ \Gamma[O(t, q, v, \ldots)])(t) \\ &= \bar{f}[O(t, q, v, \ldots)](t). \end{aligned} \tag{1.172}$$

Let $\bar{\Gamma}$ be the function that takes a path function and returns the corresponding local-tuple function:[89]

$$f = \bar{\Gamma}(\bar{f}). \tag{1.173}$$

From equation (1.172) we see that

$$\bar{\Gamma}(\bar{f})(t, q, v, \ldots) = \bar{f}[O(t, q, v, \ldots)](t). \tag{1.174}$$

The procedure `Gamma-bar` implements the function $\bar{\Gamma}$ that reconstitutes a path-dependent function into a local-tuple function:

```
(define ((Gamma-bar f-bar) local)
  ((f-bar (osculating-path local)) (time local)))
```

---

[89]The local-tuple function f is the same as the local-tuple function $\bar{\Gamma}(\bar{f})$ where $\bar{f}[q] = f \circ \Gamma[q]$. On the other hand, the path function $\bar{f}[q]$ and the path function $\bar{\Gamma}(\bar{f}) \circ \Gamma[q]$ are not necessarily the same because $\bar{f}[q]$ may require more components of the local tuple than are provided by default. For example, the Lagrange equations involve accelerations, as well as time, coordinates, and velocities. If $\Gamma[q]$ is extended to the appropriate number of components then the two are equivalent. (See footnote 90.)

The procedure `osculating-path` takes a number of local components and returns a path with these components; it is implemented
as a power series.

We can use `Gamma-bar` to construct the procedure `F->C` that
takes a coordinate transformation `F` and generates the procedure
that transforms local tuples. The procedure `F->C` constructs a
path-dependent procedure `f-bar` that takes a coordinate path in
the primed system and returns the local tuple of the corresponding
path in the unprimed coordinate system. It then uses `Gamma-bar`
to abstract `f-bar` to arbitrary local tuples in the primed coordinate system.[90]

```
(define (F->C F)
  (define (C local)
    (let ((n (vector-length local)))
      (define (f-bar q-prime)
        (define q
          (compose F (Gamma q-prime)))
        (Gamma q n))
      ((Gamma-bar f-bar) local)))
  C)
```

```
(show-expression
  ((F->C p->r)
   (up 't (up 'r 'theta) (up 'rdot 'thetadot))))
```

$$
\begin{pmatrix}
t \\
\begin{pmatrix} r\cos{(\theta)} \\ r\sin{(\theta)} \end{pmatrix} \\
\begin{pmatrix} -r\dot\theta\sin{(\theta)} + \dot r\cos{(\theta)} \\ r\dot\theta\cos{(\theta)} + \dot r\sin{(\theta)} \end{pmatrix}
\end{pmatrix}
$$

---

[90]This `F->C` is more general than the code introduced on page 46 in that it
allows computation of transformations of higher derivatives of the local tuple,
if required.

To make this work, `Gamma` is also extended to generate more elements of the
local tuple than are needed for Lagrangians that depend on time, coordinates,
and velocities. Here `Gamma` is given one more argument than it usually has.
This optional argument gives the length of the initial segment of the local
tuple needed. The default length is 3, giving components of the local tuple up
to and including the velocities.

Notice that in this definition of F->C we do not explicitly calculate any derivatives. The calculation that led up to the state transformation (1.77) is not needed.

We can also use $\bar{\Gamma}$ to make an elegant formula for computing the total time derivative $D_t F$ of the function F:

$$D_t F = \bar{\Gamma}(\bar{G}), \qquad \text{with} \qquad \bar{G}[q] = D(F \circ \Gamma[q]). \qquad (1.175)$$

The total time derivative can be expressed as a program:

```
(define (Dt F)
  (define (DtF state)
    (let ((n (vector-length state)))
      (define (DF-on-path q)
        (D (compose F (Gamma q (- n 1))))))
      ((Gamma-bar DF-on-path) state)))
  DtF)
```

Given a procedure F implementing a local-tuple function and a path q, we construct a new procedure (compose F (Gamma q)). The procedure DF-on-path implements the derivative of this function of time. We then abstract this off the path with Gamma-bar to give the total time derivative.

Exercise 1.37: Velocity transformation

Use the procedure Gamma-bar to construct a procedure that transforms velocities given a coordinate transformation. Apply this procedure to the procedure p->r to deduce (again) equation (1.67) on page 42.

Lagrange equations at a moment

Given a Lagrangian, the Lagrange equations test paths to determine whether they are realizable paths of the system. The Lagrange equations relate the path and its derivatives. The fact that the Lagrange equations must be satisfied at each moment suggests that we can abstract the Lagrange equations off the path and write them as relations among the local-tuple components of realizable paths.

Let $\bar{E}[L]$ be the path-dependent function that produces the residuals of the Lagrange equations (1.12) for the Lagrangian L:

$$\bar{E}[L][q] = D(\partial_2 L \circ \Gamma[q]) - \partial_1 L \circ \Gamma[q]. \qquad (1.176)$$

Realizable paths q satisfy the Lagrange equations

$$\bar{\mathsf{E}}[L][q] = 0. \tag{1.177}$$

The path-dependent Lagrange equations can be converted to local Lagrange equations using $\bar{\Gamma}$:

$$\mathsf{E}[L] = \bar{\Gamma}(\bar{\mathsf{E}}[L]). \tag{1.178}$$

The operator E is called the *Euler–Lagrange operator*. In terms of this operator the Lagrange equations are

$$\mathsf{E}[L] \circ \Gamma[q] = 0. \tag{1.179}$$

The Euler–Lagrange operator is explicitly

$$\mathsf{E}[L] = D_t\partial_2 L - \partial_1 L. \tag{1.180}$$

The procedure `Euler-Lagrange-operator` implements E:

```
(define (Euler-Lagrange-operator L)
  (- (Dt ((partial 2) L)) ((partial 1) L))) .
```

For example, applied to the Lagrangian for the harmonic oscillator, we have

```
((Euler-Lagrange-operator
  (L-harmonic 'm 'k))
 (up 't 'x 'v 'a))
(+ (* a m) (* k x))
```

Notice that the components of the local tuple are individually specified. Using equation (1.179), the Lagrange equations for the harmonic oscillator are

```
((compose
  (Euler-Lagrange-operator (L-harmonic 'm 'k))
  (Gamma (literal-function 'x) 4))
 't)
(+ (* k (x t)) (* m (((expt D 2) x) t)))
```

Exercise 1.38: Properties of E

Let F and G be two Lagrangian-like functions of a local tuple, C be a local-tuple transformation function, and c a constant. Demonstrate the following properties:

a. $\mathsf{E}[F + G] = \mathsf{E}[F] + \mathsf{E}[G]$

b. $\mathsf{E}[cF] = c\mathsf{E}[F]$

c. $\mathsf{E}[FG] = \mathsf{E}[F]G + F\mathsf{E}[G] + (D_tF)\partial_2 G + \partial_2 F(D_t G)$

d. $\mathsf{E}[F \circ C] = D_t(DF \circ C)\partial_2 C + DF \circ C\mathsf{E}[C]$

1.10   Constrained Motion

An advantage of the Lagrangian approach is that coordinates can often be chosen that exactly describe the freedom of the system, automatically incorporating any constraints. We may also use coordinates that have more freedom than the system actually has and consider explicit constraints among the coordinates. For example, the planar pendulum has a one-dimensional configuration space. We have formulated this problem using the angle from the vertical as the configuration coordinate. Alternatively, we may choose to represent the pendulum as a body moving in the plane, constrained to be on the circle of the correct radius around the pivot. We would like to have valid descriptions for both choices and show they are equivalent. In this section we develop tools to handle problems with explicit constraints. The constraints considered here are more general than those used in the demonstration that the Lagrangian for systems with rigid constraints can be written as the difference of kinetic and potential energies (see section 1.6.2).

Suppose the configuration of a system with n degrees of freedom is specified by $n + 1$ coordinates and that configuration paths q are constrained to satisfy some relation of the form

$$\varphi(t, q(t), Dq(t)) = 0. \tag{1.181}$$

How do we formulate the equations of motion? One approach would be to use the constraint equation to eliminate one of the coordinates in favor of the rest; then the evolution of the reduced set of generalized coordinates would be described by the usual Lagrange equations. The equations governing the evolution of coordinates that are not fully independent should be equivalent.

We can address the problem of formulating equations of motion for systems with redundant coordinates by returning to the action principle. Realizable paths are distinguished from other paths by having stationary action. Stationary refers to the fact that the action does not change with certain small variations of the path. What variations should be considered? We have seen that velocity-independent rigid constraints can be used to eliminate redundant coordinates. In the irredundant coordinates we distinguished realizable paths by using variations that by construction satisfy the constraints. Thus in the case where constraints can be used to eliminate redundant coordinates we can restrict the variations in the path to those that are consistent with the constraints.

So how does the restriction of the possible variations affect the argument that led to Lagrange's equations (refer to section 1.5)? Actually most of the calculation is unaffected. The condition that the action is stationary still reduces to the conditions (1.17) or (1.34):

$$0 = \int_{t_1}^{t_2} \left\{ (\partial_1 L \circ \Gamma[q]) - D\left(\partial_2 L \circ \Gamma[q]\right) \right\} \eta. \tag{1.182}$$

At this point we argued that because the variations η are arbitrary (except for conditions at the endpoints), the only way for the integral to be zero is for the integrand to be zero. Furthermore, the freedom in our choice of η allowed us to deduce that the factor multiplying η in the integrand must be identically zero, thereby deriving Lagrange's equations.

Now the choice of η is not completely free. We can still deduce from the arbitrariness of η that the integrand must be zero,[91] but we can no longer deduce that the factor multiplying η is zero (only that the projection of this factor onto acceptable variations is zero). So we have

$$\left\{ (\partial_1 L \circ \Gamma[q]) - D\left(\partial_2 L \circ \Gamma[q]\right) \right\} \eta = 0, \tag{1.183}$$

with η subject to the constraints.

A path q satisfies the constraint if $\bar{\varphi}[q] = \varphi \circ \Gamma[q] = 0$. The constraint must be satisfied even for the varied path, so we allow only variations η for which the variation of the constraint is zero:

$$\delta_\eta(\bar{\varphi}) = 0. \tag{1.184}$$

---

[91] Given any acceptable variation, we may make another acceptable variation by multiplying the given one by a bump function that emphasizes any particular time interval.

We can say that the variation must be "tangent" to the constraint surface. Expanding this with the chain rule, a variation η is tangent to the constraint surface φ if

$$(\partial_1\varphi \circ \Gamma[q])\,\eta + (\partial_2\varphi \circ \Gamma[q])\,D\eta = 0. \tag{1.185}$$

Note that these are functions of time; the variation at a given time is tangent to the constraint at that time.

1.10.1   Coordinate Constraints

Consider constraints that do not depend on velocities:

$$\partial_2\varphi = 0.$$

In this case the variation is tangent to the constraint surface if

$$(\partial_1\varphi \circ \Gamma)\,\eta = 0. \tag{1.186}$$

Together, equations (1.183) and (1.186) should determine the motion, but how do we eliminate η? The residual of the Lagrange equations is orthogonal[92] to any η that is orthogonal to the normal to the constraint surface. A vector that is orthogonal to all vectors orthogonal to a given vector is parallel to the given vector. Thus, the residual of Lagrange's equations is parallel to the normal to the constraint surface; the two must be proportional:

$$D\left(\partial_2 L \circ \Gamma[q]\right) - \partial_1 L \circ \Gamma[q] = \lambda(\partial_1\varphi \circ \Gamma[q]). \tag{1.187}$$

That the two vectors are parallel everywhere along the path does not guarantee that the proportionality factor is the same at each moment along the path, so the proportionality factor λ is some function of time, which may depend on the path under consideration. These equations, with the constraint equation $\varphi \circ \Gamma[q] = 0$, are the governing equations. These equations are sufficient to determine the path q and to eliminate the unknown function λ.

---

[92]We take two tuple-valued functions of time to be orthogonal if at each instant the dot product of the tuples is zero. Similarly, tuple-valued functions are considered parallel if at each moment one of the tuples is a scalar multiple of the other. The scalar multiplier is in general a function of time.

Now watch this

Suppose we form an augmented Lagrangian by treating λ as one of the coordinates:

$$L'(t; q, \lambda; \dot{q}, \dot{\lambda}) = L(t, q, \dot{q}) + \lambda \varphi(t, q, \dot{q}). \tag{1.188}$$

The Lagrange equations associated with the coordinates q are just the modified Lagrange equations (1.187), and the Lagrange equation associated with λ is just the constraint equation. (Note that $\dot{\lambda}$ does not appear in the augmented Lagrangian.) So the Lagrange equations for this augmented Lagrangian fully encapsulate the modification to the Lagrange equations that is imposed by the addition of an explicit coordinate constraint, at the expense of introducing extra degrees of freedom. Notice that this Lagrangian is of the same form as the Lagrangian (equation 1.93) that we used in the derivation of $L = T - V$ for rigid systems (section 1.6.2).

Alternatively

How do we know that we have enough information to eliminate the unknown function λ from equations (1.187), or that the extra degree of freedom introduced in Lagrangian (1.188) is purely formal?

If λ can be written as a composition of a state-dependent function with the path: $\lambda = \Lambda \circ \Gamma[q]$ then it is redundant as a degree of freedom. Consider the Lagrangian

$$L'' = L + \Lambda \varphi. \tag{1.189}$$

This new Lagrangian has no extra degrees of freedom. The Lagrange equations for L'' are the Lagrange equations for L with additional terms arising from the product $\Lambda \varphi$. Applying the Euler–Lagrange operator E (see section 1.9) to this Lagrangian gives[93]

$$\mathsf{E}[L''] = \mathsf{E}[L] + \mathsf{E}[\Lambda \varphi]$$
$$= \mathsf{E}[L] + \Lambda \,\mathsf{E}[\varphi] + \mathsf{E}[\Lambda]\,\varphi + D_t \Lambda\,\partial_2 \varphi + \partial_2 \Lambda\, D_t \varphi. \tag{1.190}$$

Composition of $\mathsf{E}[L'']$ with $\Gamma[q]$ gives the Lagrange equations for the path q. Using the fact that the constraint is satisfied on the path $\varphi \circ \Gamma[q] = 0$ and consequently $D_t \varphi \circ \Gamma[q] = 0$, we have

---

[93]Recall that the Euler–Lagrange operator E has the property

$$\mathsf{E}[FG] = F\,\mathsf{E}[G] + \mathsf{E}[F]\,G + D_t F\,\partial_2 G + \partial_2 F\,D_t G.$$

$$\mathsf{E}\,[L''] \circ \Gamma[q]$$
$$= \mathsf{E}\,[L] \circ \Gamma[q] + \lambda(\mathsf{E}\,[\varphi] \circ \Gamma[q]) + D\lambda(\partial_2\varphi \circ \Gamma[q]), \qquad (1.191)$$

where we have used $\lambda = \Lambda \circ \Gamma[q]$. If we now use the fact that we are dealing only with coordinate constraints, $\partial_2\varphi = 0$, then

$$\mathsf{E}\,[L''] \circ \Gamma[q] = \mathsf{E}\,[L] \circ \Gamma[q] + \lambda(\mathsf{E}[\varphi] \circ \Gamma[q]). \qquad (1.192)$$

The Lagrange equations are the same as those derived from the augmented Lagrangian L'. The difference is that now we see that $\lambda = \Lambda \circ \Gamma[q]$ is determined by the unaugmented state. This is the same as saying that λ can be eliminated.

Considering only the formal validity of the Lagrange equations for the augmented Lagrangian, we could not deduce that λ could be written as the composition of a state-dependent function Λ with $\Gamma[q]$. The explicit Lagrange equations derived from the augmented Lagrangian depend on the accelerations D^2q as well as λ, so we cannot deduce separately that either is the composition of a state-dependent function and $\Gamma[q]$. However, now we see that λ is such a composition. This allows us to deduce that D^2q is also a state-dependent function composed with the path. The evolution of the system is determined from the dynamical state.

The pendulum using constraints

The pendulum can be formulated as the motion of a massive particle in a vertical plane subject to the constraint that the distance to the pivot is constant (see figure 1.8).

In this formulation, the kinetic and potential energies in the Lagrangian are those of an unconstrained particle in a uniform gravitational acceleration. A Lagrangian for the unconstrained particle is

$$L(t; x, y; v_x, v_y) = \tfrac{1}{2}m(v_x^2 + v_y^2) - mgy. \qquad (1.193)$$

The constraint that the pendulum moves in a circle of radius l about the pivot is[94]

$$x^2 + y^2 - l^2 = 0. \qquad (1.194)$$

---

[94]This constraint has the same form as those used in the demonstration that $L = T - V$ can be used for rigid systems. Here it is a particular example of a more general set of constraints.

The augmented Lagrangian is

$$L'(t; x, y, \lambda; v_x, v_y, \dot{\lambda}) = \tfrac{1}{2}m(v_x^2 + v_y^2) - mgy + \lambda(x^2 + y^2 - l^2). \quad (1.195)$$

The Lagrange equations for the augmented Lagrangian are

$$mD^2x - 2\lambda x = 0 \qquad\qquad (1.196)$$
$$mD^2y + mg - 2\lambda y = 0 \qquad\qquad (1.197)$$
$$x^2 + y^2 - l^2 = 0. \qquad\qquad (1.198)$$

These equations are sufficient to solve for the motion of the pendulum.

It should not be surprising that these equations simplify if we switch to "polar" coordinates

$$x = r\sin\theta \quad y = -r\cos\theta. \qquad\qquad (1.199)$$

Substituting this into the constraint equation, we determine that $r = l$, a constant. Forming the derivatives and substituting into the other two equations, we find

$$ml(\cos\theta D^2\theta - \sin\theta(D\theta)^2) - 2\lambda\sin\theta = 0 \qquad (1.200)$$
$$ml(\sin\theta D^2\theta + \cos\theta(D\theta)^2) + mg + 2\lambda\cos\theta = 0. \qquad (1.201)$$

Multiplying the first by $\cos\theta$ and the second by $\sin\theta$ and adding, we find

$$mlD^2\theta + mg\sin\theta = 0, \qquad\qquad (1.202)$$

which we recognize as the correct equation for the pendulum. This is the same as the Lagrange equation for the pendulum using the unconstrained generalized coordinate θ. For completeness, we can find λ in terms of the other variables:

$$\lambda = \frac{mD^2x}{2x} = -\frac{1}{2l}(mg\cos\theta + ml(D\theta)^2). \qquad (1.203)$$

This confirms that λ is really the composition of a function of the state with the state path. Notice that $2l\lambda$ is a force—it is the sum of the outward component of the gravitational force and the centrifugal force. Using this interpretation in the two coordinate equations of motion, we see that the terms involving λ are the forces that must be applied to the unconstrained particle to make it move on the circle required by the constraints. Equivalently, we

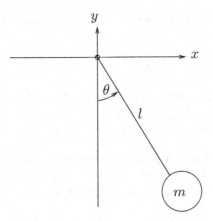

Figure 1.8   We can formulate the behavior of a pendulum as motion in the plane, constrained to a circle about the pivot.

may think of $2l\lambda$ as the tension in the pendulum rod that holds the mass.[95]

Building systems from parts

The method of using augmented Lagrangians to enforce constraints on dynamical systems provides a way to analyze a compound system by combining the results of the analysis of the parts of the system and the coupling between them.

Consider the compound spring-mass system shown at the top of figure 1.9. We could analyze this as a monolithic system with two configuration coordinates x_1 and x_2, representing the extensions of the springs from their equilibrium lengths X_1 and X_2.

An alternative procedure is to break the system into several parts. In our spring-mass system we can choose two parts: one is a spring and mass attached to the wall, and the other is a spring and mass with its attachment point at an additional configuration coordinate ξ. We can formulate a Lagrangian for each part separately. We can then choose a Lagrangian for the composite system as the sum of the two component Lagrangians with a constraint $\xi = X_1 + x_1$ to accomplish the coupling.

---

[95]Indeed, if we had scaled the constraint equations as we did in the discussion of Newtonian constraint forces, we could have identified λ with the the magnitude of the constraint force F. However, though λ will in general be related to the constraint forces it will not be one of them. We chose to leave the scaling as it naturally appeared rather than make things turn out artificially pretty.

Let's see how this works. The Lagrangian for the subsystem attached to the wall is

$$L_1(t, x_1, \dot{x}_1) = \tfrac{1}{2}m_1\dot{x}_1^2 - \tfrac{1}{2}k_1x_1^2 \tag{1.204}$$

and the Lagrangian for the subsystem that attaches to it is

$$L_2(t; \xi, x_2; \dot{\xi}, \dot{x}_2) = \tfrac{1}{2}m_2(\dot{\xi} + \dot{x}_2)^2 - \tfrac{1}{2}k_2x_2^2. \tag{1.205}$$

We construct a Lagrangian for the system composed from these parts as a sum of the Lagrangians for each of the separate parts, with a coupling term to enforce the constraint:

$$
\begin{aligned}
&L(t; x_1, x_2, \xi, \lambda; \dot{x}_1, \dot{x}_2, \dot{\xi}, \dot{\lambda}) \\
&\quad = L_1(t, x_1, \dot{x}_1) + L_2(t; \xi, x_2; \dot{\xi}, \dot{x}_2) \\
&\qquad + \lambda(\xi - (X_1 + x_1)).
\end{aligned}
\tag{1.206}
$$

Thus we can write Lagrange's equations for the four configuration coordinates, in order, as follows:

$$m_1D^2x_1 = -k_1x_1 - \lambda \tag{1.207}$$
$$m_2(D^2\xi + D^2x_2) = -k_2x_2 \tag{1.208}$$
$$m_2(D^2\xi + D^2x_2) = \lambda \tag{1.209}$$
$$0 = \xi - (X_1 + x_1). \tag{1.210}$$

Notice that in this system λ is the force of constraint holding the system together. We can now eliminate the "glue" coordinates ξ and λ to obtain the equations of motion in the coordinates x_1 and x_2:

$$m_1D^2x_1 + m_2(D^2x_1 + D^2x_2) + k_1x_1 = 0 \tag{1.211}$$
$$m_2(D^2x_1 + D^2x_2) + k_2x_2 = 0. \tag{1.212}$$

This strategy can be generalized. We can make a library of primitive components. Each component may be characterized by a Lagrangian with additional degrees of freedom for the *terminals* where that component may be attached to others. We then can construct composite Lagrangians by combining components, using constraints to glue together the terminals.

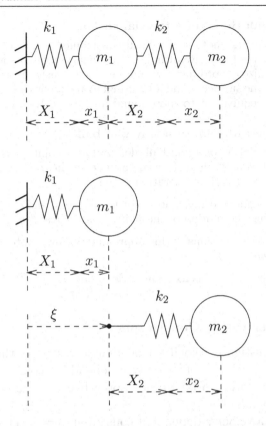

Figure 1.9 A compound spring-mass system is decomposed into two subsystems. We have two springs coupling two masses that can move horizontally. The equilibrium positions of the springs are X_1 and X_2. The systems are coupled by the coordinate constraint $\xi = X_1 + x_1$.

Exercise 1.39: Combining Lagrangians

a. Make another primitive component, compatible with the spring-mass structures described in this section. For example, make a pendulum that can attach to the spring-mass system. Build a combination and derive the equations of motion. Be careful, the algebra is horrible if you choose bad coordinates.

b. For a nice little project, construct a family of compatible mechanical parts, characterized by appropriate Lagrangians, that can be combined in a variety of ways to make interesting mechanisms. Remember that in a good language the result of combining pieces should be a piece of the same kind that can be further combined with other pieces.

Exercise 1.40: Bead on a triaxial surface

Consider again the motion of a bead constrained to move on a triaxial surface (exercise 1.18). Reformulate this using rectangular coordinates as the generalized coordinates with an explicit constraint that the bead must stay on the surface. Find a Lagrangian and show that the Lagrange equations are equivalent to those found in exercise 1.18.

Exercise 1.41: Motion of a tiny golf ball

Consider the motion of a golf ball idealized as a point mass constrained to a frictionless smooth surface of varying height $h(x, y)$ in a uniform gravitational field with acceleration g.

a. Find an augmented Lagrangian for this system, and derive the equations governing the motion of the point mass in x and y.

b. Under what conditions is this approximated by a potential function $V(x, y) = mgh(x, y)$?

c. Assume that $h(x, y)$ is axisymmetric about $x = y = 0$. Can you find such an h that yields motions with closed orbits?

1.10.2   Derivative Constraints

Here we investigate velocity-dependent constraints that are "total time derivatives" of velocity-independent constraints. The methods presented so far do not apply because the constraint is velocity-dependent.

Consider a velocity-dependent constraint $\psi = 0$. That ψ is a total time derivative means that there exists a velocity-independent function φ such that

$$\psi \circ \Gamma[q] = D(\varphi \circ \Gamma[q]). \tag{1.213}$$

That φ is velocity-independent means $\partial_2 \varphi = 0$. As state functions the relationship between ψ and φ is

$$\psi = D_t \varphi = \partial_0 \varphi + \partial_1 \varphi \dot{Q}. \tag{1.214}$$

Given a ψ we can find φ by solving this linear partial differential equation. The solution is determined up to a constant, so $\psi = 0$ implies $\varphi = K$ for some constant K. On the other hand, if we knew $\varphi = K$ then $\psi = 0$ follows. Thus the velocity-dependent constraint $\psi = 0$ is equivalent to the velocity-independent constraint $\varphi = K$, and we know how to find Lagrange equations for such systems.

If L is a Lagrangian for the unconstrained problem, the Lagrange equations with the constraint $\varphi = K$ are

$$\mathsf{E}[L] \circ \Gamma[q] + \lambda \left(\mathsf{E}[\varphi] \circ \Gamma[q] \right) = 0, \tag{1.215}$$

where λ is a function of time that will be eliminated during the solution process. The constant K does not affect the Lagrange equations. The function φ is independent of velocity, $\partial_2 \varphi = 0$, so the Lagrange equations become

$$\mathsf{E}[L] \circ \Gamma[q] - \lambda(\partial_1 \varphi \circ \Gamma[q]) = 0. \tag{1.216}$$

From equation (1.214) we see that

$$\partial_1 \varphi = \partial_2 \psi, \tag{1.217}$$

so the Lagrange equations with the constraint $\psi = 0$ are

$$\mathsf{E}[L] \circ \Gamma[q] = \lambda(\partial_2 \psi \circ \Gamma[q]). \tag{1.218}$$

The important feature is that we can write the Lagrange equations directly in terms of ψ without having to produce φ. But the validity of these Lagrange equations depends on the existence of φ.

It turns out that the augmented Lagrangian trick also works here. These Lagrange equations are given if we augment the Lagrangian with the constraint ψ multiplied by a function of time λ':

$$L' = L + \lambda' \psi. \tag{1.219}$$

The Lagrange equations for L' turn out to be

$$\mathsf{E}[L] \circ \Gamma[q] = -D\lambda'(\partial_2 \psi \circ \Gamma[q]), \tag{1.220}$$

which, with the identification $\lambda = -D\lambda'$, are the same as Lagrange equations (1.218).

Sometimes a problem can be naturally formulated in terms of velocity-dependent constraints. The formalism we have developed will handle any velocity-dependent constraint that can be written in terms of the derivative of a coordinate constraint. Such a constraint is called an *integrable constraint*. Any system for which the constraints can be put in the form of a coordinate constraint, or are already in that form, is called a *holonomic system*.

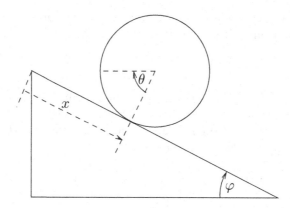

Figure 1.10   A massive hoop rolling, without slipping, down an inclined plane.

Exercise 1.42: Augmented Lagrangian

Show that the augmented Lagrangian (1.219) does lead to the Lagrange equations (1.220), taking into account the fact that ψ is a total time derivative of φ.

Goldstein's hoop

Here we consider a problem for which the constraint can be represented as a time derivative of a coordinate constraint: a hoop of mass M and radius R rolling, without slipping, down a (one-dimensional) inclined plane (see figure 1.10).[96]

We will formulate this problem in terms of the two coordinates θ, the rotation of an arbitrary point on the hoop from an arbitrary reference direction, and x, the linear progress down the inclined plane. The constraint is that the hoop does not slip. Thus a change in θ is exactly reflected in a change in x; the constraint function is

$$\psi(t; x, \theta; \dot{x}, \dot{\theta}) = R\dot{\theta} - \dot{x}. \tag{1.221}$$

This constraint is phrased as a relation among generalized velocities, but it could be integrated to get $x = R\theta + c$. We may form our augmented Lagrangian with either the integrated constraint or its derivative.

---

[96]This example appears in [20], pp. 49–51,

The kinetic energy has two parts, the energy of rotation of the hoop and the energy of the motion of its center of mass.[97] The potential energy of the hoop decreases as the height decreases. Thus we may write the augmented Lagrangian:

$$L(t; x, \theta, \lambda; \dot{x}, \dot{\theta}, \dot{\lambda})$$
$$= \tfrac{1}{2} M R^2 \dot{\theta}^2 + \tfrac{1}{2} M \dot{x}^2 + M g x \sin \varphi + \lambda (R \dot{\theta} - \dot{x}). \qquad (1.222)$$

Lagrange's equations are

$$M D^2 x - D\lambda = M g \sin \varphi \qquad (1.223)$$
$$M R^2 D^2 \theta + R \, D\lambda = 0 \qquad (1.224)$$
$$R \, D\theta - Dx = 0. \qquad (1.225)$$

And by differentiation of the third Lagrange equation we obtain

$$D^2 x = R D^2 \theta. \qquad (1.226)$$

By combining these equations we can solve for the dynamical quantities of interest. For this case of a rolling hoop the linear acceleration

$$D^2 x = \tfrac{1}{2} g \sin \varphi \qquad (1.227)$$

is just half of what it would have been if the mass had just slid down a frictionless plane without rotating. Note that for this hoop $D^2 x$ is independent of both M and R. We see from the Lagrange equations that $D\lambda$ can be interpreted as the friction force involved in enforcing the constraint. The frictional force of constraint is

$$D\lambda = \tfrac{1}{2} M g \sin \varphi \qquad (1.228)$$

and the angular acceleration is

$$D^2 \theta = \frac{1}{2} \frac{g}{R} \sin \varphi. \qquad (1.229)$$

---

[97]We will see in chapter 2 how to compute the kinetic energy of rotation, but for now the answer is $\tfrac{1}{2} M R^2 \dot{\theta}^2$.

1.10.3   Nonholonomic Systems

Systems with constraints that are not integrable are termed *non-holonomic systems*. A constraint is not integrable if it cannot be written in terms of an equivalent coordinate constraint. An example of a nonholonomic system is a ball rolling without slipping in a bowl. As the ball rolls it must turn so that its surface does not move relative to the bowl at the point of contact. This looks as if it might establish a relation between the location of the ball in the bowl and the orientation of the ball, but it doesn't. The ball may return to the same place in the bowl with different orientations depending on the intervening path it has taken. As a consequence, the constraints cannot be used to eliminate any coordinates.

What are the equations of motion governing nonholonomic systems? For the restricted set of systems with nonholonomic constraints that are linear in the velocities, it is widely reported[98] that the equations of motion are as follows. Let ψ have the form

$$\psi(t, q, v) = G_1(t, q)v + G_2(t, q), \tag{1.230}$$

a state function that is linear in the velocities. We assume ψ is not a total time derivative. If L is a Lagrangian for the unconstrained system, then the equations of motion are asserted to be

$$\mathsf{E}[L] \circ \Gamma[q] = \lambda(G_1 \circ \Gamma[q]) = \lambda(\partial_2\psi \circ \Gamma[q]). \tag{1.231}$$

With the constraint $\psi = 0$, the system is completely specified and the evolution of the system is determined. Note that these equations are identical to the Lagrange equations (1.218) for the case that ψ is a total time derivative, but here the derivation of those equations is no longer valid.

An essential step in the derivation of the Lagrange equations for coordinate constraints $\varphi = 0$ with $\partial_2\varphi = 0$ was to note that two conditions must be satisfied:

$$(\mathsf{E}[L] \circ \Gamma[q])\eta = 0, \tag{1.232}$$

and

$$(\partial_1\varphi \circ \Gamma[q])\eta = 0. \tag{1.233}$$

---

[98]For some treatments of nonholonomic systems see, for example, Whittaker [46], Goldstein [20], Gantmakher [19], or Arnold et al. [6].

Because $\mathsf{E}[L] \circ \Gamma[q]$ is orthogonal to η and η is constrained to be orthogonal to $\partial_1 \varphi \circ \Gamma[q]$, the two must be parallel at each moment:

$$\mathsf{E}[L] \circ \Gamma[q] = \lambda(\partial_1 \varphi \circ \Gamma[q]). \tag{1.234}$$

The Lagrange equations for derivative constraints were derived from this.

This derivation does not go through if the constraint function depends on velocity. In this case, for a variation η to be consistent with the velocity-dependent constraint function ψ it must satisfy (see equation 1.185)

$$(\partial_1 \psi \circ \Gamma[q])\eta + (\partial_2 \psi \circ \Gamma[q])D\eta = 0. \tag{1.235}$$

We may no longer eliminate η by the same argument, because η is no longer orthogonal to $\partial_1 \psi \circ \Gamma[q]$, and we cannot rewrite the constraint as a coordinate constraint because ψ is, by assumption, not integrable.

The following is the derivation of the nonholonomic equations from Arnold et al. [6], translated into our notation. Define a "virtual velocity" ξ to be any velocity satisfying

$$(\partial_2 \psi \circ \Gamma[q])\xi = 0. \tag{1.236}$$

The "principle of d'Alembert–Lagrange," according to Arnold, states that

$$(\mathsf{E}[L] \circ \Gamma[q])\xi = 0, \tag{1.237}$$

for any virtual velocity ξ. Because ξ is arbitrary except that it is required to be orthogonal to $\partial_2 \psi \circ \Gamma[q]$ and any such ξ is orthogonal to $\mathsf{E}[L] \circ \Gamma[q]$, then $\partial_2 \psi \circ \Gamma[q]$ must be parallel to $\mathsf{E}[L] \circ \Gamma[q]$. So

$$\mathsf{E}[L] \circ \Gamma[q] = \lambda(\partial_2 \psi \circ \Gamma[q]), \tag{1.238}$$

which are the nonholonomic equations.

To convert the stationary action equations to the equations of Arnold we must do the following. To get from equation (1.232) to equation (1.237), we must replace η by ξ. However, to get from equation (1.235) to equation (1.236), we must set $\eta = 0$ and replace $D\eta$ by ξ. All "derivations" of the nonholonomic equations have similar identifications. It comes down to this: the nonholonomic equations do not follow from the action principle. They are something else. Whether they are correct or not depends on whether or not they agree with experiment.

For systems with either coordinate constraints or derivative constraints, we have found that the Lagrange equations can be derived from a Lagrangian that is augmented with the constraint. However, if the constraints are not integrable the Lagrange equations for the augmented Lagrangian are not the same as the nonholonomic system (equations 1.231).[99] Let L' be an augmented Lagrangian with non-integrable constraint ψ:

$$L'(t; q, \lambda; \dot{q}, \dot{\lambda}) = L(t, q, \dot{q}) + \lambda\psi(t, q, \dot{q}); \tag{1.239}$$

then the Lagrange equations associated with the coordinates are

$$
\begin{aligned}
0 = {}& \mathsf{E}[L] \circ \Gamma[q] \\
& + D\lambda(\partial_2\psi \circ \Gamma[q]) + \lambda D(\partial_2\psi \circ \Gamma[q]) - \lambda(\partial_1\psi \circ \Gamma[q]).
\end{aligned} \tag{1.240}
$$

The Lagrange equation associated with λ is just the constraint equation

$$\psi \circ \Gamma[q] = 0. \tag{1.241}$$

An interesting feature of these equations is that they involve both λ and $D\lambda$. Thus the usual state variables q and Dq, with the constraint, are not sufficient to determine a full set of initial conditions for the derived Lagrange equations; we need to specify an initial value for λ as well.

In general, for any particular physical system, equations (1.231) and (1.240) are not the same, and in fact they have different solutions. It is not apparent that either set of equations accurately models the physical system. The first approach to nonholonomic systems is not justified by extension of the arguments for the holonomic case and the other is not fully determined. Perhaps this indicates that the models are inadequate, that more details of how the constraints are maintained need to be specified.

---

[99] Arnold et al. [6] call the variational mechanics with the constraints added to the Lagrangian *Vakonomic mechanics*.

1.11 Summary

To analyze a mechanical system we construct an action function
that gives us a way to distinguish realizable motions from other
conceivable motions of the system. The action function is constructed so as to be stationary only on paths describing realizable
motions, with respect to variations of the path. This is the *principle of stationary action*. The principle of stationary action is a
coordinate-independent specification of the realizable paths. For
systems with or without constraints we may choose any system
of coordinates that uniquely determines the configuration of the
system.

An action is an integral of a function, the *Lagrangian*, along
the path. For many systems an appropriate Lagrangian is the
difference of the kinetic energy and the potential energy of the
system. The choice of a Lagrangian for a system is not unique.

For any system for which we have a Lagrangian action we can
formulate a system of ordinary differential equations, the Lagrange
equations, that is satisfied by any realizable path. The method of
deriving the Lagrange equations from the Lagrangian is independent of the coordinate system used to formulate the Lagrangian.
One freedom we have in formulation is that the addition of a total time derivative to a Lagrangian for a system yields another
Lagrangian that has the same Lagrange equations.

The Lagrange equations are a set of ordinary differential equations: there is a finite state that summarizes the history of the
system and is sufficient to determine the future. There is an effective procedure for evolving the motion of the system from a
state at an instant. For many systems the state is determined by
the coordinates and the rate of change of the coordinates at an
instant.

If there are continuous symmetries in a physical system there
are conserved quantities associated with them. If the system can
be formulated in such a way that the symmetries are manifest in
missing coordinates in the Lagrangian, then there are conserved
momenta conjugate to those coordinates. If the Lagrangian is
independent of time then there is a conserved energy.

1.12   Projects

Exercise 1.43: A numerical investigation

Consider a pendulum: a mass m supported on a massless rod of length l in a uniform gravitational field. A Lagrangian for the pendulum is

$$L(t, \theta, \dot\theta) = \frac{m}{2}(l\dot\theta)^2 + mgl\cos\theta.$$

For the pendulum, the period of the motion depends on the amplitude. We wish to find trajectories of the pendulum with a given frequency. Three methods of doing this present themselves: (1) solution by the principle of least action, (2) numerical integration of Lagrange's equation, and (3) analytic solution (which requires some exposure to elliptic functions). We will carry out all three and compare the solution trajectories.

Consider the parameters $m = 1\,\mathrm{kg}$, $l = 1\,\mathrm{m}$, $g = 9.8\,\mathrm{m\,s^{-2}}$. The frequency of small-amplitude oscillations is $\omega_0 = \sqrt{g/l}$. Let's find the nontrivial solution that has the frequency $\omega_1 = \frac{4}{5}\omega_0$.

a. The angle is periodic in time, so a Fourier series representation is appropriate. We can choose the origin of time so that a zero crossing of the angle is at time zero. Since the potential is even in the angle, the angle is an odd function of time. Thus we need only a sine series. Since the angle returns to zero after one-half period, the angle is an odd function of time about the midpoint. Thus only odd terms of the series are present:

$$\theta(t) = \sum_{n=1}^{m} A_n \sin((2n-1)\omega_1 t).$$

The amplitude of the trajectory is $A = \theta_{\max} = \sum_{n=1}^{\infty}(-1)^{n+1}A_n$.

Find approximations to the first few coefficients A_n by minimizing the action. You will have to write a program similar to the `find-path` procedure in section 1.4. Watch out: there is more than one trajectory that minimizes the action.

b. Write a program to numerically integrate Lagrange's equations for the trajectories of the pendulum. The trouble with using numerical integration to solve this problem is that we do not know how the frequency of the motion depends on the initial conditions. So we have to guess, and then gradually improve our guess. Define a function $\Omega(\dot\theta)$ that numerically computes the frequency of the motion as a function of the initial angular velocity (with $\theta = 0$). Find the trajectory by solving $\Omega(\dot\theta) = \omega$ for the initial angular velocity of the desired trajectory. Methods of solving this equation include successive bisection, minimizing the squared residual, etc.—choose one.

c. Now let's formulate the analytic solution for the frequency as a function of amplitude. The period of the motion is simply

$$T = 4 \int_0^{T/4} dt = 4 \int_0^A \frac{1}{\dot{\theta}} d\theta.$$

Using the energy, solve for $\dot{\theta}$ in terms of the amplitude A and θ to write the required integral explicitly. This integral can be written in terms of elliptic functions, but in a sense this does not solve the problem—we still have to compute the elliptic functions. Let's avoid this excursion into elliptic functions and just do the integral numerically using the procedure `definite-integral`. We still have the problem that we can specify the amplitude A and get the frequency; to solve our problem we need to solve the inverse problem, but that can be done as in part **b**.

Exercise 1.44: Double pendulum behavior

Consider the ideal double pendulum shown in figure 1.11.

a. Formulate a Lagrangian to describe the dynamics. Derive the equations of motion in terms of the given angles θ_1 and θ_2. Put the equations into a form appropriate for numerical integration. Assume the following system parameters:

$$g = 9.8\,\mathrm{m\,s}^{-2}$$
$$l_1 = 1.0\,\mathrm{m}$$
$$l_2 = 0.9\,\mathrm{m}$$
$$m_1 = 1.0\,\mathrm{kg}$$
$$m_2 = 3.0\,\mathrm{kg}$$

b. Prepare graphs showing the behavior of each angle as a function of time when the system is started with the following initial conditions:

$$\theta_1(0) = \pi/2\,\mathrm{rad}$$
$$\theta_2(0) = \pi\,\mathrm{rad}$$
$$\dot{\theta}_1(0) = 0\,\mathrm{rad\,s}^{-1}$$
$$\dot{\theta}_2(0) = 0\,\mathrm{rad\,s}^{-1}$$

Make the graphs extend to 50 seconds.

c. Make a graph of the behavior of the energy of your system as a function of time. The energy should be conserved. How good is the conservation you obtained?

d. Make a new Lagrangian, for two identical uncoupled double pendulums. (Both pendulums should have the same masses and lengths.) Your new Lagrangian should have four degrees of freedom. Give initial

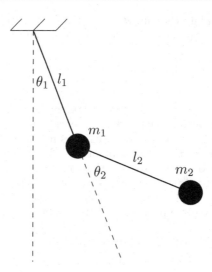

Figure 1.11   The double pendulum is pinned in two joints so that its members are free to move in a plane.

conditions for one pendulum to be the same as in the experiment of part **b** and give initial conditions for the other pendulum with the m_2 bob 10^{-10} m higher than before. The motions of the two pendulums will diverge as time progresses. Plot the logarithm of the absolute value of the difference of the positions of the m_2 bobs in your two pendulums against the time. What do you see?

e. Repeat the previous comparison, but this time use the base initial conditions:

$$\theta_1(0) = \pi/2 \,\text{rad}$$
$$\theta_2(0) = 0 \,\text{rad}$$
$$\dot{\theta}_1(0) = 0 \,\text{rad}\,\text{s}^{-1}$$
$$\dot{\theta}_2(0) = 0 \,\text{rad}\,\text{s}^{-1}$$

What do you see here?

2

Rigid Bodies

> The polhode rolls without slipping on the herpolhode lying in the invariable plane.
>
> Herbert Goldstein, *Classical Mechanics* [20], footnote, p. 207.

The motion of rigid bodies presents many surprising phenomena.

Consider the motion of a top. A top is usually thought of as an axisymmetric body, subject to gravity, with a point on the axis of symmetry that is fixed in space. The top is spun and in general executes some complicated motion. We observe that the top usually settles down into an unusual motion in which the axis of the top slowly precesses about the vertical, apparently moving perpendicular to the direction in which gravity is attempting to accelerate it.

Consider the motion of a book thrown into the air.[1] Books have three main axes. If we idealize a book as a brick with rectangular faces, the three axes are the lines through the centers of opposite faces. Try spinning the book about each axis. The motion of the book spun about the longest and the shortest axis is a simple regular rotation, perhaps with a little wobble depending on how carefully it is thrown. The motion of the book spun about the intermediate axis is qualitatively different: however carefully the book is spun about the intermediate axis, it tumbles.

The rotation of the Moon is peculiar in that the Moon always presents the same face to the Earth, indicating that the rotational period and the orbit period are the same. Considering that the orbit of the Moon is constantly changing because of interactions with the Sun and other planets, and therefore its orbital period is constantly undergoing small variations, we might expect that the face of the Moon that we see would slowly change, but it does not. What is special about the face that is presented to us?

---

[1] We put a rubber band or string around the book so that it does not open.

A rigid body may be thought of as a large number of constituent particles with rigid constraints among them. Thus the dynamical principles governing the motion of rigid bodies are the same as those governing the motion of any other system of particles with rigid constraints. What is new here is that the number of constituent particles is very large and we need to develop new tools to handle them effectively.

We have found that a Lagrangian for a system with rigid constraints can be written as the difference of the kinetic and potential energies. The kinetic and potential energies are naturally expressed in terms of the positions and velocities of the constituent particles. To write the Lagrangian in terms of the generalized coordinates and velocities we must specify functions that relate the generalized coordinates to the positions of the constituent particles. In the systems with rigid constraints considered up to now these functions were explicitly given for each of the constituent particles and individually included in the derivation of the Lagrangian. For a rigid body, however, there are too many constituent particles to handle each one of them in this way. We need to find means of expressing the kinetic and potential energies of rigid bodies in terms of the generalized coordinates and velocities, without going through the particle-by-particle details.

The strategy is to first rewrite the kinetic and potential energies in terms of quantities that characterize essential aspects of the distribution of mass in the body and the state of motion of the body. Only later do we introduce generalized coordinates. For the kinetic energy, it turns out a small number of parameters completely specify the state of motion and the relevant aspects of the distribution of mass in the body. For the potential energy, we find that for some specific problems the potential energy can be represented with a small number of parameters, but in general we have to make approximations to obtain a representation with a manageable number of parameters.

2.1   Rotational Kinetic Energy

We consider a rigid body to be made up of a large number of constituent particles with mass m_α, position $\vec{x}_\alpha$, and velocities

$\dot{\vec{x}}_\alpha$, with rigid positional constraints among them. The kinetic energy is

$$\sum_\alpha \tfrac{1}{2} m_\alpha \dot{\vec{x}}_\alpha \cdot \dot{\vec{x}}_\alpha. \tag{2.1}$$

It turns out that the kinetic energy of a rigid body can be separated into two pieces: a kinetic energy of translation and a kinetic energy of rotation. Let's see how this comes about.

The configuration of a rigid body is fully specified given the location of any point in the body and the orientation of the body. This suggests that it would be useful to decompose the position vectors for the constituent particles as the sum of the vector $\vec{X}$ to some reference position in the body and the vector $\vec{\xi}_\alpha$ from the reference position to the particular constituent element with index α:

$$\vec{x}_\alpha = \vec{X} + \vec{\xi}_\alpha. \tag{2.2}$$

Along paths, the velocities are related by

$$\dot{\vec{x}}_\alpha = \dot{\vec{X}} + \dot{\vec{\xi}}_\alpha. \tag{2.3}$$

So in terms of $\dot{\vec{X}}$ and $\dot{\vec{\xi}}_\alpha$ the kinetic energy is

$$\sum_\alpha \tfrac{1}{2} m_\alpha \left(\dot{\vec{X}} + \dot{\vec{\xi}}_\alpha \right) \cdot \left(\dot{\vec{X}} + \dot{\vec{\xi}}_\alpha \right)$$

$$= \sum_\alpha \tfrac{1}{2} m_\alpha \left(\dot{\vec{X}} \cdot \dot{\vec{X}} + 2\dot{\vec{X}} \cdot \dot{\vec{\xi}}_\alpha + \dot{\vec{\xi}}_\alpha \cdot \dot{\vec{\xi}}_\alpha \right). \tag{2.4}$$

If we select the reference position in the body to be its *center of mass*,

$$\vec{X} = \frac{1}{M} \sum_\alpha m_\alpha \vec{x}_\alpha, \tag{2.5}$$

where $M = \sum_\alpha m_\alpha$ is the total mass of the body, then

$$\sum_\alpha m_\alpha \vec{\xi}_\alpha = \sum_\alpha m_\alpha (\vec{x}_\alpha - \vec{X}) = 0. \tag{2.6}$$

So along paths the relative velocities satisfy

$$\sum_\alpha m_\alpha \dot{\vec{\xi}}_\alpha = 0. \tag{2.7}$$

The kinetic energy is then

$$\sum_\alpha \tfrac{1}{2} m_\alpha \dot{\vec{X}} \cdot \dot{\vec{X}} + \sum_\alpha \tfrac{1}{2} m_\alpha \dot{\vec{\xi}}_\alpha \cdot \dot{\vec{\xi}}_\alpha. \tag{2.8}$$

The kinetic energy is the sum of the kinetic energy of the motion of the total mass at the center of mass

$$\tfrac{1}{2} M \dot{\vec{X}} \cdot \dot{\vec{X}}, \tag{2.9}$$

and the kinetic energy of rotation about the center of mass

$$\sum_\alpha \tfrac{1}{2} m_\alpha \dot{\vec{\xi}}_\alpha \cdot \dot{\vec{\xi}}_\alpha. \tag{2.10}$$

Written in terms of appropriate generalized coordinates, the kinetic energy is a Lagrangian for a free rigid body. If we choose generalized coordinates so that the center of mass position is entirely specified by some of them and the orientation is entirely specified by others, then the Lagrange equations for a free rigid body will decouple into two groups of equations, one concerned with the motion of the center of mass and one concerned with the orientation.

Such a separation might occur in other problems, such as a rigid body moving in a uniform gravitational field, but in general, potential energies cannot be separated as the kinetic energy separates. So the motion of the center of mass and the rotational motion are usually coupled through the potential. Even in these cases, it is usually an advantage to choose generalized coordinates that separately specify the position of the center of mass and the orientation.

2.2 Kinematics of Rotation

The motion of a rigid body about a center of rotation, a reference position that is fixed with respect to the body, is characterized at each moment by a rotation axis and a rate of rotation. Let's elaborate.

We can get from any orientation of a body to any other orientation of the body by a rotation of the body. That this is true is called Euler's theorem on rotations about a point.[2] We know that rotations have the property that they do not commute: the composition of successive rotations in general depends on the order of operation. Rotating a book about the $\hat{x}$ axis and then about the $\hat{z}$ axis puts the book in a different orientation than rotating the book about the $\hat{z}$ axis and then about the $\hat{x}$ axis. Nevertheless, Euler's theorem states that however many rotations have been composed to reach a given orientation, the orientation could have been reached with a single rotation. Try it! We take a book, rotate it this way, then that, and then some other way—then find the rotation that does the job in one step. So a rotation can be specified by an axis of rotation and the angular amount of the rotation.

We can specify the orientation of a body by specifying the rotation that takes the body to its actual orientation from some reference orientation. As the body moves, the rotation that does this changes.

Let q be the coordinate path that we will use to describe the motion of the body. Let $\mathsf{M}(q(t))$ be the rotation that takes the body from the reference orientation to the orientation specified by $q(t)$ (see figure 2.1). Let $\vec{\xi}_\alpha(t)$ be the vector to some constituent particle with the body in the orientation specified by $q(t)$, and let $\vec{\xi}'_\alpha$ be the vector to the same constituent with the body in the reference orientation. Then

$$\vec{\xi}_\alpha(t) = \mathsf{M}(q(t))\vec{\xi}'_\alpha. \tag{2.11}$$

The constituent vectors $\vec{\xi}'_\alpha$ do not depend on the configuration, because they are the vectors to the positions of the constituents with the body in a fixed reference orientation.

To compute the kinetic energy we will accumulate the contributions from all of the constituent mass elements. So we need

---

[2]For an elementary geometric proof see Whittaker [46], p. 2.

the velocities of the constituents. The positions of the constituent particles, at a given time t, are

$$\vec{\xi}_\alpha(t) = \mathsf{M}(q(t))\vec{\xi}'_\alpha = M(t)\vec{\xi}'_\alpha, \tag{2.12}$$

where $M = \mathsf{M} \circ q$. The velocity is the time derivative

$$D\vec{\xi}_\alpha(t) = DM(t)\vec{\xi}'_\alpha. \tag{2.13}$$

Using equation (2.12), we can write

$$D\vec{\xi}_\alpha(t) = DM(t)(M(t))^{-1}\vec{\xi}_\alpha(t). \tag{2.14}$$

So we have a time-varying linear differential equation that describes the motion of the constituents. Let's look at the multiplier $DM(t)(M(t))^{-1}$. Since $M(t)$ is a rotation its matrix representation is an orthogonal matrix $\mathbf{M}(t)$, with the property $(\mathbf{M}(t))^{-1} = (\mathbf{M}(t))^\mathsf{T}$. Because $\mathbf{M}(t)(\mathbf{M}(t))^\mathsf{T} = \mathbf{I}$, its derivative is:

$$\mathbf{0} = D(\mathbf{MM}^\mathsf{T}) = D\mathbf{M}\,\mathbf{M}^\mathsf{T} + \mathbf{M}\,D\mathbf{M}^\mathsf{T}. \tag{2.15}$$

So

$$D\mathbf{M}\,\mathbf{M}^\mathsf{T} = -(D\mathbf{M}\,\mathbf{M}^\mathsf{T})^\mathsf{T}. \tag{2.16}$$

We can conclude that $D\mathbf{M}\mathbf{M}^\mathsf{T}$ is antisymmetric.

Let $\mathbf{u}$ have components (x, y, z). Every 3×3 antisymmetric matrix is of the following form:

$$\mathsf{A}(\mathbf{u}) = \begin{pmatrix} 0 & -z & y \\ z & 0 & -x \\ -y & x & 0 \end{pmatrix}. \tag{2.17}$$

Multiplication by this matrix can be interpreted as the operation of cross product with the vector $\vec{u}$. The vector $\vec{u}$ has a matrix representation $\mathbf{u}$.

The inverse of the function A can be applied to any skew-symmetric matrix: we can use A^{-1} to extract the components of $\mathbf{u}$.

We can interpret multiplication by $D\mathbf{M}\,\mathbf{M}^\mathsf{T}$ as a cross product with a vector that we call $\vec{\omega}$, the *angular velocity vector* with components $\boldsymbol{\omega}$. So we can write

$$\boldsymbol{\omega} = \mathsf{A}^{-1}(D\mathbf{M}\,\mathbf{M}^\mathsf{T}). \tag{2.18}$$

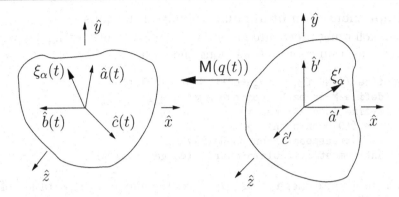

Figure 2.1 The rotation $\mathsf{M}(q(t))$ rotates the body from a reference orientation to its orientation at time t. Vectors attached to the body, such as ξ'_α are rotated with the body to the position $\xi_\alpha(t)$. Axes attached to the body, labeled by $\hat{a}'$, $\hat{b}'$, and $\hat{c}'$, specify a right-handed orthonormal coordinate system. In the reference orientation the body axes are aligned with the spatial axes, labeled by $\hat{x}$, $\hat{y}$, and $\hat{z}$. At time t the body axes are rotated to $\hat{a}(t)$, $\hat{b}(t)$, and $\hat{c}(t)$.

In terms of the angular velocity vector, the differential equations for the motion of the constituents (see equation 2.14) are

$$D\vec{\xi}_\alpha(t) = \vec{\omega}(t) \times \vec{\xi}_\alpha(t). \tag{2.19}$$

If the angular velocity vector for a body is $\vec{\omega}$ then the velocities of the constituent particles are perpendicular to the vectors to the constituent particles and proportional to the rate of rotation of the body and the distance of the constituent particle from the instantaneous rotation axis:

$$\dot{\vec{\xi}}_\alpha = \vec{\omega} \times \vec{\xi}_\alpha. \tag{2.20}$$

The components $\boldsymbol{\omega}'$ of the angular velocity vector on the body axes are $\boldsymbol{\omega}' = \mathsf{M}^\mathsf{T}\boldsymbol{\omega}$, so

$$\boldsymbol{\omega}' = \mathsf{M}^\mathsf{T}\mathsf{A}^{-1}(D\mathsf{M}\,\mathsf{M}^\mathsf{T}). \tag{2.21}$$

The relationship of the angular velocity vector to the path is a kinematic relationship; it is valid for any path. Thus we can abstract it to obtain the components of the angular velocity at a moment given the configuration and velocity at that moment.

Implementation of angular velocity functions

The following procedure gives the components of the angular velocity as a function of time along the path:

```
(define (((M-of-q->omega-of-t M-of-q) q) t)
  (define M-on-path (compose M-of-q q))
  (define (omega-cross t)
    (* ((D M-on-path) t)
       (transpose (M-on-path t))))
  (antisymmetric->column-matrix (omega-cross t)))
```

The procedure omega-cross produces the matrix representation of $\vec{\omega}\times$. The procedure antisymmetric->column-matrix, which corresponds to the function A^{-1}, is used to extract the components of the angular velocity vector from the skew-symmetric $\vec{\omega}\times$ matrix.

The components of the angular velocity vector on a basis fixed in the body, as a function of time, along the path are

```
(define (((M-of-q->omega-body-of-t M-of-q) q) t)
  (* (transpose (M-of-q (q t)))
     (((M-of-q->omega-of-t M-of-q) q) t)))
```

We can get the procedures of local state that give the angular velocity components by abstracting these procedures along arbitrary paths that have given coordinates and velocities. The abstraction of a procedure of a path to a procedure of state is accomplished by Gamma-bar (see section 1.9):

```
(define (M->omega M-of-q)
  (Gamma-bar
    (M-of-q->omega-of-t M-of-q)))

(define (M->omega-body M-of-q)
  (Gamma-bar
    (M-of-q->omega-body-of-t M-of-q)))
```

These procedures give the angular velocities as a function of state. We will see them in action after we get some M-of-q's to work with, starting in section 2.7.

2.3   Moments of Inertia

The rotational kinetic energy is the sum of the kinetic energy of each of the constituents of the rigid body. We can rewrite the rotational kinetic energy in terms of the angular velocity vector and certain aggregate quantities determined by the distribution of mass in the rigid body.

Substituting our representation of the relative velocity vectors into the rotational kinetic energy, we obtain

$$\sum_\alpha \tfrac{1}{2} m_\alpha \dot{\vec{\xi}}_\alpha \cdot \dot{\vec{\xi}}_\alpha = \sum_\alpha \tfrac{1}{2} m_\alpha \left(\vec{\omega} \times \vec{\xi}_\alpha \right) \cdot \left(\vec{\omega} \times \vec{\xi}_\alpha \right). \tag{2.22}$$

We introduce an arbitrary spatially fixed rectangular coordinate system with origin at the center of rotation and with basis vectors $\hat{e}_0$, $\hat{e}_1$, and $\hat{e}_2$, with the property that $\hat{e}_0 \times \hat{e}_1 = \hat{e}_2$. The components of $\vec{\omega}$ on this coordinate system are ω^0, ω^1, and ω^2. Rewriting $\vec{\omega}$ in terms of its components, the rotational kinetic energy becomes

$$\sum_\alpha \tfrac{1}{2} m_\alpha \left(\left(\textstyle\sum_i \hat{e}_i \omega^i \right) \times \vec{\xi}_\alpha \right) \cdot \left(\left(\textstyle\sum_j \hat{e}_j \omega^j \right) \times \vec{\xi}_\alpha \right)$$

$$= \tfrac{1}{2} \sum_{ij} \omega^i \omega^j \sum_\alpha m_\alpha \left(\hat{e}_i \times \vec{\xi}_\alpha \right) \cdot \left(\hat{e}_j \times \vec{\xi}_\alpha \right)$$

$$= \tfrac{1}{2} \sum_{ij} \omega^i \omega^j I_{ij}, \tag{2.23}$$

with

$$I_{ij} = \sum_\alpha m_\alpha \left(\hat{e}_i \times \vec{\xi}_\alpha \right) \cdot \left(\hat{e}_j \times \vec{\xi}_\alpha \right). \tag{2.24}$$

The nine time-dependent quantities I_{ij} are the components of the *inertia tensor* with respect to the chosen coordinate system.

Note what a remarkable form the kinetic energy has taken. All we have done is interchange the order of summations, but now the kinetic energy is written as a sum of products of components of the angular velocity vector, which completely specify how the orientation of the body is changing, and the quantity I_{ij}, which depends solely on the distribution of mass in the body relative to the chosen coordinate system.

We will deduce a number of properties of the inertia tensor. First, we find a somewhat simpler expression for it. The components of the vector $\vec{\xi}_\alpha$ are $(\xi_\alpha^0, \xi_\alpha^1, \xi_\alpha^2)$. If we rewrite $\vec{\xi}_\alpha$ as a sum over its components and simplify the elementary vector products of basis vectors, we can obtain the components of the inertia tensor. We can arrange the components of the inertia tensor to form the *inertia matrix*:

$$\mathbf{I} = \begin{pmatrix} I_{00} & I_{01} & I_{02} \\ I_{10} & I_{11} & I_{12} \\ I_{20} & I_{21} & I_{22} \end{pmatrix}, \tag{2.25}$$

where

$$I_{00} = \sum_\alpha m_\alpha ((\xi_\alpha^1)^2 + (\xi_\alpha^2)^2)$$

$$I_{11} = \sum_\alpha m_\alpha ((\xi_\alpha^0)^2 + (\xi_\alpha^2)^2)$$

$$I_{22} = \sum_\alpha m_\alpha ((\xi_\alpha^0)^2 + (\xi_\alpha^1)^2)$$

$$I_{ij} = -\sum_\alpha m_\alpha \xi_\alpha^i \xi_\alpha^j \quad \text{for } i \neq j \tag{2.26}$$

Note that the inertia tensor has real components and is symmetric: $I_{jk} = I_{kj}$.

We define the *moment of inertia* about a line by

$$\sum_\alpha m_\alpha (\xi_\alpha^\perp)^2, \tag{2.27}$$

where $\xi_\alpha^\perp$ is the perpendicular distance from the line to the constituent with index α. The diagonal components of the inertia tensor I_{ii} are recognized as the moments of inertia about the lines coinciding with the coordinate axes $\hat{e}_i$. The off-diagonal components of the inertia tensor are called *products of inertia*.

The rotational kinetic energy of a body depends on the distribution of mass of the body solely through the inertia tensor. Remarkably, the inertia tensor involves only second-order moments of the mass distribution with respect to the center of mass. We might have expected the kinetic energy to depend in a complicated way on all the moments of the mass distribution, interwoven in some complicated way with the components of the angular velocity vector, but this is not the case. This fact has a remarkable consequence: for the motion of a free rigid body the detailed shape of the body does not matter. If a book and a banana have the same inertia tensor, that is, the same second-order mass moments,

then if they are thrown in the same way the subsequent motion
will be the same, however complicated that motion is. The facts
that the book has corners and the banana has a stem do not affect
the motion except for their contributions to the inertia tensor. In
general, the potential energy of an extended body is not so simple
and does indeed depend on all moments of the mass distribution,
but for the kinetic energy the second moments are all that matter!

Exercise 2.1: Rotational kinetic energy

Show that the rotational kinetic energy can also be written

$$T_{\mathrm{R}} = \tfrac{1}{2} I \omega^2, \tag{2.28}$$

where I is the moment of inertia about the line through the center of
mass with direction $\hat{\omega}$, and ω is the instantaneous rate of rotation.

Exercise 2.2: Steiner's theorem

Let I be the moment of inertia of a body with respect to some given line
through the center of mass. Show that the moment of inertia I' with
respect to a second line parallel to the first is

$$I' = I + MR^2 \tag{2.29}$$

where M is the total mass of the body and R is the distance between
the lines.

Exercise 2.3: Some useful moments of inertia

Show that the moments of inertia of the following objects are as given:

a. The moment of inertia of a sphere of uniform density with mass M
and radius R about any line through the center is $\tfrac{2}{5} MR^2$.

b. The moment of inertia of a spherical shell with mass M and radius
R about any line through the center is $\tfrac{2}{3} MR^2$.

c. The moment of inertia of a cylinder of uniform density with mass M
and radius R about the axis of the cylinder is $\tfrac{1}{2} MR^2$.

d. The moment of inertia of a thin rod of uniform density per unit
length with mass M and length L about an axis perpendicular to the
rod through the center of mass is $\tfrac{1}{12} ML^2$.

Exercise 2.4: Jupiter

a. The density of a planet increases toward the center. Provide an
argument that the moment of inertia of a planet is less than that of a
sphere of uniform density of the same mass and radius.

b. The density as a function of radius inside Jupiter is well approximated by

$$\rho(r) = \frac{M}{R^3} \frac{\sin(\pi r/R)}{4r/R},$$

where M is the mass and R is the radius of Jupiter. Find the moment of inertia of Jupiter in terms of M and R.

2.4   Inertia Tensor

The representation of the rotational kinetic energy in terms of the inertia tensor was derived with the help of a rectangular coordinate system with basis vectors $\hat{e}_i$. There was nothing special about this particular rectangular basis. So, the kinetic energy must have the same form in any rectangular coordinate system. We can use this fact to derive how the inertia tensor changes if the body or the coordinate system is rotated.

Let's talk a bit about *active* and *passive* rotations. The rotation of the vector $\vec{x}$ by the rotation R produces a new vector $\vec{x}' = R\vec{x}$. We may write $\vec{x}$ in terms of its components with respect to some arbitrary rectangular coordinate system with orthonormal basis vectors $\hat{e}_i$: $\vec{x} = x^0\hat{e}_0 + x^1\hat{e}_1 + x^2\hat{e}_2$. Let $\mathbf{x}$ indicate the column matrix of components x^0, x^1, and x^2 of $\vec{x}$, and $\mathbf{R}$ be the matrix representation of R with respect to the same basis. In these terms the rotation can be written $\mathbf{x}' = \mathbf{Rx}$. The rotation matrix $\mathbf{R}$ is a real orthogonal matrix.[3] A rotation that carries vectors to new vectors is called an *active* rotation.

Alternatively, we can rotate the coordinate system by rotating the basis vectors, but leave other vectors that might be represented in terms of them unchanged. If a vector is unchanged but the basis vectors are rotated, then the components of the vector on the rotated basis vectors are not the same as the components on the original basis vectors. Denote the rotated basis vectors by $\hat{e}'_i = R\hat{e}_i$. The component of a vector along a basis vector is the dot product of the vector with the basis vector. So the components of

---

[3]Remember, an orthogonal matrix $\mathbf{R}$ satisfies $\mathbf{R}^{\mathrm{T}} = \mathbf{R}^{-1}$ and $\det \mathbf{R} = 1$.

the vector $\vec{x}$ along the rotated basis $\hat{e}_i'$ are $(x')^i = \vec{x} \cdot \hat{e}_i' = \vec{x} \cdot (R\hat{e}_i) = (R^{-1}\vec{x}) \cdot \hat{e}_i$.[4] Thus the components with respect to the rotated basis elements are the same as the components of the rotated vector $R^{-1}\vec{x}$ with respect to the original basis. In terms of components, if the vector $\vec{x}$ has components $\mathbf{x}$ with respect to the original basis vectors $\hat{e}_i$, then the components $\mathbf{x}'$ of the same vector with respect to the rotated basis vectors $\hat{e}_i'$ are $\mathbf{x}' = \mathbf{R}^{-1}\mathbf{x}$, or equivalently $\mathbf{x} = \mathbf{Rx}'$. A rotation that actively rotates the basis vectors, leaving other vectors unchanged, is called a *passive* rotation. For a passive rotation the components of a fixed vector change as if the vector were actively rotated by the inverse rotation.

With respect to the rectangular basis $\hat{e}_i$ the rotational kinetic energy is written

$$\tfrac{1}{2}\sum_{ij}\omega^i\omega^j I_{ij}. \tag{2.30}$$

In terms of matrix representations, the kinetic energy is

$$\tfrac{1}{2}\omega^{\mathrm{T}}\mathbf{I}\omega, \tag{2.31}$$

where ω is the column of components representing $\vec{\omega}$.[5] If we rotate the coordinate system by the passive rotation R about the center of rotation, the new basis vectors are $\hat{e}_i' = R\hat{e}_i$. The components ω' of the vector $\vec{\omega}$ with respect to the rotated coordinate system satisfy

$$\omega = \mathbf{R}\omega', \tag{2.32}$$

where $\mathbf{R}$ is the matrix representation of R. The kinetic energy is

$$\tfrac{1}{2}(\omega')^{\mathrm{T}}\mathbf{R}^{\mathrm{T}}\mathbf{I}\mathbf{R}\omega'. \tag{2.33}$$

However, if we had started with the basis $\hat{e}_i'$, we would have written the kinetic energy directly as

$$\tfrac{1}{2}(\omega')^{\mathrm{T}}\mathbf{I}'\omega', \tag{2.34}$$

---

[4]The last equality follows from the fact that the rotation of two vectors preserves the dot product: $\vec{x} \cdot \vec{y} = (R\vec{x}) \cdot (R\vec{y})$, or $(R^{-1}\vec{x}) \cdot \vec{y} = \vec{x} \cdot (R\vec{y})$.

[5]We take a 1-by-1 matrix as a number.

where the components are taken with respect to the $\hat{e}'_i$ basis. Comparing the two expressions, we see that

$$\mathbf{I'} = \mathbf{R}^{\mathsf{T}} \mathbf{I} \mathbf{R}. \tag{2.35}$$

Thus the inertia matrix transforms by a similarity transformation.[6]

2.5 Principal Moments of Inertia

We can use the transformation properties of the inertia tensor (2.35) to show that there are special rectangular coordinate systems for which the inertia tensor I' is diagonal, that is, $I'_{ij} = 0$ for $i \neq j$. Let's assume that $\mathbf{I'}$ is diagonal and solve for the rotation matrix $\mathbf{R}$ that does the job. Multiplying both sides of (2.35) on the left by $\mathbf{R}$, we have

$$\mathbf{R}\mathbf{I'} = \mathbf{I}\mathbf{R}. \tag{2.36}$$

We can examine pieces of this matrix equation by multiplying on the right by a trivial column vector that picks out a particular column. So we multiply on the right by the column matrix representation $\mathbf{e}_i$ of each of the coordinate unit vectors $\hat{e}_i$. These column matrices have a one in the ith row and zeros otherwise. Using $\mathbf{e}'_i = \mathbf{R}\mathbf{e}_i$, we find

$$\mathbf{R}\mathbf{I'}\mathbf{e}_i = \mathbf{I}\mathbf{R}\mathbf{e}_i = \mathbf{I}\mathbf{e}'_i. \tag{2.37}$$

On the other hand, the matrix $\mathbf{I'}$ is diagonal, so

$$\mathbf{R}\mathbf{I'}\mathbf{e}_i = \mathbf{R}\mathbf{e}_i I'_{ii} = I'_{ii}\mathbf{e}'_i. \tag{2.38}$$

So, from equations (2.37) and (2.38), we have

$$\mathbf{I}\mathbf{e}'_i = I'_{ii}\mathbf{e}'_i, \tag{2.39}$$

which we recognize as an equation for the eigenvalue I'_{ii} and $\mathbf{e}'_i$, the column matrix of components of the associated eigenvector.

---

[6] That the inertia tensor transforms in this manner could have been deduced from its definition (2.24). However, it seems that the argument based on the coordinate-system independence of the kinetic energy provides insight.

From $\mathbf{e}_i' = \mathbf{R}\mathbf{e}_i$, we see that the $\mathbf{e}_i'$ are the columns of the rotation matrix $\mathbf{R}$. Now, rotation matrices are orthogonal, so $\mathbf{R}^\mathrm{T}\mathbf{R} = 1$; thus the columns of the rotation matrix must be orthonormal—that is, $(\mathbf{e}_i')^\mathrm{T}\mathbf{e}_j' = \delta_{ij}$, where δ_{ij} is one if $i = j$ and zero otherwise. But the eigenvectors that are solutions of equation (2.39) are not necessarily even orthogonal. So we are not done yet.

If a matrix is real and symmetric then the eigenvalues are real. Furthermore, if the eigenvalues are distinct then the eigenvectors are orthogonal. However, if the eigenvalues are not distinct then the directions of the eigenvectors for the degenerate eigenvalues are not uniquely determined—we have the freedom to choose particular $\mathbf{e}_i'$ that are orthogonal.[7] The linearity of equation (2.39) implies that the $\mathbf{e}_i'$ can be normalized. Thus whether or not the eigenvalues are distinct we can obtain an orthonormal set of $\mathbf{e}_i'$. This is enough to reconstruct a rotation matrix $\mathbf{R}$ that does the job we asked of it: to rotate the coordinate system to a configuration such that the inertia tensor is diagonal. If the eigenvalues are not distinct, the rotation matrix $\mathbf{R}$ is not uniquely defined—there is more than one rotation matrix $\mathbf{R}$ that does the job.

The eigenvectors and eigenvalues are determined by the requirement that the inertia tensor be diagonal with respect to the rotated coordinate system. Thus the rotated coordinate system has a special orientation with respect to the body. The basis vectors $\hat{e}_i'$ therefore actually point along particular directions in the body. We define the axes in the body through the center of mass with these directions to be the *principal axes*. With respect to the coordinate system defined by $\hat{e}_i'$, the inertia tensor is diagonal, by construction, with the eigenvalues I_{ii}' on the diagonal. Thus the moments of inertia about the principal axes are the eigenvalues I_{ii}'. We call the moments of inertia about the principal axes the *principal moments of inertia*.

For convenience, we often label the principal moments of inertia according to their size: $A \leq B \leq C$, with principal axis unit vectors $\hat{a}$, $\hat{b}$, $\hat{c}$, respectively. The positive direction along the principal axes can be chosen so that $\hat{a}$, $\hat{b}$, $\hat{c}$ form a right-handed rectangular coordinate basis.

---

[7]If two eigenvalues are not distinct then linear combinations of the associated eigenvectors are eigenvectors. This gives us the freedom to find linear combinations of the eigenvectors that are orthonormal.

Let **x** represent the matrix of components of a vector $\vec{x}$ with respect to the basis vectors $\hat{e}_i$. Recall that the components **x'** of a vector $\vec{x}$ with respect to the principal axis unit vectors $\hat{e}'_i$ satisfy

$$\mathbf{x}' = \mathbf{R}^\mathsf{T}\mathbf{x}. \tag{2.40}$$

The components of a vector on the principal axis basis are sometimes called the *body components* of the vector.

If we choose the reference orientation of the body so that the principal axes are aligned with the spatial axes $\hat{x}, \hat{y}, \hat{z}$, then the rotation **R** that diagonalizes the inertia matrix becomes the rotation **M** shown in figure 2.1. The axes $\hat{a}'$, $\hat{b}'$, $\hat{c}'$ then become the principal axes. The rotation matrix **M** multiplies the column of components of a vector on the principal axes to make a column of components of the vector in space.

Now let's rewrite the kinetic energy in terms of the principal moments of inertia. If we choose our rectangular coordinate system so that it coincides with the principal axes then the calculation is simple. Let the components of the angular velocity vector on the principal axes be $(\omega^a, \omega^b, \omega^c)$. Then, keeping in mind that the inertia tensor is diagonal with respect to the principal axis basis, the kinetic energy is just

$$T_\mathrm{R} = \tfrac{1}{2}\left[A(\omega^a)^2 + B(\omega^b)^2 + C(\omega^c)^2\right]. \tag{2.41}$$

Or as a program:

```
(define ((T-body A B C) omega-body)
  (* 1/2
     (+ (* A (square (ref omega-body 0)))
        (* B (square (ref omega-body 1)))
        (* C (square (ref omega-body 2)))))))
```

Exercise 2.5: A constraint on the moments of inertia

Show that the sum of any two of the moments of inertia is greater than or equal to the third moment of inertia. You may assume the moments of inertia are with respect to orthogonal axes.

Exercise 2.6: Principal moments of inertia

For each of the configurations described below find the principal moments of inertia with respect to the center of mass, and find the corresponding principal axes.

a. A regular tetrahedron consisting of four equal point masses tied together with rigid massless wire.

b. A cube of uniform density.

c. Five equal point masses rigidly connected by massless stuff. The point masses are at the rectangular coordinates

$(-1, 0, 0), (1, 0, 0), (1, 1, 0), (0, 0, 0), (0, 0, 1)$.

Exercise 2.7: This book

Measure this book. You will admit that it is pretty dense. Don't worry, you will get to throw it later. Show that the principal axes are the lines connecting the centers of opposite faces of the idealized brick approximating the book. Compute the corresponding principal moments of inertia.

2.6   Vector Angular Momentum

The vector angular momentum of a particle is the cross product of its position vector and its linear momentum vector. For a rigid body the vector angular momentum is the sum of the vector angular momentum of each of the constituents. Here we find an expression for the vector angular momentum of a rigid body in terms of the inertia tensor and the angular velocity vector.

The vector angular momentum of a rigid body is

$$\sum_\alpha \vec{x}_\alpha \times (m_\alpha \dot{\vec{x}}_\alpha), \tag{2.42}$$

where $\vec{x}_\alpha$, $\dot{\vec{x}}_\alpha$, and m_α are the positions, velocities, and masses of the constituent particles. It turns out that the vector angular momentum decomposes into the sum of the angular momentum of the center of mass and the rotational angular momentum about the center of mass, just as the kinetic energy separates into the kinetic energy of the center of mass and the kinetic energy of rotation. As in the kinetic energy demonstration (section 2.1), decompose the position into the vector to the center of mass $\vec{X}$ and the vectors from the center of mass to the constituent mass elements $\vec{\xi}_\alpha$:

$$\vec{x}_\alpha = \vec{X} + \vec{\xi}_\alpha, \tag{2.43}$$

with velocities

$$\dot{\vec{x}}_\alpha = \dot{\vec{X}} + \dot{\vec{\xi}}_\alpha. \tag{2.44}$$

Substituting, the angular momentum is

$$\sum_\alpha m_\alpha (\vec{X} + \vec{\xi}_\alpha) \times (\dot{\vec{X}} + \dot{\vec{\xi}}_\alpha). \tag{2.45}$$

Multiplying out the product, and using the fact that $\vec{X}$ is the center of mass and $M = \sum_\alpha m_\alpha$ is the total mass of the body, the angular momentum is

$$\vec{X} \times (M\dot{\vec{X}}) + \sum_\alpha \vec{\xi}_\alpha \times (m_\alpha \dot{\vec{\xi}}_\alpha). \tag{2.46}$$

The angular momentum of the center of mass is

$$\vec{X} \times (M\dot{\vec{X}}), \tag{2.47}$$

and the rotational angular momentum is

$$\sum_\alpha \vec{\xi}_\alpha \times (m_\alpha \dot{\vec{\xi}}_\alpha). \tag{2.48}$$

Using $\dot{\vec{\xi}}_\alpha = \vec{\omega} \times \vec{\xi}_\alpha$, we get the rotational angular momentum vector

$$\vec{L} = \sum_\alpha m_\alpha \vec{\xi}_\alpha \times (\vec{\omega} \times \vec{\xi}_\alpha). \tag{2.49}$$

We can also reexpress the rotational angular momentum in terms of the angular velocity vector and the inertia tensor, as we did for the kinetic energy. In terms of components with respect to the basis $\{\hat{e}_0, \hat{e}_1, \hat{e}_2\}$, this is

$$L_j = \sum_k I_{jk} \omega^k, \tag{2.50}$$

where I_{jk} are the components of the inertia tensor (2.24). The angular momentum and the kinetic energy are expressed in terms of the same inertia tensor.

With respect to the principal-axis basis, the components of the angular momentum have a particularly simple form:

$$L_a = A\omega^a$$
$$L_b = B\omega^b$$
$$L_c = C\omega^c \tag{2.51}$$

Since the angular momenta are the partial derivatives of T_R (see equation 2.41) with respect to the angular velocities, they must be grouped as a down tuple (in matrix language, a row matrix): $L' = [L_a, L_b, L_c]$. As a program:

```
(define ((L-body A B C) omega-body)
  (down (* A (ref omega-body 0))
        (* B (ref omega-body 1))
        (* C (ref omega-body 2))))
```

If $\mathbf{M}$ is the matrix representation of the rotation that takes an angular-velocity vector $\vec{\omega}'$ to a rotated vector $\vec{\omega}$, the components transform as $\omega = \mathbf{M}\omega'$.

When working with matrices it is more convenient to work with a column matrix of the angular momentum components, so we introduce $\overline{\mathbf{L}} = \mathbf{L}^\mathsf{T}$. Using $\omega = \mathbf{M}\omega'$ and equation (2.35) with $\mathbf{R}$ replaced by $\mathbf{M}$ we derive an expression for the angular momentum

$$\overline{\mathbf{L}} = \mathbf{I}\omega = \mathbf{M}\mathbf{I}'\omega' = \mathbf{M}\overline{\mathbf{L}}'. \tag{2.52}$$

Transposing this result, we see that the angular momentum components must transform as $\mathbf{L} = \mathbf{L}'\mathbf{M}^\mathsf{T}$:

```
(define (((L-space M) A B C) omega-body)
  (* ((L-body A B C) omega-body)
     (transpose M)))
```

Exercise 2.8: Rotational angular momentum

Verify that expression (2.50) for the components of the rotational angular momentum (2.49) in terms of the inertia tensor is correct.

2.7   Euler Angles

To go further we must finally specify a set of generalized coordinates. We first do this using the traditional *Euler angles*. Later, we find other ways of describing the orientation of a rigid body.

We are using an intermediate representation of the orientation in terms of the function M of the generalized coordinates that gives the rotation that takes the body from some reference orientation and rotates it to the orientation specified by the generalized coordinates. Here we take the reference orientation so that principal-axis unit vectors $\hat{a}$, $\hat{b}$, $\hat{c}$ are coincident with the basis vectors $\hat{e}_i$, labeled here by $\hat{x}$, $\hat{y}$, $\hat{z}$.

We define the Euler angles in terms of simple rotations about the coordinate axes. Let $R_x(\psi)$ be a right-handed rotation about the $\hat{x}$ axis by the angle ψ, and let $R_z(\psi)$ be a right-handed rotation about the $\hat{z}$ axis by the angle ψ. The function M for Euler angles is written as a composition of three of these simple coordinate axis rotations:

$$\mathsf{M}(\theta, \varphi, \psi) = R_z(\varphi) \circ R_x(\theta) \circ R_z(\psi), \tag{2.53}$$

for the Euler angles θ, φ, ψ.

The Euler angles can specify any orientation of the body, but the orientation does not always correspond to a unique set of Euler angles. In particular, if $\theta = 0$ then the orientation is dependent only on the sum $\varphi + \psi$, so the orientation does not uniquely determine either φ or ψ.

Exercise 2.9: Euler angles
It is not immediately obvious that all orientations can be represented in terms of the Euler angles. To show that the Euler angles are adequate to represent all orientations, solve for the Euler angles that give an arbitrary rotation R. Keep in mind that some orientations do not correspond to a unique representation in terms of Euler angles.

Though the Euler angles allow us to specify all orientations and thus can be used as generalized coordinates, the definition of Euler angles is pretty arbitrary. In fact no reasoning has led us to them. This is reflected in our presentation of them by just saying "here they are." Euler angles are well suited for some problems, but cumbersome for others.

There are other ways of defining similar sets of angles. For instance, we could also take our generalized coordinates to satisfy

$$\mathsf{M}'(\theta, \varphi, \psi) = R_x(\varphi) \circ R_y(\theta) \circ R_z(\psi). \tag{2.54}$$

Such alternatives to the Euler angles sometimes come in handy.

Each of the fundamental rotations can be represented as a matrix. The rotation matrix representing a right-handed rotation about the $\hat{z}$ axis by the angle ψ is

$$\mathbf{R}_z(\psi) = \begin{pmatrix} \cos\psi & -\sin\psi & 0 \\ \sin\psi & \cos\psi & 0 \\ 0 & 0 & 1 \end{pmatrix} \tag{2.55}$$

and a right-handed rotation about the x axis by the angle ψ is represented by the matrix

$$\mathbf{R}_x(\psi) = \begin{pmatrix} 1 & 0 & 0 \\ 0 & \cos\psi & -\sin\psi \\ 0 & \sin\psi & \cos\psi \end{pmatrix}. \tag{2.56}$$

The matrix that represents the rotation that carries the body from its reference orientation to the actual orientation is

$$\mathbf{M}(\theta, \varphi, \psi) = \mathbf{R}_z(\varphi)\mathbf{R}_x(\theta)\mathbf{R}_z(\psi). \tag{2.57}$$

The rotation matrices and their product can be constructed by simple programs:

```
(define (Rz-matrix angle)
  (matrix-by-rows
    (list (cos angle) (- (sin angle))          0)
    (list (sin angle)    (cos angle)           0)
    (list          0               0           1)))
```

```
(define (Rx-matrix angle)
  (matrix-by-rows
    (list          1               0           0)
    (list          0     (cos angle) (- (sin angle)))
    (list          0     (sin angle)    (cos angle))))
```

```
(define (Euler->M angles)
  (let ((theta (ref angles 0))
        (phi   (ref angles 1))
        (psi   (ref angles 2)))
    (* (Rz-matrix phi)
       (Rx-matrix theta)
       (Rz-matrix psi))))
```

Now that we have a procedure that implements a sample M, we can find the components of the angular velocity vector and the body components of the angular velocity vector using the procedures M-of-q->omega-of-t and M-of-q->omega-body-of-t from section 2.2. For example,

```
(show-expression
 (((M-of-q->omega-body-of-t Euler->M)
   (up (literal-function 'theta)
       (literal-function 'phi)
       (literal-function 'psi)))
  't))
```

$$
\begin{pmatrix}
D\varphi\,(t)\sin\left(\theta\,(t)\right)\sin\left(\psi\,(t)\right)+\cos\left(\psi\,(t)\right)D\theta\,(t) \\[2ex]
D\varphi\,(t)\sin\left(\theta\,(t)\right)\cos\left(\psi\,(t)\right)-\sin\left(\psi\,(t)\right)D\theta\,(t) \\[2ex]
\cos\left(\theta\,(t)\right)D\varphi\,(t)+D\psi\,(t)
\end{pmatrix}
$$

To construct the kinetic energy we need the procedure of state that gives the body components of the angular velocity vector:

```
(show-expression
 ((M->omega-body Euler->M)
  (up 't
      (up 'theta 'phi 'psi)
      (up 'thetadot 'phidot 'psidot))))
```

$$
\begin{pmatrix}
\dot{\varphi}\sin\left(\psi\right)\sin\left(\theta\right)+\dot{\theta}\cos\left(\psi\right) \\[2ex]
\dot{\varphi}\sin\left(\theta\right)\cos\left(\psi\right)-\dot{\theta}\sin\left(\psi\right) \\[2ex]
\dot{\varphi}\cos\left(\theta\right)+\dot{\psi}
\end{pmatrix}
$$

We capture this result as a procedure:

```
(define (Euler-state->omega-body local)
  (let ((q (coordinate local)) (qdot (velocity local)))
    (let ((theta (ref q 0))
          (psi (ref q 2))
          (thetadot (ref qdot 0))
          (phidot (ref qdot 1))
          (psidot (ref qdot 2)))
      (let ((omega-a (+ (* thetadot (cos psi))
                        (* phidot (sin theta) (sin psi))))
            (omega-b (+ (* -1 thetadot (sin psi))
                        (* phidot (sin theta) (cos psi))))
            (omega-c (+ (* phidot (cos theta)) psidot)))
        (up omega-a omega-b omega-c)))))
```

The kinetic energy can now be written:

```
(define ((T-body-Euler A B C) local)
  ((T-body A B C)
   (Euler-state->omega-body local)))
```

We can define procedures to calculate the components of the angular momentum on the principal axes:

```
(define ((L-body-Euler A B C) local)
  ((L-body A B C)
   (Euler-state->omega-body local)))
```

We then transform the components of the angular momentum on the principal axes to the components on the fixed basis $\hat{e}_i$:

```
(define ((L-space-Euler A B C) local)
  (let ((angles (coordinate local)))
    (* ((L-body-Euler A B C) local)
       (transpose (Euler->M angles)))))
```

These procedures are local state functions, like Lagrangians.

2.8   Motion of a Free Rigid Body

The kinetic energy, expressed in terms of a suitable set of generalized coordinates, is a Lagrangian for a free rigid body. In section 2.1 we found that the kinetic energy of a rigid body can be written as the sum of the rotational kinetic energy and the translational kinetic energy. If we choose one set of coordinates to specify the position and another set to specify the orientation, the Lagrangian becomes a sum of a translational Lagrangian and a rotational Lagrangian. The Lagrange equations for translational motion are not coupled to the Lagrange equations for the rotational motion. For a free rigid body the translational motion is just that of a free particle: uniform motion. Here we concentrate on the rotational motion of the free rigid body. We can adopt the Euler angles as the coordinates that specify the orientation; the rotational kinetic energy was expressed in terms of Euler angles in the previous section.

Conserved quantities

The Lagrangian for a free rigid body has no explicit time dependence, so we can deduce that the energy, which is just the kinetic energy, is conserved by the motion.

The Lagrangian does not depend on the Euler angle φ, so we can deduce that the momentum conjugate to this coordinate is conserved. An explicit expression for the momentum conjugate to φ is

```
(define Euler-state
  (up 't
      (up 'theta 'phi 'psi)
      (up 'thetadot 'phidot 'psidot)))

(show-expression
 (ref (((partial 2) (T-body-Euler 'A 'B 'C)) Euler-state)
      1))
```

$$A\dot\varphi \left(\sin\left(\theta\right)\right)^2 \left(\sin\left(\psi\right)\right)^2 + A\dot\theta \cos\left(\psi\right)\sin\left(\theta\right)\sin\left(\psi\right)$$
$$+ B\dot\varphi \left(\cos\left(\psi\right)\right)^2 \left(\sin\left(\theta\right)\right)^2 - B\dot\theta \cos\left(\psi\right)\sin\left(\theta\right)\sin\left(\psi\right)$$
$$+ C\dot\varphi \left(\cos\left(\theta\right)\right)^2 + C\dot\psi \cos\left(\theta\right)$$

We know that this complicated quantity is conserved by the motion of the rigid body because of the symmetries of the Lagrangian.

If there are no external torques, then we expect that the vector angular momentum will be conserved. We can verify this using the Lagrangian formulation of the problem. First, we note that L_z is the same as p_φ. We can check this by direct calculation:

```
(- (ref ((L-space-Euler 'A 'B 'C) Euler-state)
        2)
   (ref (((partial 2) (T-body-Euler 'A 'B 'C)) Euler-state)
        1))
0
```

We know that p_φ is conserved because the Lagrangian for the free rigid body did not mention φ, so now we know that L_z is conserved. Since the orientation of the coordinate axes is arbitrary, we know that if any rectangular component is conserved then all of them are. So the vector angular momentum is conserved for the free rigid body.

We could have seen this with the help of Noether's theorem (see section 1.8.5). There is a continuous family of rotations that can transform any orientation into any other orientation. The orientation of the coordinate axes we used to define the Euler angles is arbitrary, and the kinetic energy (the Lagrangian) is the same for any choice of coordinate system. Thus the situation meets the requirements of Noether's theorem, which tells us that there is a conserved quantity. In particular, the family of rotations around each coordinate axis gives us conservation of the angular momentum component on that axis. We construct the vector angular momentum by combining these contributions.

Exercise 2.10: Uniformly accelerated rigid body

Show that a rigid body subject to a uniform acceleration rotates as a free rigid body, while the center of mass has a parabolic trajectory.

Exercise 2.11: Conservation of angular momentum

Fill in the details of the argument that Noether's theorem implies that vector angular momentum is conserved by the motion of the free rigid body.

2.8.1   Computing the Motion of Free Rigid Bodies

Lagrange's equations for the motion of a free rigid body in terms of Euler angles are quite disgusting, so we will not show them here. However, we will use the Lagrange equations to explore the motion of the free rigid body.

Before doing this it is worth noting that the equations of motion in Euler angles are singular for some configurations, because for these configurations the Euler angles are not uniquely defined. If we set $\theta = 0$ then an orientation does not correspond to a unique value of φ and ψ; only their sum determines the orientation.

The singularity arises in the explicit Lagrange equations when we attempt to solve for the second derivative of the generalized coordinates in terms of the generalized coordinates and the generalized velocities (see section 1.7). The isolation of the second derivative requires multiplying by the inverse of $\partial_2\partial_2 L$. The determinant of this quantity becomes zero when the Euler angle θ is zero:

```
(show-expression
 (determinant
  (((square (partial 2)) (T-body-Euler 'A 'B 'C))
   Euler-state)))
```

$$ABC \left(\sin \left(\theta\right)\right)^{2}$$

So when θ is zero, we cannot solve for the second derivatives. When θ is small, the Euler angles can move very rapidly, and thus may be difficult to compute reliably. Of course, the motion of the rigid body is perfectly well behaved for any orientation. This is a problem of the representation of that motion in Euler angles; it is a "coordinate singularity."

One solution to this problem is to use another set of Euler-like coordinates for which Lagrange's equations have singularities for different orientations, such as those defined in equation (2.54). So if as the calculation proceeds the trajectory comes close to a singularity in one set of coordinates, we can switch coordinate systems and use another set for a while until the trajectory encounters another singularity. This solves the problem, but it is cumbersome. For the moment we will ignore this problem and compute some trajectories, being careful to limit our attention to trajectories that avoid the singularities.

We will compute some trajectories by numerical integration and check our integration process by seeing how well energy and angular momentum are conserved. Then, we will investigate the evolution of the components of angular momentum on the principal axis basis. We will discover that we can learn quite a bit about the qualitative behavior of rigid bodies by combining the information we get from the energy and angular momentum.

To develop a trajectory from initial conditions we integrate the Lagrange equations, as we did in chapter 1. The system derivative is obtained from the Lagrangian:

```
(define (rigid-sysder A B C)
  (Lagrangian->state-derivative (T-body-Euler A B C)))
```

The following program monitors the errors in the energy and in the components of the angular momentum:

```
(define ((monitor-errors win A B C L0 E0) state)
  (let ((t (time state))
        (L ((L-space-Euler A B C) state))
        (E ((T-body-Euler A B C) state)))
    (plot-point win t (relative-error (ref L 0) (ref L0 0)))
    (plot-point win t (relative-error (ref L 1) (ref L0 1)))
    (plot-point win t (relative-error (ref L 2) (ref L0 2)))
    (plot-point win t (relative-error E E0))))

(define (relative-error value reference-value)
  (if (zero? reference-value)
      (error "Zero reference value -- RELATIVE-ERROR")
      (/ (- value reference-value) reference-value)))
```

We make a plot window to display the errors:

```
(define win (frame 0.0 100.0 -1.0e-12 1.0e-12))
```

The default integration method used by the system is Bulirsch–Stoer (bulirsch-stoer), but here we set the integration method to be quality-controlled Runge–Kutta (qcrk4), because the error plot is more interesting:

```
(set-ode-integration-method! 'qcrk4)
```

We use evolve to investigate the evolution:

```
(let ((A 1.0) (B (sqrt 2.0)) (C 2.0)    ; moments of inertia
      (state0 (up 0.0                    ; initial state
                  (up 1.0 0.0 0.0)
                  (up 0.1 0.1 0.1))))
  (let ((L0 ((L-space-Euler A B C) state0))
        (E0 ((T-body-Euler A B C) state0)))
    ((evolve rigid-sysder A B C)
     state0
     (monitor-errors win A B C L0 E0)
     0.1                    ; step between plotted points
     100.0                  ; final time
     1.0e-12)))             ; max local truncation error
```

The plot that is developed of the relative errors in the components of the angular momenta and the energy (see figure 2.2) shows that we have been successful in controlling the error in the conserved quantities. This should give us some confidence in the trajectory that is evolved.

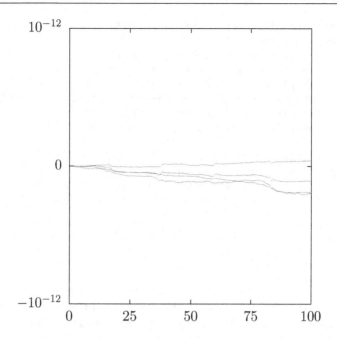

Figure 2.2   The relative error in energy and in the three spatial components of the angular momentum versus time. It is interesting to note that the energy error is one of the three falling curves.

2.8.2   Qualitative Features of Free Rigid Body Motion

The evolution of the components of the angular momentum on the principal axes has a remarkable property. For almost every initial condition the body components of the angular momentum periodically trace a simple closed curve.

We can see this by investigating a number of trajectories and plotting the components of angular momentum of the body on the principal axes (see figure 2.3). To make this figure a number of trajectories of equal energy were computed. The three-dimensional space of body components is projected onto a two-dimensional plane for display. Points on the back of this projection of the ellipsoid of constant energy are plotted with lower density than points on the front of the ellipsoid. For most initial conditions we find a one-dimensional simple closed curve. Some trajectories on the front side appear to cross trajectories on the back side, but this is an artifact of projection. There is also a family of trajectories that appear to intersect in two points, one on the front side

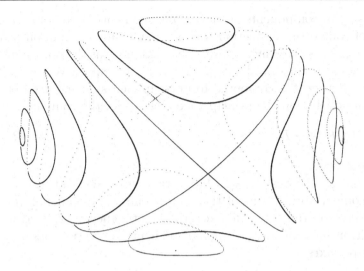

Figure 2.3   Trajectories of the components of the angular momentum vector on the principal axes, projected onto a plane. Each closed curve, except for the separatrix, is a different trajectory. All the trajectories shown here have the same energy.

and one on the back side. The curve that is the union of these trajectories is called a *separatrix*; it separates different types of motion.

What is going on? The state space for a free rigid body is six-dimensional: the three Euler angles and their time derivatives. We know four constants of the motion—the three spatial components of the angular momentum, L_x, L_y, and L_z, and the energy, E. Thus, the motion is restricted to a two-dimensional region of the state space.[8] Our experiment shows that the components of the angular momentum trace one-dimensional closed curves in the angular-momentum subspace, so there is something more going on here.

The total angular momentum is conserved if all of the components are, so we also have the constant

$$L^2 = L_x^2 + L_y^2 + L_z^2. \tag{2.58}$$

---

[8]We expect that for each constant of the motion we reduce by one the dimension of the region of the state space explored by a trajectory. This is because a constant of the motion can be used locally to solve for one of the state variables in terms of the others.

The spatial components of the angular momentum do not change, but of course the projections of the angular momentum onto the principal axes do change because the axes move as the body moves. However, the magnitude of the angular momentum vector is the same whether it is computed from components on the fixed basis or components on the principal axis basis. So, the combination

$$L^2 = L_a^2 + L_b^2 + L_c^2, \tag{2.59}$$

is conserved.

Using the expressions (2.51) for the components of the angular momentum in terms of the components of the angular velocity vector on the principal axes, the kinetic energy (2.41) can be rewritten in terms of the angular momentum components on the principal axes:

$$E = \frac{1}{2} \left(\frac{L_a^2}{A} + \frac{L_b^2}{B} + \frac{L_c^2}{C} \right). \tag{2.60}$$

The two conserved quantities (2.59 and 2.60) provide constraints on how the components of the angular momentum vector on the principal axes can change. We recognize the conservation of angular momentum constraint (2.59) as the equation of a sphere, and the conservation of kinetic energy constraint (2.60) as the equation for a triaxial ellipsoid. For every trajectory both constraints are satisfied, so the components of the angular momentum move on the intersection of these two surfaces, the energy ellipsoid and the angular momentum sphere. The intersection of an ellipsoid and a sphere with the same center is typically two closed curves, so an orbit is confined to one of these curves. This sheds light on the puzzle posed at the beginning of this section.

Because of our ordering $A \leq B \leq C$, the longest axis of this triaxial ellipsoid coincides with the $\hat{c}$ direction (all the angular momentum is along the axis of largest principal moment of inertia) and the shortest axis of the energy ellipsoid coincides with the $\hat{a}$ axis (all the angular momentum is along the smallest moment of inertia). Without actually solving the Lagrange equations, we have found strong constraints on the evolution of the components of the angular momentum on the principal axes.

To determine how the system evolves along these intersection curves we have to use the equations of motion. We observe that the evolution of the components of the angular momentum on the principal axes depends only on the components of the angular momentum on the principal axes, even though the values of these components are not enough to completely specify the dynamical state. Apparently the dynamics of these components is self-contained, and we will see that it can be described in terms of a set of differential equations whose only dynamical variables are the components of the angular momentum on the principal axes (see section 2.9).

We note that there are two axes for which the intersection curves shrink to a point if we hold the energy constant and vary the magnitude of the angular momentum. If the angular momentum starts at these points, the conserved quantities constrain the angular momentum to stay there. These points are *equilibrium* points for the body components of the angular momentum. However, they are not equilibrium points for the system as a whole. At these points the body is still rotating even though the body components of the angular momentum are not changing. This kind of equilibrium is called a *relative equilibrium*. We can also see that if the angular momentum is initially slightly displaced from one of these relative equilibria, then the angular momentum is constrained to stay near it on one of the intersection curves. The angular momentum vector is fixed in space, so the principal axis of the equilibrium point of the body rotates stably about the angular momentum vector.

At the principal axis with intermediate moment of inertia, the $\hat{b}$ axis, the intersection curves appear to cross. As we observed, the dynamics of the components of the angular momentum on the principal axes forms a self-contained dynamical system. Trajectories of a dynamical system cannot cross,[9] so the most that can happen is that if the equations of motion carry the system along the intersection curve then the system can approach the crossing point only asymptotically. So without solving any equations we can deduce that the point of crossing is another relative equilibrium. If the angular momentum is initially aligned with the intermediate axis, then it stays aligned. If the system is slightly displaced from the intermediate axis, then the evolution along the intersection curve will take the system far from the relative equilibrium. So rotation about the axis of intermediate moment of inertia is unstable—initial displacements of the angular momen-

---

[9]Systems of ODEs that satisfy a Lipschitz condition have unique solutions.

tum, however small initially, become large. Again, the angular momentum vector is fixed in space, but now the principal axis with the intermediate principal moment does not stay close to the angular momentum, so the body executes a complicated tumbling motion.

This gives some insight into the mystery of the thrown book mentioned at the beginning of the chapter. If one throws a book so that it is initially rotating about either the axis with the largest moment of inertia or the axis with the smallest moment of inertia (the shortest and longest physical axes, respectively), the book rotates regularly about that axis. However, if the book is thrown so that it is initially rotating about the axis of intermediate moment of inertia (the intermediate physical axis), then it tumbles, however carefully it is thrown. You can try it with this book (but put a rubber band or string around it first).

Before moving on, we can make some further physical deductions. Suppose a freely rotating body is subject to some sort of internal friction that dissipates energy but conserves the angular momentum. For example, real bodies flex as they spin. If the spin axis moves with respect to the body then the flexing changes with time, and this changing distortion converts kinetic energy of rotation into heat. Internal processes do not change the total angular momentum of the system. If we hold the magnitude of the angular momentum fixed but gradually decrease the energy, then the curve of intersection on which the system moves gradually deforms. For a given angular momentum there is a lower limit on the energy: the energy cannot be so low that there are no intersections. For this lowest energy the intersection of the angular momentum sphere and the energy ellipsoid is a pair of points on the axis of maximum moment of inertia. With energy dissipation, a freely rotating physical body eventually ends up with the lowest energy consistent with the given angular momentum, which is rotation about the principal axis with the largest moment of inertia (typically the shortest physical axis).

Thus, we expect that given enough time all freely rotating physical bodies will end up rotating about the axis of largest moment of inertia. You can demonstrate this to your satisfaction by twirling a small bottle containing some viscous fluid, such as correction fluid. What you will find is that, whatever spin you try to put on the bottle, it will reorient itself so that the axis of the largest moment of inertia is aligned with the spin axis. Remarkably, this is very nearly true of almost every body in the solar system for which there is enough information to decide. The deviations from principal axis rotation for the Earth are tiny: the angle between the angular momentum vector and the $\hat{c}$ axis for the Earth is less than one arc-second.[10] In fact, the evidence is that all of the planets, the Moon and all the other natural satellites, and almost all of the asteroids rotate very nearly about the largest moment of inertia. We have deduced that this is to be expected using an elementary argument. There are exceptions. Comets typically do not rotate about the largest moment. As they are heated by the sun, material spews out from localized jets, and the back reaction from these jets changes the rotation state. Among the natural satellites, the only known exception is Saturn's satellite Hyperion, which is tumbling chaotically. Hyperion is especially out of round and subject to strong gravitational torques from Saturn.

2.9   Euler's Equations

For a free rigid body we have seen that the components of the angular momentum on the principal axes comprise a self-contained dynamical system: the variation of the principal axis components depends only on the principal axis components. Here we derive equations that govern the evolution of these components.

The starting point for the derivation is the conservation of the vector angular momentum. The components of the angular momentum on the principal axes are[11]

$$\overline{\mathbf{L}}' = \mathbf{I}'\boldsymbol{\omega}', \tag{2.61}$$

where $\boldsymbol{\omega}'$ is composed of the components of the angular velocity vector on the principal axes and $\mathbf{I}'$ is the matrix representation of the inertia tensor with respect to the principal axis basis:

---

[10] The deviation of the angular momentum from the principal axis may be due to a number of effects: earthquakes, atmospheric tides, ...

[11] Here we are using the column-matrix version of the components of the angular momentum, as in equation (2.52).

$$\mathbf{I}' = \begin{pmatrix} A & 0 & 0 \\ 0 & B & 0 \\ 0 & 0 & C \end{pmatrix}. \tag{2.62}$$

The body components of the angular momentum $\mathbf{L}'$ are related to the components $\mathbf{L}$ on the fixed rectangular basis $\hat{e}_i$ by

$$\overline{\mathbf{L}} = \mathbf{M}\overline{\mathbf{L}}', \tag{2.63}$$

where $\mathbf{M}$ is the matrix representation of the rotation that carries the body and all vectors attached to the body from the reference orientation of the body to the actual orientation.

The vector angular momentum is conserved for free rigid-body motion, and so are its components on a fixed rectangular basis. So, along solution paths,

$$0 = D\overline{\mathbf{L}} = D\mathbf{M}\overline{\mathbf{L}}' + \mathbf{M}\,D\overline{\mathbf{L}}'. \tag{2.64}$$

Solving, we find

$$D\overline{\mathbf{L}}' = -\mathbf{M}^{\mathsf{T}}\,D\mathbf{M}\overline{\mathbf{L}}'. \tag{2.65}$$

In terms of ω' this is

$$\begin{aligned}
\mathbf{I}'D\omega' &= -\mathbf{M}^{\mathsf{T}}\,D\mathbf{M}\,\mathbf{I}'\omega' \\
&= -\mathbf{M}^{\mathsf{T}}\,\mathsf{A}(\mathbf{M}\omega')\,\mathbf{M}\,\mathbf{I}'\omega',
\end{aligned} \tag{2.66}$$

where we have used equation (2.21) to write $D\mathbf{M}$ in terms of A. The function A has the property[12]

$$\mathbf{R}^{\mathsf{T}}\,\mathsf{A}(\mathbf{R}\mathbf{v})\,\mathbf{R} = \mathsf{A}(\mathbf{v}) \tag{2.67}$$

for any vector with components $\mathbf{v}$ and any rotation with matrix representation $\mathbf{R}$. Using this property of A, we find *Euler's equations*:

$$\mathbf{I}'D\omega' = -\mathsf{A}(\omega')\,\mathbf{I}'\omega'. \tag{2.68}$$

Euler's equations give the time derivative of the body components of the angular velocity vector entirely in terms of the angular

---

[12] Rotating the cross product of two vectors gives the same vector as is obtained by taking the cross product of two rotated vectors: $R(\vec{u} \times \vec{v}) = (R\vec{u}) \times (R\vec{v})$.

velocity components and the principal moments of inertia. Let ω^a, ω^b, and ω^c denote the components of the angular velocity vector on the principal axes. Then Euler's equations can be written as the component equations

$$A\,D\omega^a = (B - C)\,\omega^b\omega^c$$
$$B\,D\omega^b = (C - A)\,\omega^c\omega^a$$
$$C\,D\omega^c = (A - B)\,\omega^a\omega^b. \tag{2.69}$$

Alternatively, we can rewrite Euler's equations in terms of the components of the angular momentum on the principal axes

$$DL_a = \left(\frac{1}{C} - \frac{1}{B}\right) L_b L_c$$
$$DL_b = \left(\frac{1}{A} - \frac{1}{C}\right) L_a L_c$$
$$DL_a = \left(\frac{1}{B} - \frac{1}{A}\right) L_a L_b. \tag{2.70}$$

These equations confirm that the time derivatives of the components of the angular momentum on the principal axes depend only on the components of the angular momentum on the principal axes.

Euler's equations are very simple, but they do not completely determine the evolution of a rigid body—they do not give the spatial orientation of the body. However, equation (2.21) and property (2.67) can be used to relate the derivative of the orientation matrix to the body components of the angular velocity vector:

$$D\mathbf{M} = \mathbf{M}A(\boldsymbol{\omega}'). \tag{2.71}$$

A straightforward method of using these equations is to integrate them componentwise as a set of nine first-order ordinary differential equations, with initial conditions determining the initial configuration matrix. Together with Euler's equations, which describe how the body components of the angular velocity vector change with time, this system of equations governing the motion of a rigid body is complete. However, the reader will no doubt have noticed that this approach is rather wasteful. The fact that the orientation matrix can be specified with only three parameters has not been taken into account. We should be integrating three equations for the orientation, given $\boldsymbol{\omega}'$, not nine. To accomplish this we once again need to parameterize the configuration matrix.

For example, we can use Euler angles to parameterize the orientation:

$$\mathsf{M}(\theta, \varphi, \psi) = \mathbf{R}_z(\varphi)\mathbf{R}_x(\theta)\mathbf{R}_z(\psi). \tag{2.72}$$

We form $\mathbf{M}$ by composing M with an Euler coordinate path. Equation (2.71) can then be used to solve for $D\theta$, $D\varphi$, and $D\psi$. We find

$$\begin{pmatrix} D\theta \\ D\varphi \\ D\psi \end{pmatrix} = \frac{1}{\sin\theta} \begin{pmatrix} \cos\psi\sin\theta & -\sin\psi\sin\theta & 0 \\ \sin\psi & \cos\psi & 0 \\ -\sin\psi\cos\theta & -\cos\psi\cos\theta & \sin\theta \end{pmatrix} \begin{pmatrix} \omega^a \\ \omega^b \\ \omega^c \end{pmatrix}. \tag{2.73}$$

This gives us the desired equation for the orientation. Note that it is singular for $\theta = 0$, as are Lagrange's equations. So Euler's equations using Euler angles for the configuration have the same problem as did the Lagrange equations using Euler angles. Again, this is a manifestation of the fact that for $\theta = 0$ the orientation depends only on $\varphi + \psi$. The singularity in the equations of motion for $\theta = 0$ does not correspond to anything funny in the motion of the rigid body. A practical solution to the singularity problem is to choose another set of Euler-like angles that have a singularity in a different place, and switch from one to the other when the going gets tough.

Exercise 2.12:
Fill in the details of the derivation of equation (2.73). You may want to use the computer to help with the algebra.

Euler's equations for forced rigid bodies
Euler's equations were derived for a free rigid body. In general, we must be able to deal with external forcing. How do we do this? First, we derive expressions for the vector torque. Then we include the vector torque in the Euler equations.

We derive the vector torque in a manner analogous to the derivation of the vector angular momentum. That is, we derive one component and then argue that since the coordinate system is arbitrary, all components have the same form.

Suppose we have a rigid body subject to some potential energy that depends only on time and the configuration. A Lagrangian is $L = T - V$. If we use the Euler angles as generalized coordinates, the last of the three active Euler rotations that define the orientation is a rotation about the $\hat{z}$ axis by the angle φ. The Lagrange equation for φ gives[13]

$$Dp_\varphi(t) = -\partial_{1,1}V(t; \theta(t), \varphi(t), \psi(t)). \qquad (2.74)$$

If we define T_z, the component of the torque about the z axis, to be minus the derivative of the potential energy with respect to the angle of rotation of the body about the z axis,

$$T_z(t) = -\partial_{1,1}V(t; \theta(t), \varphi(t), \psi(t)), \qquad (2.75)$$

then we see that

$$Dp_\varphi(t) = T_z(t). \qquad (2.76)$$

We have already identified the momentum conjugate to φ as one component, L_z, of the vector angular momentum $\vec{L}$ (see section 2.8), so

$$DL_z(t) = T_z. \qquad (2.77)$$

Since the orientation of the reference rectangular basis vectors is arbitrary, we may choose them any way that we please. Thus if we want any component of the vector torque, we may choose the z-axis so that we can compute it in this way. We can conclude that the vector torque gives the rate of change of the vector angular momentum

$$D\vec{L} = \vec{T}. \qquad (2.78)$$

Having obtained a general prescription for the vector torque, we address how the vector torque may be included in Euler's equations. Euler's equations expressed the fact that the vector angular

---

[13]In this equation we have a partial derivative with respect to a component of the coordinate argument of the potential energy function. The first subscript on the ∂ symbol indicates the coordinate argument. The second one selects the φ component.

momentum is conserved. Let's return to that calculation, but now include a torque with components $\mathbf{T}$ arranged as a column matrix:

$$D\overline{\mathbf{L}} = \overline{\mathbf{T}} = D\mathbf{M}\overline{\mathbf{L}}' + \mathbf{M}D\overline{\mathbf{L}}'. \tag{2.79}$$

Carrying out the same steps as before, we find

$$T_a = DL_a - \left(\frac{1}{C} - \frac{1}{B}\right) L_b L_c$$

$$T_b = DL_b - \left(\frac{1}{A} - \frac{1}{C}\right) L_a L_c$$

$$T_c = DL_a - \left(\frac{1}{B} - \frac{1}{A}\right) L_a L_b, \tag{2.80}$$

where the components of the torque on the principal axes are

$$\overline{\mathbf{T}}' = \mathbf{M}^{-1}\overline{\mathbf{T}}. \tag{2.81}$$

In terms of $\boldsymbol{\omega}'$ this is

$$\mathbf{I}'D\boldsymbol{\omega}' + \mathsf{A}(\boldsymbol{\omega}')\,\mathbf{I}'\boldsymbol{\omega}' = \overline{\mathbf{T}}'; \tag{2.82}$$

in components,

$$A\,D\omega^a - (B - C)\,\omega^b\omega^c = T_a$$
$$B\,D\omega^b - (C - A)\,\omega^c\omega^a = T_b$$
$$C\,D\omega^c - (A - B)\,\omega^a\omega^b = T_c. \tag{2.83}$$

Note that the torque entered only the equations for the body angular momentum and for the body angular velocity vector. The equations that relate the derivative of the orientation to the angular velocity vector are not modified by the torque. In a sense, Euler's equations contain the dynamics, and the equations governing the orientation are kinematic. Of course, Lagrange's equations must be modified by the potential that gives rise to the torques; in this sense Lagrange's equations contain both dynamics and kinematics.

Exercise 2.13: Bicycle wheel

a. Imagine that you are holding a bicycle wheel by the axle (in both hands) and the wheel is spinning so that the top edge is going away from your face. If you torque the wheel by pushing down with your right hand and pulling up with your left hand the wheel will precess. Which way does it try to turn?

b. A free bicycle wheel rolls on a horizontal surface. If it starts to tilt, the torque from gravity will cause the wheel to turn. Which way will it turn? The reasoning that applied to part **a** does not directly apply to the rolling bicycle wheel, which is not a holonomic system. However, it is interesting to think about whether the behavior of the two systems is related.

2.10   Axisymmetric Tops

We have all played with a top at one time or another. For the purposes of analysis we will consider an idealized top that does not wander around. Thus, an ideal top is a rotating rigid body, one point of which is fixed in space. Furthermore, the center of mass of the top is not at the fixed point, which is the center of rotation, and there is a uniform gravitational acceleration.

For our top we can take the Lagrangian to be the difference of the kinetic energy and the potential energy. We already know how to write the kinetic energy—what is new here is that we must express the potential energy in terms of the configuration. In the case of a body in a uniform gravitational field this is easy. The potential energy is the sum of "mgh" for all the constituent particles:

$$\sum_{\alpha} m_{\alpha} g h_{\alpha}, \tag{2.84}$$

where g is the gravitational acceleration, $h_{\alpha} = \vec{x}_{\alpha} \cdot \hat{z}$, and the unit vector $\hat{z}$ indicates which way is up. Rewriting the vector to the constituents in terms of the vector $\vec{X}$ to the center of mass, the potential energy is

$$\sum_{\alpha} m_{\alpha} g \left(\vec{X} + \vec{\xi}_{\alpha} \right) \cdot \hat{z}$$

$$= gM\vec{X} \cdot \hat{z} + g \left(\sum_{\alpha} m_{\alpha} \vec{\xi}_{\alpha} \right) \cdot \hat{z}$$

$$= gM\vec{X} \cdot \hat{z}, \tag{2.85}$$

where the last sum is zero because the center of mass is the origin of $\vec{\xi}_{\alpha}$. So the potential energy of a body in a gravitational field with uniform acceleration is very simple: it is just Mgh, where M is the total mass and $h = \vec{X} \cdot \hat{z}$ is the height of the center of mass.

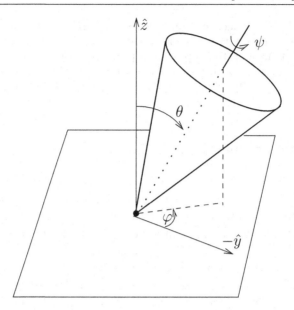

Figure 2.4   An axisymmetric top is a symmetrical rigid body in a uniform gravitational field with one point of the body fixed in space. The Euler angles used to specify the configuration are indicated.

Here we consider an axisymmetric top (see figure 2.4). Such a top has an axis of symmetry of the mass distribution, so the center of mass is on the symmetry axis and the fixed point is also on the axis of symmetry.

In order to write the Lagrangian we need to choose a set of generalized coordinates. If we choose them well we can take advantage of the symmetries of the problem. If the Lagrangian does not depend on a particular coordinate, the conjugate momentum is conserved, and the complexity of the system is reduced.

The axisymmetric top has two apparent symmetries. The fact that the mass distribution is axisymmetric implies that neither the kinetic nor the potential energy is sensitive to the orientation of the top about that symmetry axis. Additionally, the kinetic and potential energy are insensitive to a rotation of the physical system about the vertical axis, because the gravitational field is uniform.

We can take advantage of these symmetries by choosing appropriate coordinates, and we already have a coordinate system that does the job—the Euler angles.[14] We choose the reference orientation so that the symmetry axis is vertical. The first Euler angle, ψ, expresses a rotation about the symmetry axis. The next Euler angle, θ, is the tilt of the symmetry axis of the top from the vertical. The third Euler angle, φ, expresses a rotation of the top about the $\hat{z}$ axis. The symmetries of the problem imply that the first and third Euler angles do not appear in the Lagrangian. As a consequence the momenta conjugate to these angles are conserved quantities. Let's work out the details.

First, we develop the Lagrangian explicitly. The general form of the kinetic energy about a fixed point is given by equation 2.41. The top is constrained so that it pivots about a fixed point that is not at the center of mass. So the moments of inertia that enter the kinetic energy are the moments of inertia of the top with respect to the pivot point, not with respect to the center of mass. If we know the moments of inertia about the center of mass we can write the moments of inertia about the pivot in terms of them (see exercise 2.2 on Steiner's theorem). So let's assume the principal moments of inertia of the top about the pivot are A, B, and C, and $A = B$ because of the symmetry.[15] We can use the computer to help us figure out the Lagrangian for this special case:

```
(show-expression
 ((T-body-Euler 'A 'A 'C)
  (up 't
      (up 'theta 'phi 'psi)
      (up 'thetadot 'phidot 'psidot))))
```

$$\frac{1}{2}\left(\sin\left(\theta\right)\right)^2 A\dot{\varphi}^2 + \cos\left(\theta\right)\left(\frac{1}{2}\cos\left(\theta\right)C\dot{\varphi}^2 + C\dot{\varphi}\dot{\psi}\right) + \frac{1}{2}A\dot{\theta}^2 + \frac{1}{2}C\dot{\psi}^2$$

We can rearrange this a bit to get

$$T(t; \theta, \varphi, \psi; \dot{\theta}, \dot{\varphi}, \dot{\psi})$$
$$= \tfrac{1}{2}A\left(\dot{\theta}^2 + \dot{\varphi}^2\sin^2\theta\right) + \tfrac{1}{2}C\left(\dot{\psi} + \dot{\varphi}\cos\theta\right)^2. \qquad (2.86)$$

---

[14]That the axisymmetric top can be solved in Euler angles is, no doubt, the reason for the traditional choice of the definition of these. For other problems, the Euler angles may offer no particular advantage.

[15]Here, we do not require that C be larger than $A = B$, because they are not measured with respect to the center of mass.

In terms of Euler angles, the potential energy is

$$V(t; \theta, \varphi, \psi; \dot{\theta}, \dot{\varphi}, \dot{\psi}) = MgR \cos \theta, \tag{2.87}$$

where R is the distance of the center of mass from the pivot. The Lagrangian is $L = T - V$. We see that the Lagrangian is indeed independent of ψ and φ, as expected.

There is no particular reason to look at the Lagrange equations. We can assign that job to the computer when needed. However, we have already seen that it may be useful to examine the conserved quantities associated with the symmetries.

The energy is conserved, because the Lagrangian has no explicit time dependence. Also, the energy is the sum of the kinetic and potential energy $E = T + V$, because the kinetic energy is a homogeneous quadratic form in the generalized velocities. The energy is

$$E = \tfrac{1}{2} A \left(\dot{\theta}^2 + \dot{\varphi}^2 \sin^2 \theta \right) + \tfrac{1}{2} C \left(\dot{\psi} + \dot{\varphi} \cos \theta \right)^2 + MgR \cos \theta. \tag{2.88}$$

Two of the generalized coordinates do not appear in the Lagrangian, so there are two conserved momenta. The momentum conjugate to φ is

$$p_\varphi = \left(A(\sin \theta)^2 + C(\cos \theta)^2 \right) \dot{\varphi} + C \dot{\psi} \cos \theta. \tag{2.89}$$

The momentum conjugate to ψ is

$$p_\psi = C(\dot{\psi} + \dot{\varphi} \cos \theta). \tag{2.90}$$

The state of the system at a moment is specified by the tuple $(t; \theta, \varphi, \psi; \dot{\theta}, \dot{\varphi}, \dot{\psi})$. Because the two coordinates φ and ψ do not appear in the Lagrangian, they do not appear in the Lagrange equations or the conserved momenta. So the evolution of the remaining four state variables, θ, $\dot{\theta}$, $\dot{\varphi}$, and $\dot{\psi}$, depends only on those remaining state variables. This subsystem for the top has a four-dimensional state space. The variables that did not appear in the Lagrangian can be determined by integrating the derivatives of these variables, which are determined separately by solving the independent subsystem.

The evolution of the top is described by a four-dimensional subsystem and two auxiliary quadratures.[16] This subdivision is a consequence of choosing generalized coordinates that incorporate the symmetries. However, the choice of generalized coordinates that incorporate the symmetries also gives conserved momenta. We can make use of these momenta to simplify the formulation of the problem further. Each conserved quantity can be used to locally eliminate one dimension of the subsystem. In this case the subsystem has four dimensions and there are three conserved quantities, so the system can be completely reduced to quadratures. For the top, this can be done analytically, but we think it is a waste of time to do so. Rather, we are interested in extracting interesting features of the motion. We concentrate on the energy and use the two conserved momenta to eliminate $\dot{\varphi}$ and $\dot{\psi}$. After a bit of algebra we find:

$$E = \frac{1}{2}A\dot{\theta}^2 + \frac{(p_\varphi - p_\psi \cos\theta)^2}{2A(\sin\theta)^2} + \frac{p_\psi^2}{2C} + MgR\cos\theta. \tag{2.91}$$

Along a path θ, where $D\theta(t)$ is substituted for $\dot{\theta}$, this is an ordinary differential equation for θ. This differential equation involves various constants, some of which are set by the initial conditions of the other state variables. The solution of the differential equation for θ involves no more than ordinary integrals. So the top is essentially solved. We could continue this argument to obtain the qualitative behavior of θ: Using the energy (2.91), we can plot the trajectories in the plane of $\dot{\theta}$ versus θ and see that the motion of θ is simply periodic. However, we will defer this until chapter 3, when we have developed more tools for analysis.

Let's get real. Let's make a top out of an aluminum disk with a steel rod through the center to make the pivot. Measuring the top very carefully, we find that the moment of inertia of the top about the symmetry axis is about $1.32 \times 10^{-4}\,\mathrm{kg\,m^2}$, and the moment of inertia about the pivot point is about $6.96 \times 10^{-4}\,\mathrm{kg\,m^2}$. The combination gMR is about $0.112\mathrm{kg\,m^2\,s^{-2}}$. We spin the top up with an initial angular velocity of $\dot{\psi} = 200\,\mathrm{rad\,s^{-1}}$ (about 1910 rpm).

---

[16]Traditionally, evaluating a definite integral is known as performing a quadrature.

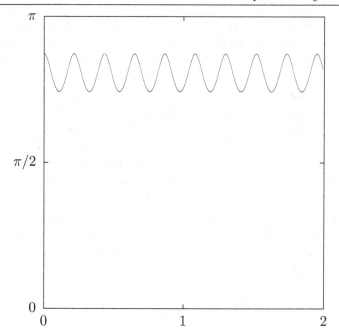

Figure 2.5   The tilt angle $\pi - \theta$ of the top versus time. The tilt of the top varies periodically. This motion is called *nutation*.

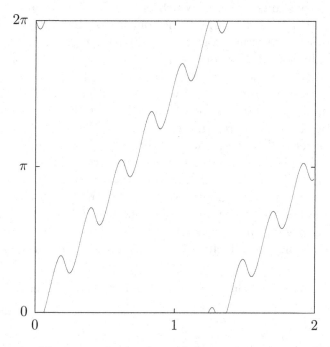

Figure 2.6   The precession angle φ of the top versus time. The top precesses nonuniformly—the rate of precession varies as the tilt varies.

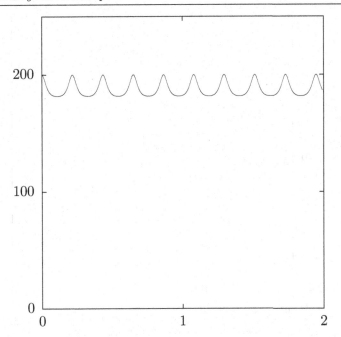

Figure 2.7   The rate of rotation $\dot{\psi}$ of the top versus time. The rate of rotation of the top changes periodically, as the tilt of the top varies.

The top initially has $\dot{\theta} = \varphi = \psi = 0$ and is initially tilted with $\theta = 0.4$ rad. We then kick it so that $\dot{\varphi} = -10$ rad s^{-1}. Figures 2.5–2.8 display aspects of the evolution of the top for 2 seconds. The tilt of the top (measured by θ) varies in a periodic manner. The orientation about the vertical is measured by φ: we see that the top also precesses, and the rate of precession varies with θ. We also see that as the top bobs up and down the rate of rotation of the top oscillates—the top spins faster when it is more vertical. The plot of tilt versus precession angle shows that in this case the top executes a looping motion. If we do not kick it but just let it drop, then the loop disappears, leaving just a cusp. If we kick it in the other direction, then there is no cusp nor any looping motion.

Exercise 2.14: Kinetic energy of the top

The rotational kinetic energy of the top can be written in terms of the principal moments of inertia with respect to the pivot point and the angular velocity vector of rotation with respect to the pivot point. Show that this formulation of the kinetic energy yields the same value that one would obtain by computing the sum of the rotational kinetic energy about its center of mass and the kinetic energy of the motion of the center of mass.

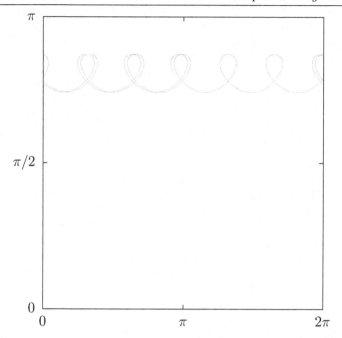

Figure 2.8   An idea of the actual motion of the top is obtained by plotting the tilt angle $\pi - \theta$ versus the precession angle φ. This is a "latitude-longitude" map showing the path of the center of mass of the top. We see that, though the top has a net precession, it executes a looping motion as it precesses.

Exercise 2.15: Nutation of the top

a. Carry out the algebra to obtain the energy (2.91) in terms of θ and $\dot{\theta}$.

b. Numerically integrate the Lagrange equations for the top to obtain figure 2.5, θ versus time.

c. Note that the energy is a differential equation for $\dot{\theta}$ in terms of θ, with conserved quantities p_φ, p_ψ, and E determined by initial conditions. Can we use this differential equation to obtain θ as a function of time? Explain.

Exercise 2.16: Precession of the top

Consider a top that is rotating so that θ is constant.

a. Using conservation of angular momentum, compute the rate of precession $\dot{\varphi}$ as a function of the conserved angular momenta and the equilibrium value of θ.

b. For θ to be at an equilibrium the acceleration $D^2\theta$ must be zero. Use the Lagrange equation for θ to find the rate of precession $\dot{\varphi}$ at the equilibrium in terms of the equilibrium θ and $\dot{\psi}$.

c. Find an approximate expression for the precession rate in the limit that $\dot{\psi}$ is large.

d. The Newtonian rule is that the rate of change of the angular momentum is the torque. Assume the top is spinning so fast that the angular momentum is nearly the same as the angular momentum of rotation about the symmetry axis. By equating the rate of change of this vector angular momentum to the gravitational torque on the center of mass develop an approximate formula for the precession rate.

e. Numerically integrate the top to check your deductions.

2.11 Spin-Orbit Coupling

The rotation of planets and natural satellites is affected by the gravitational forces from other celestial bodies. As an extended application of the Lagrangian method for forced rigid bodies, we consider the rotation of celestial objects subject to gravitational forces.

We first develop the form of the potential energy for the gravitational interaction of an extended body with an external point mass. With this potential energy and the usual rigid-body kinetic energy we can form Lagrangians that model a number of systems. We will take an initial look at the rotation of the Moon and Mercury; later, after we have developed more tools, we will return to study these systems.

2.11.1 Development of the Potential Energy

The first task is to develop convenient expressions for the gravitational potential energy of the interaction of a rigid body with a distant point mass. A rigid body can be thought of as made of a large number of mass elements, subject to rigid coordinate constraints. We have seen that the kinetic energy of a rigid body is conveniently expressed in terms of the moments of inertia of the body and the angular velocity vector, which in turn can be represented in terms of a suitable set of generalized coordinates. The potential energy can be developed in a similar manner. We first represent the potential energy in terms of moments of the mass distribution and later introduce generalized coordinates as particular parameters of the potential energy.

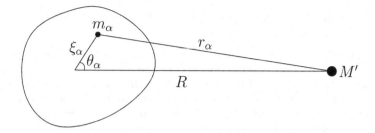

Figure 2.9   The gravitational potential energy of a point mass and a rigid body is the sum of the gravitational potential energy of the point mass with each constituent mass element of the rigid body.

The gravitational potential energy of a point mass and a rigid body (see figure 2.9) is the sum of the potential energy of the point mass with each mass element of the body:

$$-\sum_{\alpha} \frac{GM'm_{\alpha}}{r_{\alpha}}, \tag{2.92}$$

where M' is the mass of the external point mass, r_{α} is the distance between the point mass and the constituent mass element with index α, m_{α} is the mass of this constituent element, and G is the gravitational constant. Let R be the distance of the center of mass of the rigid body from the point mass. The distance from the center of mass to the constituent with index α is ξ_{α}. The distance r_{α} is then given by the law of cosines as $r_{\alpha}^2 = R^2 + \xi_{\alpha}^2 - 2\xi_{\alpha}R\cos\theta_{\alpha}$, where θ_{α} is the angle between the lines from the center of mass to the constituent and to the point mass.

Because this is a three-dimensional body the distance ξ_{α} and angle θ_{α} do not completely specify the position of the constituent mass element; to do that one must also specify the angle of rotation about the line between the center of mass and the external point mass. But the potential energy does not depend on this angle.

The potential energy is then

$$-GM'\sum_{\alpha} \frac{m_{\alpha}}{\left(R^2 + \xi_{\alpha}^2 - 2\xi_{\alpha}R\cos\theta_{\alpha}\right)^{1/2}}. \tag{2.93}$$

This is complete, but we need to find a representation that does not mention each constituent.

Typically, the size of celestial bodies is small compared to the distance between them. We can make use of this to find a more compact representation of the potential energy. If we expand the potential energy in the small ratio ξ_α/R we find

$$-GM' \sum_\alpha m_\alpha \frac{1}{R} \sum_l \frac{\xi_\alpha^l}{R^l} P_l(\cos\theta_\alpha), \tag{2.94}$$

where P_l is the lth Legendre polynomial.[17]   Interchanging the order of the summations yields:

$$-\frac{GM'}{R} \sum_l \sum_\alpha m_\alpha \frac{\xi_\alpha^l}{R^l} P_l(\cos\theta_\alpha). \tag{2.95}$$

Successive terms in this expansion of the potential energy typically decrease very rapidly because celestial bodies are small compared to the distance between them. We can compute an upper bound to the size of these terms by replacing each factor in the sum over α by an upper bound. The Legendre polynomials all have magnitudes less than one for arguments in the range -1 to 1. The distances ξ_α are all less than some maximum extent of the body $\xi_{\max}$. The sum over m_α times these upper bounds is just the total mass M times the upper bounds. Thus

$$\left| \sum_\alpha m_\alpha \frac{\xi_\alpha^l}{R^l} P_l(\cos\theta_\alpha) \right| \leq M \frac{\xi_{\max}^l}{R^l}. \tag{2.96}$$

We see that the upper bound on successive terms decreases by a factor $\xi_{\max}/R$. Successive terms may be smaller still. For large bodies the gravitational force is strong enough to overcome the

---

[17]The Legendre polynomials P_l may be obtained by expanding the expression $(1 + y^2 - 2yx)^{-1/2}$ as a power series in y. The coefficient of y^l is $P_l(x)$. The first few Legendre polynomials are: $P_0(x) = 1$, $P_1(x) = x$, $P_2(x) = \frac{3}{2}x^2 - \frac{1}{2}$, and so on. The rest satisfy the recurrence relation

$$lP_l(x) = (2l - 1)xP_{l-1}(x) - (l - 1)P_{l-2}(x).$$

internal material strength of the body, so the body, over time, becomes nearly spherical. Successive terms in the expansion of the potential are measures of the deviation of the mass distribution from a spherical mass distribution. Thus for large bodies the higher-order terms are small because the bodies are nearly spherical.

Consider the first few terms in l. For $l = 0$ the sum over α just gives the total mass M of the rigid body. For $l = 1$ the sum over α is zero, as a consequence of choosing the origin of the $\vec{\xi}_\alpha$ to be the center of mass. For $l = 2$ we have to do a little more work. The sum involves second moments of the mass distribution, and can be written in terms of moments of inertia of the rigid body:

$$\sum_\alpha m_\alpha \xi_\alpha^2 P_2(\cos\theta_\alpha) = \sum_\alpha m_\alpha \xi_\alpha^2 \left(\frac{3}{2}(\cos\theta_\alpha)^2 - \frac{1}{2}\right)$$

$$= \sum_\alpha m_\alpha \xi_\alpha^2 \left(1 - \frac{3}{2}(\sin\theta_\alpha)^2\right)$$

$$= \frac{1}{2}(A + B + C - 3I), \tag{2.97}$$

where A, B, and C are the principal moments of inertia, and I is the moment of inertia of the rigid body about the line between the center of mass of the body and the external point mass. The moment I depends on the orientation of the rigid body relative to the line between the bodies. The contributions to the potential energy up to $l = 2$ are then[18]

$$-\frac{GMM'}{R} - \frac{GM'}{2R^3}(A + B + C - 3I). \tag{2.98}$$

Let $c_a = \cos\theta_a$, $c_b = \cos\theta_b$, and $c_c = \cos\theta_c$ be the direction cosines of the angles θ_a, θ_b and θ_c between the principal axes $\hat{a}$, $\hat{b}$, and $\hat{c}$ and the line between the center of mass and the point mass. (See figure 2.10.) A little algebra shows that $I = c_a^2 A + c_b^2 B + c_c^2 C$. The potential energy is then

$$-\frac{GMM'}{R} - \frac{GM'}{2R^3}[(1 - 3c_a^2)A + (1 - 3c_b^2)B + (1 - 3c_c^2)C]. \tag{2.99}$$

---

[18]This approximate representation of the potential energy is sometimes called MacCullagh's formula.

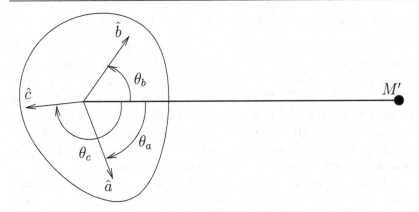

Figure 2.10   The orientation of the rigid body is specified by the three angles from the line between the centers and the principal axes.

This is a good first approximation to the potential energy of interaction for most situations in the solar system; if we intended to land on the Moon we probably would want to take into account higher-order terms in the expansion.

Exercise 2.17:

a. Fill in the details that show that the sum over constituents in equation (2.97) can be expressed as written in terms of moments of inertia. In particular, show that

$$\sum_\alpha m_\alpha \xi_\alpha \cos\theta_\alpha = 0,$$

$$2\sum_\alpha m_\alpha \xi_\alpha^2 = A + B + C,$$

and that

$$\sum_\alpha m_\alpha \xi_\alpha^2 (\sin\theta_\alpha)^2 = I.$$

b. Show that if the principal moments of inertia of a rigid body are A, B, and C, then the moment of inertia about an axis that goes through the center of mass of the body with angles θ_a, θ_b, and θ_c to the principal axes is

$$I = (\cos\theta_a)^2 A + (\cos\theta_b)^2 B + (\cos\theta_c)^2 C.$$

2.11.2 Rotation of the Moon and Hyperion

The approximation to the potential energy that we have derived can be used for a number of different problems. It can be used to investigate the effect of oblateness on the motion of an artificial satellite about the Earth, or to incorporate the effect of planetary oblateness on the evolution of the orbits of natural satellites, such as the Moon or the Galilean satellites of Jupiter. However, as the principal application here, we will use it to investigate the rotational dynamics of natural satellites and planets.

The potential energy depends on the position of the point mass relative to the rigid body and on the orientation of the rigid body. Thus the changing orientation is coupled to the orbital evolution; each affects the other. However, in many situations the effect of the orientation of the body on the evolution of the orbit may be ignored. One way to see this is to look at the relative magnitudes of the two terms in the potential energy (2.99). We already know that the second term is guaranteed to be smaller than the first by a factor of $(\xi_{max}/R)^2$, but often it is much smaller still because the body involved is nearly spherical. For example, the radius of the Moon is about a third the radius of the Earth and the distance to the Moon is about 60 Earth-radii. So the second term is smaller than the first by a factor of order 10^{-4} due to the size factors. In addition, the Moon is roughly spherical and for any orientation the combination $A + B + C - 3I$ is of order $10^{-4}C$. Now C is itself of order $\frac{2}{5}MR^2$, because the density of the Moon does not vary strongly with radius. So for the Moon the second term is of order 10^{-8} relative to the first. Even radical changes in the orientation of the Moon would have little dynamical effect on its orbit.

We can learn some important qualitative aspects of the orientation dynamics by studying a simplified model problem. First, we assume that the body is rotating about its largest moment of inertia. This is a natural assumption. Remember that for a free rigid body the loss of energy while conserving angular momentum leads to rotation about the largest moment of inertia. This is observed for most bodies in the solar system. Next, we assume that the spin axis is perpendicular to the orbital motion. This is a good approximation for the rotation of natural satellites, and is a natural consequence of tidal friction—dissipative solid-body tides raised on the satellite by the gravitational interaction with the planet. Finally, for simplicity we take the rigid body to be mov-

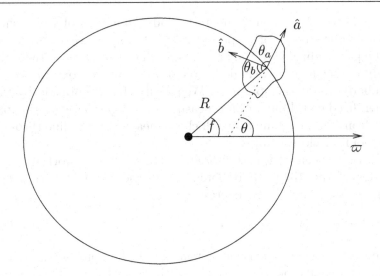

Figure 2.11 The spin-orbit model problem in which the spin axis is constrained to be perpendicular to the orbit plane has a single degree of freedom, the orientation of the body in the orbit plane. Here the orientation is specified by the generalized coordinate θ.

ing on a fixed elliptic orbit. This may approximate the motion of some physical systems, provided the time scale of the evolution of the orbit is large compared to any time scale associated with the rotational dynamics we are investigating. So we have a nice toy problem, one that has been used to investigate the rotational dynamics of Mercury, the Moon, and other natural satellites. It makes specific predictions concerning the rotation of Phobos, a satellite of Mars, that can be compared with observations. It provides a basic explanation of the fact that Mercury rotates precisely three times for every two orbits it completes, and is the starting point for understanding the chaotic tumbling of Saturn's satellite Hyperion.

We are assuming that the orbit does not change or precess. The orbit is an ellipse with the point mass at a focus of the ellipse. The angle f (see figure 2.11) measures the position of the rigid body in its orbit relative to the point in the orbit at which the two bodies are closest.[19] We assume the orbit is a fixed ellipse, so the

---

[19]Traditionally, the point in the orbit at which the two bodies are closest is called the *pericenter* and the angle f is called the *true anomaly*.

angle f and the distance R are periodic functions of time, with period equal to the orbit period. With the spin axis constrained to be perpendicular to the orbit plane, the orientation of the rigid body is specified by a single degree of freedom: the orientation of the body about the spin axis. We specify this orientation by the generalized coordinate θ that measures the angle to the $\hat{a}$ principal axis from the same line from which we measure f, the line through the point of closest approach.

Having specified the coordinate system, we can work out the details of the kinetic and potential energies, and thus find the Lagrangian. The kinetic energy is

$$T(t, \theta, \dot{\theta}) = \tfrac{1}{2}C\dot{\theta}^2, \tag{2.100}$$

where C is the moment of inertia about the spin axis and the angular velocity of the body about the $\hat{c}$ axis is $\dot{\theta}$. There is no component of angular velocity on the other principal axes.

To get an explicit expression for the potential energy, write the direction cosines in terms of θ and f: $\cos\theta_a = -\cos(\theta - f)$, $\cos\theta_b = \sin(\theta - f)$, and $\cos\theta_c = 0$ because the $\hat{c}$ axis is perpendicular to the orbit plane. The potential energy is then

$$-\frac{GMM'}{R}$$
$$-\frac{1}{2}\frac{GM'}{R^3}\left[(1 - 3\cos^2(\theta - f))A + (1 - 3\sin^2(\theta - f))B + C\right].$$

Since we are assuming that the orbit is given, we need keep only terms that depend on θ. Expanding the squares of the cosine and the sine in terms of the double angles and dropping all the terms that do not depend on θ, we find the potential energy for the orientation[20]

$$V(t, \theta, \dot{\theta}) = -\frac{3}{4}\frac{GM'}{R^3(t)}(B - A)\cos 2(\theta - f(t)). \tag{2.101}$$

---

[20]The given potential energy differs from the actual potential energy in that non-constant terms that do not depend on θ and consequently do not affect the evolution of θ have been dropped.

A Lagrangian for the model spin-orbit coupling problem is then
$L = T - V$:

$$L(t, \theta, \dot\theta) = \frac{1}{2}C\dot\theta^2 + \frac{3}{4}\frac{GM'}{R^3(t)}(B - A)\cos 2(\theta - f(t)). \qquad (2.102)$$

We introduce the dimensionless "out-of-roundness" parameter

$$\epsilon = \sqrt{\frac{3(B - A)}{C}}, \qquad (2.103)$$

and use the fact that the orbital frequency n and the semimajor
axis a satisfy Kepler's third law, $n^2a^3 = G(M + M')$, which is
approximately $n^2a^3 = GM'$ for a small body in orbit around a
much more massive one $(M \ll M')$. In terms of ϵ and n the
spin-orbit Lagrangian is

$$L(t, \theta, \dot\theta) = \frac{1}{2}C\dot\theta^2 + \frac{n^2\epsilon^2 C}{4}\frac{a^3}{R^3(t)}\cos 2(\theta - f(t)). \qquad (2.104)$$

This is a problem with one degree of freedom with terms that vary
periodically with time.

The Lagrange equations are derived in the usual manner:

$$CD^2\theta(t) = -\frac{n^2\epsilon^2 C}{2}\frac{a^3}{R^3(t)}\sin 2(\theta(t) - f(t)). \qquad (2.105)$$

The equation of motion is very similar to that of the periodically
driven pendulum. The main difference here is that not only is the
strength of the acceleration changing periodically, but in the spin-
orbit problem the center of attraction is also varying periodically.

We can give a physical interpretation of this equation of motion.
It states that the rate of change of the angular momentum is equal
to the applied torque. The torque on the body arises because the
body is out of round and the gravitational force varies as the
inverse square of the distance. Thus the force per unit mass on
the near side of the body is a little more than the acceleration
of the body as a whole, and the force per unit mass on the far
side of the body is a little less than the acceleration of the body
as a whole. Thus, relative to the acceleration of the body as a

whole, the far side is forced outward while the inner part of the
body is forced inward. The net effect is a torque on the body
that tries to align the long axis of the body with the line to the
external point mass. If θ is a bit larger than f then there is a
negative torque, and if θ is a bit smaller than f then there is
a positive torque, both of which would align the long axis with
the point mass if given a fair chance. The torque arises because
of the difference of the inverse R^2 force across the body, so the
torque is proportional to R^{-3}. There is a torque only if the body
is out of round, for otherwise there is no handle to pull on. This
is reflected in the factor $B - A$ in the expression for the torque.
The potential depends on the mass distribution as described by
the moments of inertia, and thus the body has the same dynamics
if it is rotated by 180°. The factor of 2 in the argument of sine
reflects this symmetry. This torque is called the "gravity gradient
torque."

To compute the evolution requires a lot of detailed prepara-
tion similar to what has been done for other problems. There are
many interesting phenomena to explore. We can take parameters appropriate for the Moon and find that Mr. Moon does not
steadily point the same face to the Earth, but instead constantly
shakes his head in dismay at what goes on here. If we nudge the
Moon a bit, say by hitting it with an asteroid, we find that the
long axis oscillates back and forth with respect to the direction
that points to the Earth. For the Moon, the orbital eccentricity is
currently about 0.05, and the out-of-roundness parameter is about
$\epsilon = 0.026$. Figure 2.12 shows the angle $\theta - f$ as a function of time
for two different values of the "lunar" eccentricity. The plot spans
50 lunar orbits, or a little under four years. This Moon has been
kicked by a large asteroid and has initial rotational angular velocity $\dot{\theta}$ equal to 1.01 times the orbit frequency. The initial orientation is $\theta = 0$. The smooth trace shows the evolution if the orbital
eccentricity is set to zero. We see an oscillation with a period
of about 40 lunar orbits or about three years. The more wiggly
trace shows the evolution of $\theta - f$ with an orbital eccentricity of
0.05, near the current lunar eccentricity. The lunar eccentricity
superimposes an apparent shaking of the face of the Moon back
and forth with the period of the lunar orbit. Though the Moon
does slightly change its rate of rotation during the course of its
orbit, most of this shaking is due to the nonuniform motion of the
Moon in its elliptical orbit. This oscillation, called the "optical
libration of the Moon," allows us to see a bit more than half of

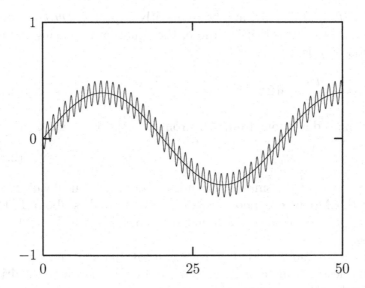

Figure 2.12   The angle $\theta - f$ versus time for 50 orbit periods. The ordinate scale is ± 1 radian. The Moon has been kicked so that the initial rotational angular velocity is 1.01 times the orbital frequency. The trace with fewer wiggles was computed with zero lunar orbital eccentricity; the other trace was computed with lunar orbital eccentricity of 0.05. The period of the rapid oscillations is the lunar orbit period. These oscillations are due mostly to the nonuniform motion of f.

the Moon's surface. The longer-period oscillation induced by the kick is called the "free libration of the Moon." It is "free" because we are free to excite it by choosing appropriate initial conditions. The mismatch of the orientation of the Moon caused by the optical libration actually produces a periodic torque on the Moon, which slightly speeds it up and slows it down during every orbit. The resulting oscillation is called the "forced libration of the Moon," but it is too small to see in this plot.

   The oscillation period of the free libration is easily calculated. We see that the eccentricity of the orbit does not substantially affect the period, so we consider the special case of zero eccentricity. In this case $R = a$, a constant, and $f(t) = nt$, where n is the orbital frequency.[21] The equation of motion becomes

$$D^2\theta(t) = -\frac{n^2\epsilon^2}{2}\sin 2(\theta(t) - nt). \qquad (2.106)$$

---

[21] Traditionally, the orbital angular frequency is called the *mean motion*.

Let $\varphi(t) = \theta(t) - nt$, and consequently $D\varphi(t) = D\theta(t) - n$, and $D^2\varphi = D^2\theta$. Substituting these, the equation governing the evolution of φ is

$$D^2\varphi = -\frac{n^2\epsilon^2}{2}\sin 2\varphi. \tag{2.107}$$

For small deviations from synchronous rotation (small φ) this is

$$D^2\varphi = -n^2\epsilon^2\varphi, \tag{2.108}$$

so we see that the small-amplitude oscillation frequency of φ is $n\epsilon$. For the Moon, ϵ is about 0.026, so the period is about $1/0.026$ orbit periods or about 40 lunar orbit periods, which is what we observed.

It is perhaps more fun to see what happens if the out-of-roundness parameter is large. After our experience with the driven pendulum it is no surprise that we find abundant chaos in the spin-orbit problem when the system is strongly driven by having large ϵ and significant orbital eccentricity e. There is indeed one body in the solar system that exhibits chaotic rotation—Hyperion, a small satellite of Saturn. Though our toy model is not adequate for a complete account of Hyperion, we can show that it exhibits chaotic behavior for parameters appropriate for Hyperion. We take $\epsilon = 0.89$ and $e = 0.1$. Figure 2.13 shows $\theta - f$ for 50 orbits, starting with $\theta = 0$ and $\dot{\theta} = 1.05$. We see that sometimes one face of the body oscillates facing the planet, sometimes the other face oscillates facing the planet, and sometimes the body rotates relative to the planet in either direction.

If we relax our restriction that the spin axis be fixed perpendicular to the orbit, then we find that the Moon maintains this orientation of the spin axis even if nudged a bit, but for Hyperion the spin axis almost immediately falls away from this configuration. The state in which Hyperion on average points one face to Saturn is dynamically unstable to chaotic tumbling. Observations of Hyperion are consistent with the deduction that it is chaotically tumbling.

Exercise 2.18: Precession of the equinox

The Earth spins very nearly about the largest moment of inertia, and the spin axis is tilted by about 23° to the orbit normal. There is a gravity-gradient torque on the Earth from the Sun that causes the spin axis of the Earth to precess. Investigate this precession in the approximation

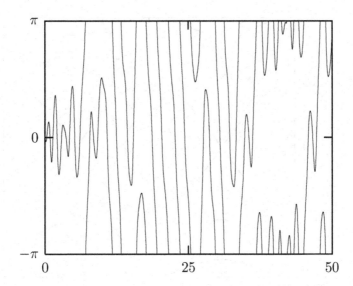

Figure 2.13   The angle $\theta - f$ versus time for 50 orbit periods. The ordinate scale is $\pm\pi$ radian. The out-of-roundness parameter is large $\epsilon = 0.89$, with an orbital eccentricity of $e = 0.1$. The system is strongly driven. The rotation is apparently chaotic.

that the orbit of the Earth is circular and the Earth is axisymmetric. Determine the rate of precession in terms of the moments of inertia of the Earth.

2.11.3   Spin-Orbit Resonances

Consider the motion of the Moon in synchronous rotation. We have seen that if we give the Moon a kick so that it is not exactly pointing one face to the Earth, then the face will oscillate back and forth relative to the direction to the Earth. If we give it a really big kick, then instead of oscillating it will spin relative to the direction to the Earth. How do we understand this?

Let's look again at the equations of motion for the rotation of the Moon when the orbit is circular (equation 2.106):

$$CD^2\theta(t) = -\frac{C}{2}n^2\epsilon^2 \sin 2(\theta(t) - n)t). \tag{2.109}$$

Changing variables to $\varphi(t) = \theta(t) - nt$ this equation becomes

$$CD^2\varphi(t) = -\frac{C}{2}n^2\epsilon^2 \sin 2\varphi(t). \tag{2.110}$$

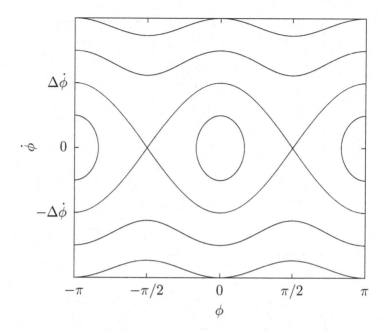

Figure 2.14   Trajectories of φ and $\dot{\varphi}$ in the spin-orbit problem when the orbital eccentricity is zero.

This equation can be solved; it has an "energy-like" conserved quantity

$$E(\varphi, \dot{\varphi}) = \frac{C}{2}\dot{\varphi}^2 - \frac{C}{4}n^2\epsilon^2\cos(2\varphi). \tag{2.111}$$

The solutions are just contours of this conserved quantity (see figure 2.14). There are two centers of oscillation corresponding to the two different faces of the Moon that could point towards Earth. There are also trajectories that rotate relative to the Earth. And there are separating trajectories that divided the oscillating trajectories from the circulating trajectories. Where these separating trajectories appear to cross, the system is at an unstable equilibrium. The separating trajectories are asymptotic to the unstable equilibria (a system on that trajectory takes an infinite time to get to the equilibrium point). These asymptotic trajectories are analogous to the trajectories of the simple pendulum that are asymptotic to the vertical.

The extent of the oscillation region can be evaluated with the help of the conserved quantity E. Let's evaluate it on the separating trajectory:

$$E(\pi/2, 0) = \frac{C}{4}n^2\epsilon^2,$$
$$E(0, \Delta\dot\varphi) = \frac{C}{2}\Delta\dot\varphi^2 - \frac{C}{4}n^2\epsilon^2. \tag{2.112}$$

Equating these and solving, we find the maximum extent of the oscillating region:

$$\Delta\dot\varphi = n\epsilon. \tag{2.113}$$

So we see that the out-of-roundness parameter ϵ not only gives the frequency of small-amplitude oscillations, but also gives the extent of the oscillation region.

Mercury rotates exactly three times for every two times it goes around the Sun, as discovered by Pettengill and Dyce in 1965, using Arecibo radar. We can understand this spin-orbit resonance using our simple spin-orbit model problem.

Let's first use qualitative reasoning to understand how the resonance comes about. The spin-orbit equation of motion, equation (2.105), equates the rate of change of the angular momentum to the gravity gradient torque. The torque is proportional to the inverse cube of the distance, so it is largest when the distance is smallest, at pericenter. For the purpose of qualitative reasoning, consider the torque only at pericenter. Suppose Mercury is rotating exactly three times for every two orbits. Then if the long axis of Mercury is pointed at the Sun at one pericenter, it will point the other end of this long axis the next time it passes pericenter (it will have rotated one and a half times). Now, suppose Mercury is rotating a little faster. Then if the long axis is aligned at one pericenter passage, on the next pericenter passage it will have rotated a little too much and the long axis will no longer point to the Sun. In this case $\theta - f$ will be positive and there will be a negative torque, slowing down the rotation a bit. Over many orbits the rotation of Mercury is reduced. A similar argument shows that if Mercury is rotating slower than three times for every two orbits, then the torques at succesive pericenter passages will tend to increase the rotation rate. An oscillation ensues.

We can also understand this spin-orbit resonance analytically. The right-hand side of the equation of motion (equation 2.105) has factors that vary periodically with the orbit period:

$$CD^2\theta(t) = -\frac{C}{2}n^2\epsilon^2 \left(\frac{a}{R(t)}\right)^3 \sin 2(\theta(t) - f(t)), \qquad (2.114)$$

where both $f(t)$ and $R(t)$ are periodic with period $2\pi/n$. We can expand this as a Fourier series:

$$CD^2\theta(t) = -\frac{C}{2}n^2\epsilon^2 \sum_{m=-\infty}^{\infty} A_m(e)\sin(2\theta(t) - mnt), \qquad (2.115)$$

where the coefficients $A_m(e)$ are functions of the orbital eccentricity e. The coefficients are proportional to $e^{|m-2|}$ and so for small eccentricity we need to consider only a few of them.[22] We have

$$A_1(e) = -\frac{e}{2} + o(e^3)$$

$$A_2(e) = 1 - \frac{5e^2}{2} + o(e^4)$$

$$A_3(e) = \frac{7e}{2} + o(e^3). \qquad (2.116)$$

All other terms are of higher order in e. With just the terms of order e or less, the equation of motion becomes

$$CD^2\theta(t) = -\frac{C}{2}n^2\epsilon^2 \left[\sin(2\theta(t) - 2nt)\right.$$
$$+ \frac{7e}{2}\sin(2\theta(t) - 3nt)$$
$$- \frac{e}{2}\sin(2\theta(t) - nt)$$
$$\left. + \cdots\right]. \qquad (2.117)$$

Suppose we are close to the 3:2 Mercury resonance. Then $\dot\theta$ is close to $(3/2)n$. So the combination $\theta(t) - (3/2)nt$ is slowly varying compared to the other two arguments: $\theta(t) - nt$ and $\theta(t) - (1/2)nt$. The rapidly varying torques due to these other terms

---

[22]Deriving the coefficients is a matter of celestial mechanics; $A_m(e) = X_2^{-3,2}(e)$ where $X_k^{i,j}(e)$ are called Hansen functions. They are written as power series in eccentricity.

tend to average out, leaving a slowly varying torque that controls the motion.[23] The averaged equation of motion for motion near the 3:2 resonance is then

$$CD^2\theta(t) = -\frac{C}{2}n^2\epsilon^2\left(\frac{7e}{2}\right)\sin(2\theta(t) - 3nt). \qquad (2.118)$$

We can solve this by changing variables to

$$\varphi(t) = \theta(t) - (3/2)nt. \qquad (2.119)$$

The equation of motion becomes

$$CD^2\varphi(t) = \frac{C}{2}n^2\epsilon^2\left(\frac{7e}{2}\right)\sin(2\varphi(t)). \qquad (2.120)$$

This has the "energy-like" conserved quantity

$$E(\varphi, \dot\varphi) = \frac{C}{2}\dot\varphi^2 - \frac{C}{4}n^2\epsilon^2\left(\frac{7e}{2}\right)\cos(2\varphi), \qquad (2.121)$$

which is very similar to the conserved quantity we found for the zero-eccentricity synchronous rotation case considered earlier; see equation (2.111). Indeed the trajectories are contours of the conserved quantity and look just like those in figure 2.14. Using analogous reasoning we can determine the extent of the oscillation region and find

$$\Delta\dot\varphi = n\epsilon\sqrt{\frac{7e}{2}}. \qquad (2.122)$$

This gives the approximate range of rotation rate over which Mercury can oscillate stably in the 3:2 resonance.

2.12   Nonsingular Coordinates and Quaternions

The Euler angles provide a convenient way to parameterize the orientation of a rigid body. However, the equations of motion derived for them have singularities. Though we can avoid the singularities by using other Euler-like combinations with different

---

[23]This method of *averaging* is rather vague; we will justify it later when we study perturbation theory.

singularities, this kludge is not very satisfying. Let's brainstorm a bit and see if we can come up with something better.

What does it take to specify an orientation? Perhaps we can take a hint from Euler's theorem. Recall that Euler's theorem states that any orientation can be reached with a single rotation. So one idea to specify the orientation of a body is to parameterize this single rotation that does the job. To specify this rotation we need to specify the rotation axis and the amount of rotation. We contrast this with the Euler angles, which specify three successive rotations. These three rotations need not have any relation to the single composite rotation that gives the orientation. Isn't it curious that the Euler angles make no use of Euler's theorem?

We can think of several ways of specifying a rotation. One way would be to specify the rotation axis by the latitude and the longitude at which the rotation axis pierces a sphere. The amount of rotation needed to take the body from the reference position could be specified by one more angle. We can predict, though, that this choice of coordinates will have similar problems to those of the Euler angles: if the amount of rotation is zero, then the latitude and longitude of the rotation axis are undefined. So the Lagrange equations for these angles are probably singular. Another idea, without this defect, is to represent the rotation by the rectangular components of an orientation vector $\vec{o}$; we take the direction of the orientation vector to be the same as the axis of rotation that takes the body from the reference orientation to the present orientation, and the length of the orientation vector to be the angle by which the body must be rotated, in a right-hand sense, about the orientation vector. With this choice of coordinates, if the angle of rotation is zero then the length of the vector is zero and has no unwanted direction. This choice looks promising, but there is another problem: a rotation by 2π is equivalent to no rotation at all, so it will not have a well-defined rotation vector. This can be fixed by making the length of the orientation vector be the sine of half of the angle of rotation rather than the angle of rotation. With this choice a rotation by zero angle will have the same orientation vector as a rotation by 2π. But there is still another problem: rotations by θ and $2\pi - \theta$ are not distinguished. We can solve this by keeping track of the cosine of half the angle of rotation. (Actually we need to know only the sign of the cosine, but the cosine is convenient.) Wrapping this all up into 4-tuples gives us Hamilton's *quaternions*.

Let θ be the angle of rotation about the axis $\hat{n}$. The components of a quaternion representing this rotation are:

$$(\cos(\theta/2),\ \sin(\theta/2)\,\hat{n}_x,\ \sin(\theta/2)\,\hat{n}_y,\ \sin(\theta/2)\,\hat{n}_z)\,, \tag{2.123}$$

where $(\hat{n}_x, \hat{n}_y, \hat{n}_z)$ are rectangular components of $\hat{n}$. The sum of the squares of the components of this quaternion is 1: it is a *unit quaternion*. So there is a unit quaternion associated with every rotation.

We can invert this: given a quaternion we can compute the angle and the axis. Let (r, x, y, z) be the components of a quaternion q. We separate the first component (called the *real part*) and the tuple $v = (x, y, z)$ (called the *3-vector*) of the remaining components. The Euclidean norm of the tuple $\|v\| = |\sin(\theta/2)|$. The first component $r = \cos(\theta/2)$. So the angle $\theta = 2\arctan(\|v\|, r)$ and the axis is $v/\|v\|$. This process is independent of the scale of the quaternion.

By taking the absolute value of $\sin(\theta/2)$ we have lost information about the quadrant, but this is not a real problem because the rotation represented by a quaternion is not changed by reversing the sign of all its components: changing the sign of v reverses the axis but does not change the angle; changing the sign of the first component changes the angle θ to $2\pi - \theta$, so the actual rotation is unchanged.

Given the four elements of a quaternion, we need to find the corresponding rotation matrix. We can get the angle and axis given a quaternion. We can get a rotation matrix given the angle θ and the axis given by a unit vector $\hat{n}$. We rotate by θ around the $\hat{z}$ axis, and then transform this rotation to the axis specified by colatitude φ and longitude λ:

$$\mathbf{R}(\theta, \hat{n}) = \mathbf{R}_z(\lambda)\mathbf{R}_y(\varphi)\mathbf{R}_z(\theta)(\mathbf{R}_y(\varphi))^{\mathsf{T}}(\mathbf{R}_z(\lambda))^{\mathsf{T}}, \tag{2.124}$$

where $\varphi = \arccos(\hat{n}_z)$ and $\lambda = \arctan(\hat{n}_y, \hat{n}_x)$.

A procedure for making the rotation matrix is:

```
(define (angle-axis->rotation-matrix theta n)
  (let ((nx (ref n 0)) (ny (ref n 1)) (nz (ref n 2)))
    (let ((colatitude (acos nz)) (longitude (atan ny nx)))
      (* (Rz-matrix longitude)
         (Ry-matrix colatitude)
         (Rz-matrix theta)
         (transpose (Ry-matrix colatitude))
         (transpose (Rz-matrix longitude))))))
```

And a procedure for obtaining the angle and axis of a quaternion is

```
(define (quaternion->angle-axis q)
  (let* ((v (quaternion->3vector q))
         (theta (* 2 (atan (euclidean-norm v)
                           (quaternion->real-part q))))
         (axis (/ v (euclidean-norm v))))
    (list theta axis)))
```

Combining these, we can compute the rotation matrix associated with a quaternion:

```
(define (quaternion->RM q)
  (let ((aa (quaternion->angle-axis q)))
    (let ((theta (ref aa 0)) (n (ref aa 1)))
      (angle-axis->rotation-matrix theta n))))
```

The resulting matrix has the square of the magnitude of the quaternion dividing each term. For a unit quaternion this denominator has no effect, but the expression looks simpler if we multiply through:

```
(show-expression
  (let ((v (up 'q_0 'q_1 'q_2 'q_3)))
    (let ((m^2 (dot-product v v)))
      (* m^2 (quaternion->RM (make-quaternion v))))))
```

$$
\begin{pmatrix}
q_0^2 + q_1^2 - q_2^2 - q_3^2 & -2q_0q_3 + 2q_1q_2 & 2q_0q_2 + 2q_1q_3 \\
2q_0q_3 + 2q_1q_2 & q_0^2 - q_1^2 + q_2^2 - q_3^2 & -2q_0q_1 + 2q_2q_3 \\
-2q_0q_2 + 2q_1q_3 & 2q_0q_1 + 2q_2q_3 & q_0^2 - q_1^2 - q_2^2 + q_3^2
\end{pmatrix}
$$

We then capture this result as a useful procedure (dividing through by the square of the magnitude). The resulting matrix is homoge-

neous of degree zero in the quaternion components, so the result
is insensitive to the scale.

```
(define (quaternion->rotation-matrix q)
  (let ((q0 (quaternion-ref q 0)) (q1 (quaternion-ref q 1))
        (q2 (quaternion-ref q 2)) (q3 (quaternion-ref q 3)))
    (let ((m^2
           (+ (expt q0 2) (expt q1 2)
              (expt q2 2) (expt q3 2))))
      (/ (matrix-by-rows
          (list (- (+ (expt q0 2) (expt q1 2))
                   (+ (expt q2 2) (expt q3 2)))
                (* 2 (- (* q1 q2) (* q0 q3)))
                (* 2 (+ (* q1 q3) (* q0 q2))))
          (list (* 2 (+ (* q1 q2) (* q0 q3)))
                (- (+ (expt q0 2) (expt q2 2))
                   (+ (expt q1 2) (expt q3 2)))
                (* 2 (- (* q2 q3) (* q0 q1))))
          (list (* 2 (- (* q1 q3) (* q0 q2)))
                (* 2 (+ (* q2 q3) (* q0 q1)))
                (- (+ (expt q0 2) (expt q3 2))
                   (+ (expt q1 2) (expt q2 2)))))
         m^2))))
```

Next we determine the components of the angular velocity on
the body using this result and the `M->omega-body` of section 2.2:

```
(show-expression
  ((M->omega-body
    (compose quaternion->rotation-matrix make-quaternion))
   (up 't
       (up 'q_0 'q_1 'q_2 'q_3)
       (up 'qdot_0 'qdot_1 'qdot_2 'qdot_3))))
```

$$
\begin{pmatrix}
\dfrac{2q_0\dot{q}_1 - 2q_1\dot{q}_0 - 2q_2\dot{q}_3 + 2q_3\dot{q}_2}{q_0^2 + q_1^2 + q_2^2 + q_3^2} \\[2mm]
\dfrac{2q_0\dot{q}_2 + 2q_1\dot{q}_3 - 2q_2\dot{q}_0 - 2q_3\dot{q}_1}{q_0^2 + q_1^2 + q_2^2 + q_3^2} \\[2mm]
\dfrac{2q_0\dot{q}_3 - 2q_1\dot{q}_2 + 2q_2\dot{q}_1 - 2q_3\dot{q}_0}{q_0^2 + q_1^2 + q_2^2 + q_3^2}
\end{pmatrix}
$$

The result is simple (ignoring the denominators, which have value
1 for unit quaternions). Note that this result is not, on the surface, independent of the scale of the quaternion. But since the

quaternion is representing the orientation of a rotating body it is a function of time. So the time derivative of the quaternion must scale as the quaternion scales: in this sense the formula is independent of scale.

But we can write this in an even simpler way. Notice that the numerators are linear in both the $\dot{q}_i$ and q_j. We can invent matrices that perform the relevant combinations. Introduce

$$\mathbf{i} = \begin{pmatrix} 0 & +1 & 0 & 0 \\ -1 & 0 & 0 & 0 \\ 0 & 0 & 0 & -1 \\ 0 & 0 & +1 & 0 \end{pmatrix}$$

$$\mathbf{j} = \begin{pmatrix} 0 & 0 & +1 & 0 \\ 0 & 0 & 0 & +1 \\ -1 & 0 & 0 & 0 \\ 0 & -1 & 0 & 0 \end{pmatrix}$$

$$\mathbf{k} = \begin{pmatrix} 0 & 0 & 0 & +1 \\ 0 & 0 & -1 & 0 \\ 0 & +1 & 0 & 0 \\ -1 & 0 & 0 & 0 \end{pmatrix}. \tag{2.125}$$

In terms of these matrices, we can write the angular velocity on the body more simply, given a unit quaternion

$$\omega^a = 2\mathbf{q}^T \mathbf{i}\,\dot{\mathbf{q}}/\|\mathbf{q}\|^2$$
$$\omega^b = 2\mathbf{q}^T \mathbf{j}\,\dot{\mathbf{q}}/\|\mathbf{q}\|^2$$
$$\omega^c = 2\mathbf{q}^T \mathbf{k}\,\dot{\mathbf{q}}/\|\mathbf{q}\|^2, \tag{2.126}$$

where q is a column matrix of the components of q. As a program:

```
(define (quaternion-state->omega-body s)
  (let ((q (coordinates s)) (qdot (velocities s)))
    (let ((m^2 (dot-product q q)))
      (let ((omega^a
             (/ (* 2 (dot-product q (* q:i qdot))) m^2))
            (omega^b
             (/ (* 2 (dot-product q (* q:j qdot))) m^2))
            (omega^c
             (/ (* 2 (dot-product q (* q:k qdot))) m^2)))
        (up omega^a omega^b omega^c)))))
```

where q:i, q:j, and q:k implement **i**, **j**, and **k**.

The antisymmetric matrices **i**, **j**, and **k** have interesting algebraic properties:

$$\mathbf{i}^2 = \mathbf{j}^2 = \mathbf{k}^2 = \mathbf{ijk} = -1, \tag{2.127}$$

where **1** is the 4×4 unit matrix.

If we forget that these are matrices, and just use the algebraic properties of **i**, **j**, and **k** we get the "imaginary number" representation invented by Hamilton.

Composition of rotations

What is the quaternion that represents the composition of two rotations?

```
(let ((q (quaternion 'q_0 'q_1 'q_2 'q_3))
      (p (quaternion 'p_0 'p_1 'p_2 'p_3)))
  (let ((Mq (quaternion->rotation-matrix q))
        (Mp (quaternion->rotation-matrix p)))
    (rotation-matrix->quaternion (* Mq Mp))))
```

Unfortunately, the result is messy because each component is scaled by a factor of $\|q\|\|p\|$, which is 1 for unit quaternions. For each rotation matrix there are many quaternions that can represent it. Indeed, a quaternion scaled by any nonzero number represents the same rotation matrix. So the process of choosing a quaternion to represent that matrix picks a unit quaternion. Here are the components of the chosen quaternion, eliminating the normalizing factor $\|q\|\|p\|$:

$$\begin{pmatrix} p_0 q_0 - p_1 q_1 - p_2 q_2 - p_3 q_3 \\ p_0 q_1 + p_1 q_0 - p_2 q_3 + p_3 q_2 \\ p_0 q_2 + p_1 q_3 + p_2 q_0 - p_3 q_1 \\ p_0 q_3 - p_1 q_2 + p_2 q_1 + p_3 q_0 \end{pmatrix}. \tag{2.128}$$

The first component, the real part of the resulting quaternion, can be interpreted as

$$r_0 = q_0 p_0 - v_q \cdot v_p, \tag{2.129}$$

where $v_p = (p_1, p_2, p_3)$ and $v_q = (q_1, q_2, q_3)$ are the 3-vector parts of the quaternions p and q. The remaining three components can be interpreted as:

$$v_r = q_0 v_p + p_0 v_q + v_q \times v_p. \tag{2.130}$$

We take this to specify the product of two quaternions, whether or not they are unit quaternions. This extension of multiplicaton to non-unit quaternions works because we did not include the normalization factors.

Each quaternion has a matrix representation in terms of the matrices **i**, **j**, **k**, and **1**, the *quaternion units*:

$$\mathbf{q} = q_0\mathbf{1} + q_1\mathbf{i} + q_2\mathbf{j} + q_3\mathbf{k}. \tag{2.131}$$

We can use this representation to write our result as a matrix product:

$$\mathbf{r} = \mathbf{qp}. \tag{2.132}$$

The elements of the top row of the matrix **r** are the components of the quaternion r.

It turns out that the rotation matrix corresponding to a unit quaternion can also be written in terms of **i**, **j**, and **k**. Let **M** be a rotation matrix corresponding to the unit quaternion p. A vector with component 3-tuple **w** can be rotated by multiplication by **M** on the left. We can perform the same operation using the quaternion units. Let q_w be the quaternion whose real part is 0 and whose 3-vector part is **w**, then the product pq_wp^* is a quaternion whose 3-vector part is the rotated vector and whose real part is zero. The conjugate p^* is obtained from p by reversing the sign of the 3-vector part. As an equation: $\mathbf{Mw} = v_{pq_wp^*}$.

Exercise 2.19: Quaternions

Verify equations (2.129) and (2.130) using only the algebraic properties given in equation (2.127).

2.12.1   Motion in Terms of Quaternions

Quaternions give us nice coordinates that do not suffer from the singularities of Euler angles. So we can make use of them to compute motions of a rigid body without needing to worry about the singularities.

We have computed the body components of the angular velocities from a state consisting of quaternion components and the rates of change of those components (see equation 2.126). We can invert these to find the rates of change of the quaternion components in terms of the angular velocities and the quaternion components. The result of this inversion is:

$$\dot{q} = -\tfrac{1}{2}(\omega^a \mathbf{i} + \omega^b \mathbf{j} + \omega^c \mathbf{k})q. \tag{2.133}$$

This set of differential equations is driven by Euler's equations for the motion of the body components of the angular velocity (see equations 2.69 on page 153).[24]

We construct a system derivative for the free rigid body with mixed coordinates. The configuration is represented by a quaternion that specifies the rotation that takes the body from the reference orientation to the actual orientation. The rate of change of the configuration is specified by the components of the angular velocities on the body.

```
(define (qw-sysder A B C)
  (let ((B-C/A (/ (- B C) A))
        (C-A/B (/ (- C A) B))
        (A-B/C (/ (- A B) C)))
    (define (the-deriv qw-state)
      (let ((t (time qw-state))
            (q (coordinates qw-state))
            (omega-body (ref qw-state 2)))
        (let ((omega^a (ref omega-body 0))
              (omega^b (ref omega-body 1))
              (omega^c (ref omega-body 2)))
          (let ((tdot 1)
                (qdot        ;driven quaternion
                 (* -1/2
                    (+ (* omega^a q:i)
                       (* omega^b q:j)
                       (* omega^c q:k))
                    q))
                (omegadot  ;Euler's equations
                 (up (* B-C/A omega^b omega^c)
                     (* C-A/B omega^c omega^a)
                     (* A-B/C omega^a omega^b))))
            (up tdot qdot omegadot)))))
    the-deriv))
```

---

[24]We could incorporate external torques by using the augmented Euler's equations (2.83).

Note that this system derivative is not constructed by an automatic process: this was not derived from a Lagrangian. This is part of the price we pay for using redundant coordinates (the four quaternion components) to represent the configuration of a system with only three degrees of freedom. By using Euler's equations we avoid having to eliminate the constraint. However, the computations with quaternions are easier than the ones using Euler angles, because they do not involve evaluating transcendental functions or avoiding the singularities.

Since we will monitor the errors in the conserved quantities, angular momentum and energy, we need to compute these quantites from the state. The kinetic energy and the angular momentum components on the body are exactly the same as we used before, because they depend on only the components of the angular velocities on the body. However, to get the components of the angular momentum on spatial axes we need the rotation computed from the quaternion coordinates:

```
(define ((qw-state->L-space A B C) qw-state)
  (let ((q (coordinates qw-state)))
    (let ((Lbody ((L-body A B C) (ref qw-state 2)))
          (M (quaternion->rotation-matrix
              (make-quaternion q))))
      (* Lbody (transpose M)))))
```

From the initial angular momentum and energy we can compute the relative error of these quantities, as we did in section 2.8.1:

```
(define ((monitor-errors win A B C L0 E0) qw-state)
  (let ((t (time qw-state))
        (L ((qw-state->L-space A B C) qw-state))
        (E ((T-body A B C) (ref qw-state 2))))
    (plot-point win t (relative-error (ref L 0) (ref L0 0)))
    (plot-point win t (relative-error (ref L 1) (ref L0 1)))
    (plot-point win t (relative-error (ref L 2) (ref L0 2)))
    (plot-point win t (relative-error E E0))
    qw-state))
```

Below we set up the initial conditions and use monitor-errors to plot the errors. We use the same initial conditions that we did for Euler angles. We get the rotation matrix M that transforms the reference position to the initial Euler state and use that to construct the equivalent quaternion state for this evolution.

```
(define win (frame 0.0 100.0 -1.0e-13 1.0e-13))

(let* ((A 1.0) (B (sqrt 2.0)) (C 2.0)    ; moments of inertia
       (Euler-state (up 0.0               ; initial state
                        (up 1.0 0.0 0.0)
                        (up 0.1 0.1 0.1)))
       (M (Euler->M (coordinates Euler-state)))
       (q (quaternion->vector (rotation-matrix->quaternion M)))
       (qw-state0
        (up (time Euler-state)

            q
            (Euler-state->omega-body Euler-state))))
  (let ((L0 ((qw-state->L-space A B C) qw-state0))
        (E0 ((T-body A B C) (ref qw-state0 2))))
    ((evolve qw-sysder A B C)
     qw-state0
     (monitor-errors win A B C L0 E0)
     0.1                    ; step between plotted points
     100.0                  ; final time
     1.0e-12)))
```

Figure 2.15 shows the relative errors in the energy and the spatial components of the angular momentum that arise in this integration. It is interesting to note that the errors incurred by integrating using Euler angles and quaternions are about a factor of ten smaller than the ones that appear when the coordinates are Euler angles.

2.13 Summary

A rigid body is an example of a mechanical system with constraints. Thus, in a sense this chapter on rigid bodies was nothing but an extended example of the application of the ideas developed in the first chapter. The equations of motion are just the Lagrange equations.

The kinetic energy for a rigid body separates into a translational kinetic energy and a rotational kinetic energy. The center of mass plays a special role in this separation. The rotational kinetic energy is simply expressed in terms of the inertia tensor and the angular velocity vector. We developed the expressions for the kinetic energy that take into account the body constraints, and we expressed the remaining degrees of freedom in terms of suitable generalized coordinates.

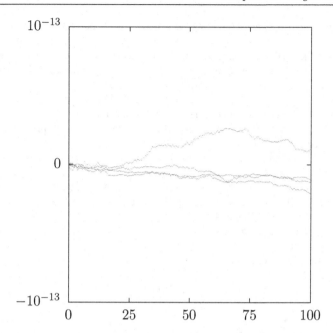

Figure 2.15   The relative errors in energy and in the three spatial components of the angular momentum versus time. The equations were integrated with quality-controlled 4th-order Runge–Kutta.

The vector angular momentum is conserved if there are no external torques. The time derivative of the body components of the angular momentum can be written entirely in terms of the body components of the angular momentum, and the three principal moments of inertia. The body components of angular momentum form a self-contained dynamical subsystem.

One choice for generalized coordinates is the Euler angles. They form suitable generalized coordinates, but are otherwise not special or well motivated. The Lagrange equations for the Euler angles are singular for some Euler angles. Other choices of generalized coordinates like the Euler angles have similar problems. Equations of motion using quaternions are nonsingular.

In general the potential energy depends on the details of the mass distribution, and does not separate as the kinetic energy separated into center of mass and relative contributions.

For an axisymmetric top with uniform gravitational acceleration, the potential energy is exactly the potential energy due to elevation of the center of mass. Aspects of the motion of the top are deduced from the conserved quantities. Euler angles are just the right thing for this problem.

For other problems, such as the rotational motion of an out-of-round satellite near a planet, the potential energy cannot be written in finite terms, and judicious approximations must be made. The essential character of such diverse systems as the rotation of the Moon, Hyperion, and Mercury are captured by a simple model problem.

2.14   Projects

Exercise 2.20: Free rigid body

Write and demonstrate a program that reproduces diagrams like figure 2.3 (section 2.8.2). Can you find trajectories that are asymptotic to the unstable relative equilibrium on the intermediate principal axis?

Exercise 2.21: Rotation of Mercury

In the '60s it was discovered that Mercury has a rotation period that is precisely 2/3 times its orbital period. We can see this resonant behavior in the spin-orbit model problem, and we can also play with nudging Mercury a bit to see how far off the rotation rate can be and still be trapped in this spin-orbit resonance. If the mismatch in angular velocity is too great, Mercury's rotation is no longer resonantly locked to its orbit. Set $\epsilon = 0.026$ and $e = 0.2$.

a. Write a program for the spin-orbit problem so this resonance dynamics can be investigated numerically. You will need to know (or, better, show!) that f satisfies the equation

$$Df(t) = n(1 - e^2)^{1/2} \left(\frac{a}{R(t)} \right)^2, \tag{2.134}$$

with

$$\frac{a}{R(t)} = \frac{1 + e \cos f(t)}{1 - e^2}. \tag{2.135}$$

Notice that n disappears from the equations if they are written in terms of a new independent variable $\tau = nt$. Also notice that a and $R(t)$ appear only in the combination $a/R(t)$.

b. Show that the 3:2 resonance is stable by numerically integrating the system when the rotation is not exactly in resonance and observing that the angle $\theta - \frac{3}{2}nt$ oscillates.

c. Find the range of initial $\dot{\theta}$ for which this resonance angle oscillates.

3
Hamiltonian Mechanics

Numerical experiments are just what their name
implies: experiments. In describing and evaluating
them, one should enter the state of mind of the
experimental physicist, rather than that of the
mathematician. Numerical experiments cannot be
used to prove theorems; but, from the physicist's
point of view, they do often provide convincing
evidence for the existence of a phenomenon. We
will therefore follow an informal, descriptive and
non-rigorous approach. Briefly stated, our aim will
be to *understand* the fundamental properties of
dynamical systems rather than to *prove* them.

Michel Hénon, "Numerical Exploration of
Hamiltonian Systems," in *Chaotic Behavior of
Deterministic Systems* [21], p. 57.

The formulation of mechanics with generalized coordinates and
momenta as dynamical state variables is called the Hamiltonian
formulation. The Hamiltonian formulation of mechanics is equivalent to the Lagrangian formulation; however, each presents a useful
point of view. The Lagrangian formulation is especially useful in
the initial formulation of a system. The Hamiltonian formulation
is especially useful in understanding the evolution of a system,
especially when there are symmetries and conserved quantities.

For each continuous symmetry of a mechanical system there
is a conserved quantity. If the generalized coordinates can be
chosen to reflect a symmetry, then, by the Lagrange equations,
the momenta conjugate to the cyclic coordinates are conserved.
We have seen that such conserved quantities allow us to deduce
important properties of the motion. For instance, consideration
of the energy and angular momentum allowed us to deduce that
rotation of a free rigid body about the axis of intermediate moment
of inertia is unstable, and that rotation about the other principal
axes is stable. For the axisymmetric top, we used two conserved
momenta to reexpress the equations governing the evolution of the
tilt angle so that they involve only the tilt angle and its derivative.

The evolution of the tilt angle can be determined independently and has simply periodic solutions. Consideration of the conserved momenta has provided key insight. The Hamiltonian formulation is motivated by the desire to focus attention on the momenta.

In the Lagrangian formulation the momenta are, in a sense, secondary quantities: the momenta are functions of the state space variables, but the evolution of the state space variables depends on the state space variables and not on the momenta. To make use of any conserved momenta requires fooling around with the specific equations. The momenta can be rewritten in terms of the coordinates and the velocities, so, locally, we can solve for the velocities in terms of the coordinates and momenta. For a given mechanical system, and a Lagrangian describing its dynamics in a given coordinate system, the momenta and the velocities can be deduced from each other. Thus we can represent the dynamical state of the system in terms of the coordinates and momenta just as well as with the coordinates and the velocities. If we use the coordinates and momenta to represent the state and write the associated state derivative in terms of the coordinates and momenta, then we have a self-contained system. This formulation of the equations governing the evolution of the system has the advantage that if some of the momenta are conserved, the remaining equations are immediately simplified.

The Lagrangian formulation of mechanics has provided the means to investigate the motion of complicated mechanical systems. We have found that dynamical systems exhibit a bewildering variety of possible motions. The motion is sometimes rather simple and sometimes very complicated. Sometimes the evolution is very sensitive to the initial conditions, and sometimes it is not. And sometimes there are orbits that maintain resonance relationships with a drive. Consider the periodically driven pendulum: it can behave more or less as an undriven pendulum with extra wiggles, it can move in a strongly chaotic manner, or it can move in resonance with the drive, oscillating once for every two cycles of the drive or looping around once per drive cycle. Or consider the Moon. The Moon rotates synchronously with its orbital motion, always pointing roughly the same face to the Earth. However, Mercury rotates three times every two times it circles the Sun, and Hyperion rotates chaotically.

How can we make sense of this? How do we put the possible motions of these systems in relation to one another? What other

motions are possible? The Hamiltonian formulation of dynamics
provides a convenient framework in which the possible motions
may be placed and understood. We will be able to see the range
of stable resonance motions and the range of states reached by
chaotic trajectories, and discover other unsuspected possible motions. So the Hamiltonian formulation gives us much more than
the stated goal of expressing the system derivative in terms of
potentially conserved quantities.

3.1 Hamilton's Equations

The momenta are given by momentum state functions of the time,
the coordinates, and the velocities.[1] Locally, we can find inverse
functions that give the velocities in terms of the time, the coordinates, and the momenta. We can use this inverse function
to represent the state in terms of the coordinates and momenta
rather than the coordinates and velocities. The equations of motion when recast in terms of coordinates and momenta are called
Hamilton's canonical equations.

We present three derivations of Hamilton's equations. The first
derivation is guided by the strategy outlined above and uses nothing more complicated than implicit functions and the chain rule.
The second derivation (section 3.1.1) first abstracts a key part of
the first derivation and then applies the more abstract machinery
to derive Hamilton's equations. The third (section 3.1.2) uses the
action principle.

Lagrange's equations give us the time derivative of the momentum p on a path q:

$$Dp(t) = \partial_1 L(t, q(t), Dq(t)), \tag{3.1}$$

where

$$p(t) = \partial_2 L(t, q(t), Dq(t)). \tag{3.2}$$

To eliminate Dq we need to solve equation (3.2) for Dq in terms
of p.

---

[1] Here we restrict our attention to Lagrangians that depend only on the time,
the coordinates, and the velocities.

Let $\mathcal{V}$ be the function that gives the velocities in terms of the time, coordinates, and momenta. Defining $\mathcal{V}$ is a problem of functional inverses. To prevent confusion we use names for the variables that have no mnemonic significance. Let

$$a = \partial_2 L(b, c, d); \tag{3.3}$$

then $\mathcal{V}$ satisfies

$$d = \mathcal{V}(b, c, a). \tag{3.4}$$

So $\mathcal{V}$ and $\partial_2 L$ are inverses on the third argument position:

$$d = \mathcal{V}(b, c, \partial_2 L(b, c, d)) \tag{3.5}$$
$$a = \partial_2 L(b, c, \mathcal{V}(b, c, a)). \tag{3.6}$$

The Lagrange equation (3.1) can be rewritten in terms of p using $\mathcal{V}$:

$$Dp(t) = \partial_1 L(t, q(t), \mathcal{V}(t, q(t), p(t))). \tag{3.7}$$

We can also use $\mathcal{V}$ to rewrite equation (3.2) as an equation for Dq in terms of t, q and p:

$$Dq(t) = \mathcal{V}(t, q(t), p(t)). \tag{3.8}$$

Equations (3.7) and (3.8) give the rate of change of q and p along realizable paths as functions of t, q, and p along the paths.

Though these equations fulfill our goal of expressing the equations of motion entirely in terms of coordinates and momenta, we can find a better representation. Define the function

$$\widetilde{L}(t, q, p) = L(t, q, \mathcal{V}(t, q, p)), \tag{3.9}$$

which is the Lagrangian reexpressed as a function of time, coordinates, and momenta.[2] For the equations of motion we need $\partial_1 L$ evaluated with the appropriate arguments. Consider

---

[2]Here we are using mnemonic names t, q, p for formal parameters of the function being defined. We could have used names like a, b, c as above, but this would have made the argument harder to read.

$$\partial_1 \tilde{L}(t, q, p) = \partial_1 L(t, q, \mathcal{V}(t, q, p)) + \partial_2 L(t, q, \mathcal{V}(t, q, p)) \partial_1 \mathcal{V}(t, q, p)$$
$$= \partial_1 L(t, q, \mathcal{V}(t, q, p)) + p \partial_1 \mathcal{V}(t, q, p), \qquad (3.10)$$

where we used the chain rule in the first step and the inverse property (3.6) of $\mathcal{V}$ in the second step. Introducing the momentum selector[3] $P(t, q, p) = p$, and using the property $\partial_1 P = 0$, we have

$$\partial_1 L(t, q, \mathcal{V}(t, q, p)) = \partial_1 \tilde{L}(t, q, p) - P(t, q, p) \partial_1 \mathcal{V}(t, q, p)$$
$$= \partial_1 (\tilde{L} - P\mathcal{V})(t, q, p)$$
$$= -\partial_1 H(t, q, p), \qquad (3.11)$$

where the *Hamiltonian* H is defined by[4]

$$H = P\mathcal{V} - \tilde{L}. \qquad (3.12)$$

Using the algebraic result (3.11), the Lagrange equation (3.7) for Dp becomes

$$Dp(t) = -\partial_1 H(t, q(t), p(t)). \qquad (3.13)$$

The equation for Dq can also be written in terms of H. Consider

$$\partial_2 H(t, q, p) = \partial_2 (P\mathcal{V} - \tilde{L})(t, q, p)$$
$$= \mathcal{V}(t, q, p) + p\partial_2 \mathcal{V}(t, q, p) - \partial_2 \tilde{L}(t, q, p). \qquad (3.14)$$

To carry out the derivative of $\tilde{L}$ we write it out in terms of L:

$$\partial_2 \tilde{L}(t, q, p) = \partial_2 L(t, q, \mathcal{V}(t, q, p)) \partial_2 \mathcal{V}(t, q, p) = p\partial_2 \mathcal{V}(t, q, p), \quad (3.15)$$

again using the inverse property (3.6) of $\mathcal{V}$. So, putting equations (3.14) and (3.15) together, we obtain

$$\partial_2 H(t, q, p) = \mathcal{V}(t, q, p). \qquad (3.16)$$

Using the algebraic result (3.16), equation (3.8) for Dq becomes

$$Dq(t) = \partial_2 H(t, q(t), p(t)). \qquad (3.17)$$

---

[3] $P = I_2$. See equations (9.7) in the appendix on notation.

[4] The overall minus sign in the definition of the Hamiltonian is traditional.

Equations (3.13) and (3.17) give the derivatives of the coordinate and momentum path functions at each time in terms of the time, and the coordinates and momenta at that time. These equations are known as *Hamilton's equations*:[5]

$$Dq(t) = \partial_2 H(t, q(t), p(t))$$
$$Dp(t) = -\partial_1 H(t, q(t), p(t)). \tag{3.18}$$

The first equation is just a restatement of the relationship of the momenta to the velocities in terms of the Hamiltonian and holds for any path, whether or not it is a realizable path. The second equation holds only for realizable paths.

Hamilton's equations have an especially simple and symmetrical form. Just as Lagrange's equations are constructed from a real-valued function, the Lagrangian, Hamilton's equations are constructed from a real-valued function, the Hamiltonian. The Hamiltonian function is[6]

$$H(t, q, p) = p\mathcal{V}(t, q, p) - L(t, q, \mathcal{V}(t, q, p)). \tag{3.19}$$

The Hamiltonian has the same value as the energy function $\mathcal{E}$ (see equation 1.142), except that the velocities are expressed in terms of time, coordinates, and momenta by $\mathcal{V}$:

$$H(t, q, p) = \mathcal{E}(t, q, \mathcal{V}(t, q, p)). \tag{3.20}$$

Illustration

Let's try something simple: the motion of a particle of mass m with potential energy $V(x, y)$. A Lagrangian is

$$L(t; x, y; v_x, v_y) = \tfrac{1}{2}m(v_x^2 + v_y^2) - V(x, y). \tag{3.21}$$

---

[5]In traditional notation, Hamilton's equations are written as a separate equation for each component:

$$\frac{dq^i}{dt} = \frac{\partial H}{\partial p_i} \quad \text{and} \quad \frac{dp_i}{dt} = -\frac{\partial H}{\partial q^i}.$$

[6]Traditionally, the Hamiltonian is written

$$H = p\dot{q} - L.$$

This way of writing the Hamiltonian confuses the values of functions with the functions that generate them: both $\dot{q}$ and L must be reexpressed as functions of time, coordinates, and momenta.

To form the Hamiltonian we find the momenta $p = \partial_2 L(t, q, v)$: $p_x = mv_x$ and $p_y = mv_y$. Solving for the velocities in terms of the momenta is easy here: $v_x = p_x/m$ and $v_y = p_y/m$. The Hamiltonian is $H(t, q, p) = pv - L(t, q, v)$, with v reexpressed in terms of (t, q, p):

$$H(t; x, y; p_x, p_y) = \frac{p_x^2 + p_y^2}{2m} + V(x, y). \tag{3.22}$$

The kinetic energy is a homogeneous quadratic form in the velocities, so the energy is $T + V$ and the Hamiltonian is the energy expressed in terms of momenta rather than velocities. Hamilton's equations for Dq are

$$Dx(t) = p_x(t)/m$$
$$Dy(t) = p_y(t)/m. \tag{3.23}$$

Note that these equations merely restate the relation between the momenta and the velocities. Hamilton's equations for Dp are

$$Dp_x(t) = -\partial_0 V(x(t), y(t))$$
$$Dp_y(t) = -\partial_1 V(x(t), y(t)). \tag{3.24}$$

The rate of change of the linear momentum is minus the gradient of the potential energy.

Exercise 3.1: Deriving Hamilton's equations

For each of the following Lagrangians derive the Hamiltonian and Hamilton's equations. These problems are simple enough to do by hand.

a. A Lagrangian for a planar pendulum: $L(t, \theta, \dot{\theta}) = \frac{1}{2}ml^2\dot{\theta}^2 + mgl\cos\theta$.

b. A Lagrangian for a particle of mass m with a two-dimensional potential energy $V(x, y) = (x^2 + y^2)/2 + x^2y - y^3/3$ is $L(t; x, y; \dot{x}, \dot{y}) = \frac{1}{2}m(\dot{x}^2 + \dot{y}^2) - V(x, y)$.

c. A Lagrangian for a particle of mass m constrained to move on a sphere of radius R: $L(t; \theta, \varphi; \dot{\theta}, \dot{\varphi}) = \frac{1}{2}mR^2(\dot{\theta}^2 + (\dot{\varphi}\sin\theta)^2)$, where θ is the colatitude and φ is the longitude on the sphere.

Exercise 3.2: Sliding pendulum

For the pendulum with a sliding support (see exercise 1.20), derive a Hamiltonian and Hamilton's equations.

Hamiltonian state

Given a coordinate path q and a Lagrangian L, the corresponding momentum path p is given by equation (3.2). Equation (3.17) expresses the same relationship in terms of the corresponding Hamiltonian H. That these relations are valid for any path, whether or not it is a realizable path, allows us to abstract to arbitrary velocity and momentum at a moment. At a moment, the momentum p for the state tuple (t, q, v) is $p = \partial_2 L(t, q, v)$. We also have $v = \partial_2 H(t, q, p)$. In the Lagrangian formulation the state of the system at a moment can be specified by the local state tuple (t, q, v) of time, generalized coordinates, and generalized velocities. Lagrange's equations determine a unique path emanating from this state. In the Hamiltonian formulation the state can be specified by the tuple (t, q, p) of time, generalized coordinates, and generalized momenta. Hamilton's equations determine a unique path emanating from this state. The Lagrangian state tuple (t, q, v) encodes exactly the same information as the Hamiltonian state tuple (t, q, p); we need a Lagrangian or a Hamiltonian to relate them. The two formulations are equivalent in that the same coordinate path emanates from them for equivalent initial states.

The Lagrangian state derivative is constructed from the Lagrange equations by solving for the highest-order derivative and abstracting to arbitrary positions and velocities at a moment.[7] The Lagrangian state path is generated by integration of the Lagrangian state derivative given an initial Lagrangian state (t, q, v). Similarly, the Hamiltonian state derivative can be constructed from Hamilton's equations by abstracting to arbitrary positions and momenta at a moment. Hamilton's equations are a set of first-order differential equations in explicit form. The Hamiltonian state derivative can be directly written in terms of them. The Hamiltonian state path is generated by integration of the Hamiltonian state derivative given an initial Hamiltonian state (t, q, p). If these state paths are obtained by integrating the state derivatives with equivalent initial states, then the coordinate path components of these state paths are the same and satisfy the Lagrange

---

[7] In the construction of the Lagrangian state derivative from the Lagrange equations we must solve for the highest-order derivative. The solution process requires the inversion of $\partial_2 \partial_2 L$. In the construction of Hamilton's equations, the construction of $\mathcal{V}$ from the momentum state function $\partial_2 L$ requires the inverse of the same structure. If the Lagrangian formulation has singularities, they cannot be avoided by going to the Hamiltonian formulation.

equations. The coordinate path and the momentum path components of the Hamiltonian state path satisfy Hamilton's equations.
The Hamiltonian formulation and the Lagrangian formulation are
equivalent.

Given a path q, the Lagrangian state path and the Hamiltonian
state paths can be deduced from it. The Lagrangian state path
$\Gamma[q]$ can be constructed from a path q simply by taking derivatives.
The Lagrangian state path satisfies:

$$\Gamma[q](t) = (t, q(t), Dq(t)).\qquad(3.25)$$

The Lagrangian state path is uniquely determined by the path q.
The Hamiltonian state path $\Pi_L[q]$ can also be constructed from
the path q but the construction requires a Lagrangian. The Hamiltonian state path satisfies

$$\Pi_L[q](t) = (t, q(t), \partial_2 L(t, q(t), Dq(t))) = (t, q(t), p(t)).\qquad(3.26)$$

The Hamiltonian state tuple is not uniquely determined by the
path q because it depends upon our choice of Lagrangian, which
is not unique.

The $2n$-dimensional space whose elements are labeled by the
n generalized coordinates q^i and the n generalized momenta p_i is
called the *phase space*. The components of the generalized coordinates and momenta are collectively called the *phase-space components*.[8] The dynamical state of the system is completely specified by the phase-space state tuple (t, q, p), given a Lagrangian or
Hamiltonian to provide the map between velocities and momenta.

Computing Hamilton's equations

Hamilton's equations are a system of first-order ordinary differential equations. A procedural formulation of Lagrange's equations
as a first-order system was presented in section 1.7. The following
formulation of Hamilton's equations is analogous:

```
(define ((Hamilton-equations Hamiltonian) q p)
  (let ((state-path (qp->H-state-path q p)))
    (- (D state-path)
       (compose (Hamiltonian->state-derivative Hamiltonian)
                state-path))))
```

---

[8]The term *phase space* was introduced by Josiah Willard Gibbs in his formulation of statistical mechanics. The Hamiltonian plays a fundamental role
in the Boltzmann–Gibbs formulation of statistical mechanics and in both the
Heisenberg and Schrödinger approaches to quantum mechanics.

The Hamiltonian state derivative is computed as follows:

```
(define ((Hamiltonian->state-derivative Hamiltonian) H-state)
  (up 1
      (((partial 2) Hamiltonian) H-state)
      (- (((partial 1) Hamiltonian) H-state))))
```

The state in the Hamiltonian formulation is composed of the time, the coordinates, and the momenta. We call this an H-state, to distinguish it from the state in the Lagrangian formulation. We can select the components of the Hamiltonian state with the selectors time, coordinate, momentum. We construct Hamiltonian states from their components with up. The first component of the state is time, so the first component of the state derivative is one, the time rate of change of time. Given procedures q and p implementing coordinate and momentum path functions, the Hamiltonian state path can be constructed with the following procedure:

```
(define ((qp->H-state-path q p) t)
  (up t (q t) (p t)))
```

The Hamilton-equations procedure returns the residuals of Hamilton's equations for the given paths.

For example, a procedure implementing the Hamiltonian for a point mass with potential energy $V(x, y)$ is

```
(define ((H-rectangular m V) state)
  (let ((q (coordinate state))
        (p (momentum state)))
    (+ (/ (square p) (* 2 m))
       (V (ref q 0) (ref q 1)))))
```

Hamilton's equations are

```
(show-expression
  (let ((V (literal-function 'V (-> (X Real Real) Real)))
        (q (up (literal-function 'x)
               (literal-function 'y)))
        (p (down (literal-function 'p_x)
                 (literal-function 'p_y))))
    (((Hamilton-equations (H-rectangular 'm V)) q p) 't)))
```

$$\begin{pmatrix} 0 \\ \begin{pmatrix} Dx\left(t\right) - \dfrac{p_x\left(t\right)}{m} \\ Dy\left(t\right) - \dfrac{p_y\left(t\right)}{m} \end{pmatrix} \\ \begin{bmatrix} Dp_x\left(t\right) + \partial_0 V\left(x\left(t\right), y\left(t\right)\right) \\ Dp_y\left(t\right) + \partial_1 V\left(x\left(t\right), y\left(t\right)\right) \end{bmatrix} \end{pmatrix}$$

The zero in the first element of the structure of Hamilton's equation residuals is just the tautology that time advances uniformly: the time function is just the identity, so its derivative is one and the residual is zero. The equations in the second element of the structure relate the coordinate paths and the momentum paths. The equations in the third element give the rate of change of the momenta in terms of the applied forces.

Exercise 3.3: Computing Hamilton's equations

Check your answers to exercise 3.1 with the `Hamilton-equations` procedure.

3.1.1   The Legendre Transformation

The Legendre transformation abstracts a key part of the process of transforming from the Lagrangian to the Hamiltonian formulation of mechanics—the replacement of functional dependence on generalized velocities with functional dependence on generalized momenta. The momentum state function is defined as a partial derivative of the Lagrangian, a real-valued function of time, coordinates, and velocities. The Legendre transformation provides an inverse that gives the velocities in terms of the momenta: we are able to write the velocities as a partial derivative of a different real-valued function of time, coordinates, and momenta.[9]

Given a real-valued function F, if we can find a real-valued function G such that $DF = (DG)^{-1}$, then we say that F and G are related by a Legendre transform.

---

[9]The Legendre transformation is more general than its use in mechanics in that it captures the relationship between conjugate variables in systems as diverse as thermodynamics, circuits, and field theory.

Locally, we can define the inverse function[10] $\mathcal{V}$ of DF so that $DF \circ \mathcal{V} = I$, where I is the identity function $I(w) = w$. Consider the composite function $\widetilde{F} = F \circ \mathcal{V}$. The derivative of $\widetilde{F}$ is

$$D\widetilde{F} = (DF \circ \mathcal{V})D\mathcal{V} = ID\mathcal{V}. \tag{3.27}$$

Since

$$D(I\mathcal{V}) = \mathcal{V} + ID\mathcal{V}, \tag{3.28}$$

we have

$$D\widetilde{F} = D(I\mathcal{V}) - \mathcal{V}, \tag{3.29}$$

or

$$\mathcal{V} = D(I\mathcal{V}) - D\widetilde{F} = D(I\mathcal{V} - \widetilde{F}). \tag{3.30}$$

The integral is determined up to a constant of integration. If we define

$$G = I\mathcal{V} - \widetilde{F}, \tag{3.31}$$

then we have

$$\mathcal{V} = DG. \tag{3.32}$$

The function G has the desired property that DG is the inverse function $\mathcal{V}$ of DF. The derivation just given applies equally well if the arguments of F and G have multiple components.[11]

Given a relation $w = DF(v)$ for some given function F, then $v = DG(w)$ for $G = I\mathcal{V} - F \circ \mathcal{V}$, where $\mathcal{V}$ is the inverse function of DF, provided it exists.

A picture may help (see figure 3.1). The curve is the graph of the function DF. Turned sideways, it is also the graph of the function DG, because DG is the inverse function of DF. The integral of DF from v_0 to v is $F(v) - F(v_0)$; this is the area below the curve from v_0 to v. Likewise, the integral of DG from w_0 to

---

[10]This can be done so long as the derivative is not zero.

[11]Equation (3.28) looks like an application of the product rule for derivatives, $D(I\mathcal{V}) = DI\mathcal{V} + ID\mathcal{V}$. Although this works for real-valued functions, it is inadequate for functions with structured outputs. The result $D(I\mathcal{V}) = \mathcal{V} + ID\mathcal{V}$ is correct, but to verify it the computation must be done after the structures are multiplied out. See page 522.

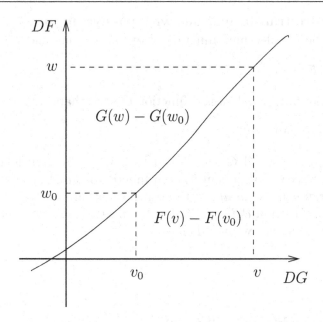

Figure 3.1   The Legendre transform can be interpreted in terms of geometric areas. The curve is the graph of DF, and viewed sideways is the graph of $DG = (DF)^{-1}$. This figure should remind you of the geometric interpretation of the product rule for derivatives, or alternatively integration by parts.

w is $G(w) - G(w_0)$; this is the area to the left of the curve from w_0 to w. The union of these two regions has area $wv - w_0v_0$. So

$$wv - w_0v_0 = F(v) - F(v_0) + G(w) - G(w_0), \tag{3.33}$$

which is the same as

$$wv - F(v) - G(w) = w_0v_0 - G(w_0) - F(v_0). \tag{3.34}$$

The left-hand side depends only on the point labeled by w and v and the right-hand side depends only on the point labeled by w_0 and v_0, so these must be constant, independent of the variable endpoints. So as the point is changed the combination $G(w) + F(v) - wv$ is invariant. Thus

$$G(w) = wv - F(v) + C, \tag{3.35}$$

with constant C. The requirement for G depends only on DG so we can choose to define G with $C = 0$.

Legendre transformations with passive arguments

Let F be a real-valued function of two arguments and

$$w = \partial_1 F(x, v). \tag{3.36}$$

If we can find a real-valued function G such that

$$v = \partial_1 G(x, w) \tag{3.37}$$

we say that F and G are related by a Legendre transformation, that the second argument in each function is *active*, and that the first argument is *passive* in the transformation.

If the function $\partial_1 F$ can be locally inverted with respect to the second argument we can define

$$v = \mathcal{V}(x, w), \tag{3.38}$$

giving

$$w = \partial_1 F(x, \mathcal{V}(x, w)) = W(x, w), \tag{3.39}$$

where $W = I_1$ is the selector function for the second argument.

For the active arguments the derivation goes through as before. The first argument to F and G is just along for the ride—it is a passive argument. Let

$$\widetilde{F}(x, w) = F(x, \mathcal{V}(x, w)), \tag{3.40}$$

then define

$$G = W\mathcal{V} - \widetilde{F}. \tag{3.41}$$

We can check that G has the property $\mathcal{V} = \partial_1 G$ by carrying out the derivative:

$$\begin{aligned}
\partial_1 G &= \partial_1 (W\mathcal{V} - \widetilde{F}) \\
&= \mathcal{V} + W\partial_1 \mathcal{V} - \partial_1 \widetilde{F},
\end{aligned} \tag{3.42}$$

but

$$\begin{aligned}
\partial_1 \widetilde{F}(x, w) &= \partial_1 F(x, \mathcal{V}(x, w)) \partial_1 \mathcal{V}(x, w) \\
&= W(x, w)\partial_1 \mathcal{V}(x, w),
\end{aligned} \tag{3.43}$$

or

$$\partial_1 \widetilde{F} = W\partial_1 \mathcal{V}. \tag{3.44}$$

So, from equation (3.42),

$$\partial_1 G = \mathcal{V}, \tag{3.45}$$

as required. The active argument may have many components.

The partial derivatives with respect to the passive arguments are related in a remarkably simple way. Let's calculate the derivative $\partial_0 G$ in pieces. First,

$$\partial_0 (W\mathcal{V}) = W\partial_0 \mathcal{V} \tag{3.46}$$

because $\partial_0 W = 0$. We calculate $\partial_0 \widetilde{F}$:

$$\begin{aligned}
\partial_0 \widetilde{F}(x, w) &= \partial_0 F(x, \mathcal{V}(x, w)) + \partial_1 F(x, \mathcal{V}(x, w))\partial_0 \mathcal{V}(x, w) \\
&= \partial_0 F(x, \mathcal{V}(x, w)) + W(x, w)\partial_0 \mathcal{V}(x, w).
\end{aligned} \tag{3.47}$$

Putting these together, we find

$$\partial_0 G(x, w) = -\partial_0 F(x, \mathcal{V}(x, w)) = -\partial_0 F(x, v). \tag{3.48}$$

The calculation is unchanged if the passive argument has many components.

We can write the Legendre transformation more symmetrically:

$$\begin{aligned}
w &= \partial_1 F(x, v) \\
wv &= F(x, v) + G(x, w) \\
v &= \partial_1 G(x, w) \\
0 &= \partial_0 F(x, v) + \partial_0 G(x, w).
\end{aligned} \tag{3.49}$$

The last relation is not as trivial as it looks, because x enters the equations connecting w and v. With this symmetrical form, we see that the Legendre transform is its own inverse.

Exercise 3.4: Simple Legendre transforms

For each of the following functions, find the function that is related to the given function by the Legendre transform on the indicated active argument. Show that the Legendre transform relations hold for your solution, including the relations among passive arguments, if any.

a. $F(x) = ax + bx^2$, with no passive arguments.

b. $F(x, y) = a \sin x \cos y$, with x active.

c. $F(x, y, \dot{x}, \dot{y}) = x\dot{x}^2 + 3\dot{x}\dot{y} + y\dot{y}^2$, with $\dot{x}$ and $\dot{y}$ active.

Hamilton's equations from the Legendre transformation
We can use the Legendre transformation with the Lagrangian
playing the role of F and with the generalized velocity slot playing
the role of the active argument. The Hamiltonian plays the role
of G with the momentum slot active. The coordinate and time
slots are passive arguments.

The Lagrangian L and the Hamiltonian H are related by a
Legendre transformation:

$$e = (\partial_2 L)(a, b, c) \tag{3.50}$$

$$ec = L(a, b, c) + H(a, b, e) \tag{3.51}$$

and

$$c = (\partial_2 H)(a, b, e), \tag{3.52}$$

with passive equations

$$0 = \partial_0 L(a, b, c) + \partial_0 H(a, b, e), \tag{3.53}$$
$$0 = \partial_1 L(a, b, c) + \partial_1 H(a, b, e). \tag{3.54}$$

Presuming it exists, we can define the inverse of $\partial_2 L$ with respect
to the last argument:

$$c = \mathcal{V}(a, b, e), \tag{3.55}$$

and write the Hamiltonian

$$H(a, b, c) = c\mathcal{V}(a, b, c) - L(a, b, \mathcal{V}(a, b, c)). \tag{3.56}$$

These relations are purely algebraic in nature.

On a path q we have the momentum p:

$$p(t) = \partial_2 L(t, q(t), Dq(t)), \tag{3.57}$$

and from the definition of $\mathcal{V}$ we find

$$Dq(t) = \mathcal{V}(t, q(t), p(t)). \tag{3.58}$$

The Legendre transform gives

$$Dq(t) = \partial_2 H(t, q(t), p(t)). \tag{3.59}$$

This relation is purely algebraic and is valid for any path. The passive equation (3.54) gives

$$\partial_1 L(t, q(t), Dq(t)) = -\partial_1 H(t, q(t), p(t)), \tag{3.60}$$

but the left-hand side can be rewritten using the Lagrange equations, so

$$Dp(t) = -\partial_1 H(t, q(t), p(t)). \tag{3.61}$$

This equation is valid only for realizable paths, because we used the Lagrange equations to derive it. Equations (3.59) and (3.61) are Hamilton's equations.

The remaining passive equation is

$$\partial_0 L(t, q(t), Dq(t)) = -\partial_0 H(t, q(t), p(t)). \tag{3.62}$$

This passive equation says that the Lagrangian has no explicit time dependence ($\partial_0 L = 0$) if and only if the Hamiltonian has no explicit time dependence ($\partial_0 H = 0$). We have found that if the Lagrangian has no explicit time dependence, then energy is conserved. So if the Hamiltonian has no explicit time dependence then it is a conserved quantity.

Exercise 3.5: Conservation of the Hamiltonian
Using Hamilton's equations, show directly that the Hamiltonian is a conserved quantity if it has no explicit time dependence.

Legendre transforms of quadratic functions
We cannot implement the Legendre transform in general because it involves finding the functional inverse of an arbitrary function. However, many physical systems can be described by Lagrangians that are quadratic forms in the generalized velocities. For such functions the generalized momenta are linear functions of the generalized velocities, and thus explicitly invertible.

More generally, we can compute a Legendre transformation for polynomial functions where the leading term is a quadratic form:

$$F(v) = \frac{1}{2}v^{\mathsf{T}}Mv + bv + c. \tag{3.63}$$

Because the first term is a quadratic form only the symmetric part of M contributes to the result, so we can assume M is symmetric.[12] Let $w = DF(v)$, then

$$w = DF(v) = Mv + b. \tag{3.64}$$

So if M is invertible we can solve for v in terms of w. Thus we may define a function $\mathcal{V}$ such that

$$v = \mathcal{V}(w) = M^{-1}(w - b) \tag{3.65}$$

and we can use this to compute the value of the function G:

$$G(w) = w\mathcal{V}(w) - F(\mathcal{V}(w)). \tag{3.66}$$

Computing Hamiltonians

We implement the Legendre transform for quadratic functions by the procedure[13]

```
(define (Legendre-transform F)
  (let ((w-of-v (D F)))
    (define (G w)
      (let ((zero (compatible-zero w)))
        (let ((M ((D w-of-v) zero))
              (b (w-of-v zero)))
          (let ((v (solve-linear-left M (- w b))))
            (- (* w v) (F v)))))))
    G))
```

The procedure `Legendre-transform` takes a procedure of one argument and returns the procedure that is associated with it by the Legendre transform. If $w = DF(v)$, $wv = F(v) + G(w)$, and $v = DG(w)$ specifies a one-argument Legendre transformation, then G is the function associated with F by the Legendre transform: $G = I\mathcal{V} - F \circ \mathcal{V}$, where $\mathcal{V}$ is the functional inverse of DF.

We can use the `Legendre-transform` procedure to compute a Hamiltonian from a Lagrangian:

---

[12]If $\mathbf{M}$ is the matrix representation of M, then $\mathbf{M} = \mathbf{M}^{\mathsf{T}}$.

[13]The procedure `solve-linear-left` was introduced in footnote 75 on page 71.

```
(define ((Lagrangian->Hamiltonian Lagrangian) H-state)
  (let ((t (time H-state))
        (q (coordinate H-state))
        (p (momentum H-state)))
    (define (L qdot)
      (Lagrangian (up t q qdot)))
    ((Legendre-transform L) p)))
```

Notice that the one-argument `Legendre-transform` procedure is sufficient. The passive variables are given no special attention, they are just passed around.

The Lagrangian may be obtained from the Hamiltonian by the procedure:

```
(define ((Hamiltonian->Lagrangian Hamiltonian) L-state)
  (let ((t (time L-state))
        (q (coordinate L-state))
        (qdot (velocity L-state)))
    (define (H p)
      (Hamiltonian (up t q p)))
    ((Legendre-transform H) qdot)))
```

Notice that the two procedures `Hamiltonian->Lagrangian` and `Lagrangian->Hamiltonian` are identical, except for the names.

For example, the Hamiltonian for the motion of the point mass with the potential energy $V(x, y)$ may be computed from the Lagrangian:

```
(define ((L-rectangular m V) local)
  (let ((q (coordinate local))
        (qdot (velocity local)))
    (- (* 1/2 m (square qdot))
       (V (ref q 0) (ref q 1)))))
```

And the Hamiltonian is, as we saw in equation (3.22):

```
(show-expression
 ((Lagrangian->Hamiltonian
   (L-rectangular
    'm
    (literal-function 'V (-> (X Real Real) Real))))
  (up 't (up 'x 'y) (down 'p_x 'p_y))))
```

$$V\left(x, y\right) + \frac{\frac{1}{2}p_x^2}{m} + \frac{\frac{1}{2}p_y^2}{m}$$

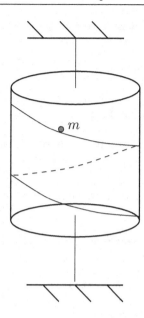

Figure 3.2   A point mass on a helical track.

Exercise 3.6: On a helical track

A uniform cylinder of mass M, radius R, and height h is mounted so as to rotate freely on a vertical axis. A point mass of mass m is constrained to move on a uniform frictionless helical track of pitch β (measured in radians per meter of drop along the cylinder) mounted on the surface of the cylinder (see figure 3.2). The mass is acted upon by standard gravity ($g = 9.8$ ms^{-2}).

a. What are the degrees of freedom of this system? Pick and describe a convenient set of generalized coordinates for this problem. Write a Lagrangian to describe the dynamical behavior. It may help to know that the moment of inertia of a cylinder around its axis is $\frac{1}{2}MR^2$. You may find it easier to do the algebra if various constants are combined and represented as single symbols.

b. Make a Hamiltonian for the system. Write Hamilton's equations for the system. Are there any conserved quantities?

c. If we release the point mass at time $t = 0$ at the top of the track with zero initial speed and let it slide down, what is the motion of the system?

Exercise 3.7: An ellipsoidal bowl

Consider a point particle of mass m constrained to move in a bowl and acted upon by a uniform gravitational acceleration g. The bowl is ellipsoidal, with height $z = ax^2 + by^2$. Make a Hamiltonian for this system. Can you make any immediate deductions about this system?

3.1.2 Hamilton's Equations from the Action Principle

The previous two derivations of Hamilton's equations made use of the Lagrange equations. Hamilton's equations can also be derived directly from the action principle.

The action is the integral of the Lagrangian along a path:

$$S[q](t_1, t_2) = \int_{t_1}^{t_2} L \circ \Gamma[q].$$ (3.67)

The action is stationary with respect to variations of a realizable path that preserve the configuration at the endpoints (for Lagrangians that are functions of time, coordinates, and velocities).

We can rewrite the integrand in terms of the Hamiltonian

$$L(t, q(t), p(t)) = p(t)Dq(t) - H(t, q(t), p(t)),$$ (3.68)

with $p(t) = \partial_2 L(t, q(t), Dq(t))$. The Legendre transformation construction gives

$$Dq(t) = \partial_2 H(t, q(t), p(t)),$$ (3.69)

which is one of Hamilton's equations, the one that does not depend on the path being a realizable path.

In order to vary the action we should make the dependences on the path explicit. We introduce

$$\tilde{p}[q](t) = \partial_2 L(t, q(t), Dq(t)),$$ (3.70)

and[14]

$$\Pi[q](t) = (t, q(t), \tilde{p}[q](t)) = (t, q(t), p(t)).$$ (3.71)

The integrand of the action integral is then

$$L \circ \Gamma[q] = \tilde{p}[q]Dq - H \circ \Pi[q].$$ (3.72)

---

[14]The function $\Pi[q]$ is the same as $\Pi_L[q]$ introduced on page 203. Indeed, the Lagrangian is needed to define momentum in every case, but we are suppressing the dependency here because it does not matter in this argument.

Using the shorthand δp for $\delta \tilde{p}[q]$,[15] and noting that $p = \tilde{p}[q]$, the variation of the action is

$$\delta S[q](t_1, t_2)$$
$$= \int_{t_1}^{t_2} (\delta p \, Dq + p \, \delta Dq - (DH \circ \Pi[q])\delta\Pi[q])$$
$$= \int_{t_1}^{t_2} \{\delta p \, Dq + p \, D\delta q \qquad\qquad\qquad (3.73)$$
$$-(\partial_1 H \circ \Pi[q])\delta q - (\partial_2 H \circ \Pi[q])\delta p\} \, .$$

Integrating the second term by parts, using $D(p\delta q) = Dp\delta q + pD\delta q$, we get

$$\delta S[q](t_1, t_2) = p\delta q|_{t_1}^{t_2}$$
$$+ \int_{t_1}^{t_2} \{\delta p \, Dq - Dp \, \delta q$$
$$-(\partial_1 H \circ \Pi[q])\delta q - (\partial_2 H \circ \Pi[q])\delta p\} \, . \qquad (3.74)$$

The variations are constrained so that $\delta q(t_1) = \delta q(t_2) = 0$, so the integrated part vanishes. Rearranging terms, the variation of the action is

$$\delta S[q](t_1, t_2) \qquad\qquad\qquad\qquad\qquad\qquad\qquad (3.75)$$
$$= \int_{t_1}^{t_2} ((Dq - \partial_2 H \circ \Pi[q]) \, \delta p - (Dp + \partial_1 H \circ \Pi[q]) \, \delta q) \, .$$

As a consequence of equation (3.69), the factor multiplying δp is zero. We are left with

$$\delta S[q](t_1, t_2) = - \int_{t_1}^{t_2} (Dp + \partial_1 H \circ \Pi[q]) \, \delta q. \qquad (3.76)$$

For the variation of the action to be zero for arbitrary variations, except for the endpoint conditions, we must have

$$Dp = -\partial_1 H \circ \Pi[q], \qquad\qquad\qquad\qquad\qquad (3.77)$$

---

[15]The variation of the momentum $\delta \tilde{p}[q]$ need not be further expanded in this argument because it turns out that the factor multiplying it is zero. However, it is handy to see how it is related to the variations in the coordinate path δq:

$$\delta p = \delta \tilde{p}[q](t) = \partial_1 \partial_2 L(t, q(t), Dq(t))\delta q(t) + \partial_2 \partial_2 L(t, q(t), Dq(t))D\delta q(t).$$

or

$$Dp = -\partial_1 H(t, q(t), p(t)), \qquad\qquad (3.78)$$

which is the "dynamical" Hamilton equation.[16]

3.1.3 A Wiring Diagram

Figure 3.3 shows a summary of the functional relationship between the Lagrangian and the Hamiltonian descriptions of a dynamical system. The diagram shows a "circuit" interconnecting some "devices" with "wires." The devices represent the mathematical functions that relate the quantities on their terminals. The wires represent identifications of the quantities on the terminals that they connect. For example, there is a box that represents the Lagrangian function. Given values t, q, and $\dot{q}$, the value of the Lagrangian $L(t, q, \dot{q})$ is on the terminal labeled L, which is wired to an addend terminal of an adder. Other terminals of the Lagrangian carry the values of the partial derivatives of the Lagrangian function.

The upper part of the diagram summarizes the relationship of the Hamiltonian to the Lagrangian. For example, the sum of the values on the terminals L of the Lagrangian and H of the Hamiltonian is the product of the value on the $\dot{q}$ terminal of the Lagrangian and the value on the p terminal of the Hamiltonian. This is the active part of the Legendre transform. The passive variables are related by the corresponding partial derivatives being negations of each other. In the lower part of the diagram the equations of motion are indicated by the presence of the integrators, relating the dynamical quantities to their time derivatives.

One can use this diagram to help understand the underlying unity of the Lagrangian and Hamiltonian formulations of mechanics. Lagrange's equations are just the connection of the $\dot{p}$ wire to the $\partial_1 L$ terminal of the Lagrangian device. One of Hamilton's equations is just the connection of the $\dot{p}$ wire (through the nega-

---

[16]It is sometimes asserted that the momenta have a different status in the Lagrangian and Hamiltonian formulations: that in the Hamiltonian framework the momenta are "independent" of the coordinates. From this it is argued that the variations δq and δp are arbitrary and independent, therefore implying that the factor multiplying each of them in the action integral (3.75) must independently be zero, apparently deriving both of Hamilton's equations. The argument is fallacious: we can write δp in terms of δq (see footnote 15).

tion device) to the $\partial_1 H$ terminal of the Hamiltonian device. The other is just the connection of the $\dot{q}$ wire to the $\partial_2 H$ terminal of the Hamiltonian device. We see that the two formulations are consistent. One does not have to abandon any part of the Lagrangian formulation to use the Hamiltonian formulation: there are deductions that can be made using both simultaneously.

3.2   Poisson Brackets

Here we introduce the Poisson bracket, in terms of which Hamilton's equations have an elegant and symmetric expression. Consider a function F of time, coordinates, and momenta. The value of F along the path $\sigma(t) = (t, q(t), p(t))$ is $(F \circ \sigma)(t) = F(t, q(t), p(t))$. The time derivative of $F \circ \sigma$ is

$$
\begin{aligned}
D(F \circ \sigma) &= (DF \circ \sigma)D\sigma \\
&= \partial_0 F \circ \sigma + (\partial_1 F \circ \sigma)Dq + (\partial_2 F \circ \sigma)Dp. \quad (3.79)
\end{aligned}
$$

If the phase-space path is a realizable path for a system with Hamiltonian H, then Dq and Dp can be reexpressed using Hamilton's equations:

$$
\begin{aligned}
D(F \circ \sigma) &= \partial_0 F \circ \sigma + (\partial_1 F \circ \sigma)(\partial_2 H \circ \sigma) - (\partial_2 F \circ \sigma)(\partial_1 H \circ \sigma) \\
&= \partial_0 F \circ \sigma + (\partial_1 F \partial_2 H - \partial_2 F \partial_1 H) \circ \sigma \\
&= \partial_0 F \circ \sigma + \{F, H\} \circ \sigma \quad (3.80)
\end{aligned}
$$

where the *Poisson bracket* $\{F, H\}$ of F and H is defined by[17]

$$
\{F, H\} = \partial_1 F \partial_2 H - \partial_2 F \partial_1 H. \quad (3.81)
$$

Note that the Poisson bracket of two functions on the phase-state space is also a function on the phase-state space.

---

[17]In traditional notation the Poisson bracket is written

$$
\{F, H\} = \sum_i \left(\frac{\partial F}{\partial q^i} \frac{\partial H}{\partial p_i} - \frac{\partial F}{\partial p_i} \frac{\partial H}{\partial q^i} \right).
$$

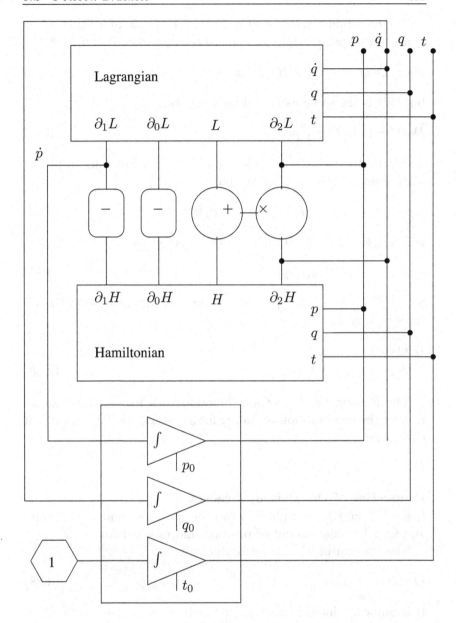

Figure 3.3   A "wiring diagram" describing the relationships among the dynamical quantities occurring in Lagrangian and Hamiltonian mechanics.

The coordinate selector $Q = I_1$ is an example of a function on phase-state space: $Q(t, q, p) = q$. According to equation (3.80),

$$Dq = D(Q \circ \sigma) = \{Q, H\} \circ \sigma = \partial_2 H \circ \sigma, \tag{3.82}$$

but this is the same as Hamilton's equation

$$Dq(t) = \partial_2 H(t, q(t), p(t)). \tag{3.83}$$

Similarly, the momentum selector $P = I_2$ is a function on phase-state space: $P(t, q, p) = p$. We have

$$Dp = D(P \circ \sigma) = \{P, H\} \circ \sigma = -\partial_1 H \circ \sigma, \tag{3.84}$$

which is the same as Hamilton's other equation

$$Dp(t) = -\partial_1 H(t, q(t), p(t)). \tag{3.85}$$

So the Poisson bracket provides a uniform way of writing Hamilton's equations:

$$\begin{aligned} D(Q \circ \sigma) &= \{Q, H\} \circ \sigma \\ D(P \circ \sigma) &= \{P, H\} \circ \sigma. \end{aligned} \tag{3.86}$$

The Poisson bracket of any function with itself is zero, so we recover the conservation of energy for a system that has no explicit time dependence:

$$DE = D(H \circ \sigma) = (\partial_0 H + \{H, H\}) \circ \sigma = \partial_0 H \circ \sigma. \tag{3.87}$$

Properties of the Poisson bracket

Let F, G, and H be functions of time, position, and momentum, and let c be independent of position and momentum.

The Poisson bracket is antisymmetric:

$$\{F, G\} = -\{G, F\}. \tag{3.88}$$

It is bilinear (linear in each argument):

$$\{F, G + H\} = \{F, G\} + \{F, H\} \tag{3.89}$$
$$\{F, cG\} = c\{F, G\} \tag{3.90}$$
$$\{F + G, H\} = \{F, H\} + \{G, H\} \tag{3.91}$$
$$\{cF, G\} = c\{F, G\}. \tag{3.92}$$

The Poisson bracket satisfies Jacobi's identity:

$$0 = \{F, \{G, H\}\} + \{H, \{F, G\}\} + \{G, \{H, F\}\}. \tag{3.93}$$

All but the last of (3.88–3.93) can immediately be verified from the definition. Jacobi's identity requires a little more effort to verify. We can use the computer to avoid some work. Define some literal phase-space functions of Hamiltonian type:

```
(define F
  (literal-function 'F
    (-> (UP Real (UP Real Real) (DOWN Real Real)) Real)))

(define G
  (literal-function 'G
    (-> (UP Real (UP Real Real) (DOWN Real Real)) Real)))

(define H
  (literal-function 'H
    (-> (UP Real (UP Real Real) (DOWN Real Real)) Real)))
```

Then we check the Jacobi identity:

```
((+ (Poisson-bracket F (Poisson-bracket G H))
    (Poisson-bracket G (Poisson-bracket H F))
    (Poisson-bracket H (Poisson-bracket F G)))
 (up 't (up 'x 'y) (down 'px 'py)))
0
```

The residual is zero, so the Jacobi identity is satisfied for any three phase-space state functions with two degrees of freedom.

Poisson brackets of conserved quantities

The Poisson bracket of conserved quantities is conserved. Let F and G be time-independent phase-space state functions: $\partial_0 F = \partial_0 G = 0$. If F and G are conserved by the evolution under H then

$$0 = D(F \circ \sigma) = \{F, H\} \circ \sigma$$
$$0 = D(G \circ \sigma) = \{G, H\} \circ \sigma. \tag{3.94}$$

So the Poisson brackets of F and G with H are zero: $\{F, H\} = \{G, H\} = 0$. The Jacobi identity then implies

$$\{\{F, G\}, H\} = 0, \tag{3.95}$$

and thus

$$D(\{F, G\} \circ \sigma) = 0, \tag{3.96}$$

so $\{F, G\}$ is a conserved quantity. The Poisson bracket of two conserved quantities is also a conserved quantity.

3.3   One Degree of Freedom

The solutions of time-independent systems with one degree of freedom can be found by quadrature. Such systems conserve the Hamiltonian: the Hamiltonian has a constant value on each realizable trajectory. We can use this constraint to eliminate the momentum in favor of the coordinate, obtaining the single equation $Dq(t) = f(q(t))$.[18]

A geometric view reveals more structure. Time-independent systems with one degree of freedom have a two-dimensional phase space. Energy is conserved, so all orbits are level curves of the Hamiltonian. The possible orbit types are restricted to curves that are contours of a real-valued function. The possible orbits are paths of constant altitude in the mountain range on the phase plane described by the Hamiltonian.

Only a few characteristic features are possible. There are points that are stable equilibria of the dynamical system. These are the peaks and pits of the Hamiltonian mountain range. These equilibria are stable in the sense that neighboring trajectories on nearby contours stay close to the equilibrium point. There are orbits that trace simple closed curves on contours that surround a peak or pit, or perhaps several peaks. There are also trajectories lying on contours that cross at a saddle point. The crossing point is an unstable equilibrium, unstable in the sense that neighboring trajectories leave the vicinity of the equilibrium point. Such contours that cross at saddle points are called *separatrices* (singular: *separatrix*), contours that "separate" two regions of distinct behavior.

---

[18]For systems with kinetic energy that is quadratic in velocity, this equation does not satisfy the Lipschitz condition at isolated points where the velocity is zero. However the solution for q can be extracted using a definite integral.

At every point Hamilton's equations give a unique rate of evolution and direct the system to move perpendicular to the gradient of the Hamiltonian. At the peaks, pits, and saddle points, the gradient of the Hamiltonian is zero, so according to Hamilton's equations these are equilibria. At other points, the gradient of the Hamiltonian is nonzero, so according to Hamilton's equations the rate of evolution is nonzero. Trajectories evolve along the contours of the Hamiltonian. Trajectories on simple closed contours periodically trace the contour. At a saddle point, contours cross. The gradient of the Hamiltonian is zero at the saddle point, so a system started at the saddle point does not leave the saddle point. On the separatrix away from the saddle point the gradient of the Hamiltonian is not zero, so trajectories evolve along the contour. Trajectories on the separatrix are asymptotic forward or backward in time to a saddle point. Going forward or backward in time, such trajectories forever approach an unstable equilibrium but never reach it. If the phase space is bounded, asymptotic trajectories that lie on contours of a smooth Hamiltonian are always asymptotic to unstable equilibria at both ends (but they may be different equilibria).

These orbit types are all illustrated by the prototypical phase plane of the pendulum (see figure 3.4). The solutions lie on contours of the Hamiltonian. There are three regions of the phase plane; in each the motion is qualitatively different. In the central region the pendulum oscillates; above this there is a region in which the pendulum circulates in one direction; below the oscillation region the pendulum circulates in the other direction. In the center of the oscillation region there is a stable equilibrium, at which the pendulum is hanging motionless. At the boundaries between these regions, the pendulum is asymptotic to the unstable equilibrium, at which the pendulum is standing upright.[19] There are two asymptotic trajectories, corresponding to the two ways the equilibrium can be approached. Each of these is also asymptotic to the unstable equilibrium going backward in time.

---

[19]The pendulum has only one unstable equilibrium. Remember that the coordinate is an angle.

3.4   Phase Space Reduction

Our motivation for the development of Hamilton's equations was
to focus attention on the quantities that can be conserved—the
momenta and the energy. In the Hamiltonian formulation the
generalized configuration coordinates and the conjugate momenta
comprise the state of the system at a given time. We know from
the Lagrangian formulation that if the Lagrangian does not depend on some coordinate then the conjugate momentum is conserved. This is also true in the Hamiltonian formulation, but there
is a distinct advantage to the Hamiltonian formulation. In the Lagrangian formulation the knowledge of the conserved momentum
does not lead immediately to any simplification of the problem,
but in the Hamiltonian formulation the fact that momenta are
conserved gives an immediate reduction in the dimension of the
system to be solved. In fact, if a coordinate does not appear in the
Hamiltonian then the dimension of the system of coupled equations that remain to be solved is reduced by two—the coordinate
does not appear and the conjugate momentum is constant.

Let $H(t, q, p)$ be a Hamiltonian for some problem with an n-
dimensional configuration space and $2n$-dimensional phase space.
Suppose the Hamiltonian does not depend upon the ith coordinate
q^i: $(\partial_1 H)_i = 0$.[20] According to Hamilton's equations, the conjugate momentum p_i is conserved. Hamilton's equations of motion
for the remaining $2n - 2$ phase-space variables do not involve q^i
(because it does not appear in the Hamiltonian), and p_i is a constant. Thus the dimension of the difficult part of the problem,
the part that involves the solution of coupled ordinary differential
equations, is reduced by two. The remaining equation governing
the evolution of q^i in general depends on all the other variables,
but once the reduced problem has been solved, the equation of
motion for q^i can be written so as to give Dq^i explicitly as a function of time. We can then find q^i as a definite integral of this
function.[21]

---

[20] If a Lagrangian does not depend on a particular coordinate then neither does
the corresponding Hamiltonian, because the coordinate is a passive variable
in the Legendre transform. Such a Hamiltonian is said to be cyclic in that
coordinate.

[21] Traditionally, when a problem has been reduced to the evaluation of a definite integral it is said to be reduced to a "quadrature." Thus, the determi-

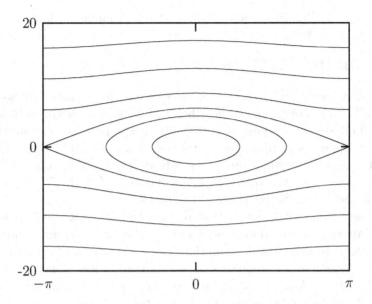

Figure 3.4   Contours of the Hamiltonian for the undriven pendulum on the phase plane. The horizontal axis is the angle θ and the vertical axis is the conjugate angular momentum p_θ. All realizable trajectories lie on contours of the Hamiltonian. There are three regions in this contour graph, displaying two distinct kinds of behavior. For small energy the pendulum oscillates, producing trajectories that are ovoid curves around the stable equilibrium point at the center. For larger energy the pendulum circulates, producing wavy tracks outside the eye-shaped region of oscillation. The oscillation region is separated from the circulation regions by the separatrix, which emanates from the unstable equilibrium at $(\pm\pi, 0)$. The pendulum has length 1m and a bob of mass 1kg. The acceleration of gravity is $9.8\,\mathrm{m\,s^{-2}}$.

Contrast this result with analogous results for more general systems of differential equations. There are two independent situations. One situation is that we know a constant of the motion. In general, constants of the motion can be used to reduce by one the dimension of the unsolved part of the problem. To see this, let the system of equations be

$$Dz^i(t) = F^i(z^0(t), z^1(t), \ldots, z^{m-1}(t)), \tag{3.97}$$

nation of the evolution of a cyclic coordinate q^i is reduced to a problem of quadrature.

where m is the dimension of the system. Assume we know some constant of the motion

$$C(z^0(t), z^1(t), \ldots, z^{m-1}(t)) = 0. \tag{3.98}$$

At least locally, we expect that we can use this equation to solve for $z^{m-1}(t)$ in terms of all the other variables, and use this solution to eliminate the dependence on $z^{m-1}(t)$. The first $m-1$ equations then depend only upon the first $m-1$ variables. The dimension of the system of equations to be solved is reduced by one. After the solution for the other variables has been found, $z^{m-1}(t)$ can be found using the constant of the motion.

The second situation is that one of the variables, say z^i, does not appear in the equations of motion (but there is an equation for Dz^i). In this case the equations for the other variables form an independent set of equations of one dimension less than the original system. After these are solved, then the remaining equation for z^i can be solved by definite integration.

In both situations the dimension of the system of coupled equations is reduced by one. Hamilton's equations are different in that these two situations come together. If a Hamiltonian for a system does not depend on a particular coordinate, then the equations of motion for the other coordinates and momenta do not depend on that coordinate. Furthermore, the momentum conjugate to that coordinate is a constant of the motion. An added benefit is that the use of this constant of the motion to reduce the dimension of the remaining equations is automatic in the Hamiltonian formulation. The conserved momentum is a state variable and just a parameter in the remaining equations.

So if there is a continuous symmetry it will probably be to our advantage to choose a coordinate system that explicitly incorporates the symmetry, making the Hamiltonian independent of a coordinate. Then the dimension of the phase space of the coupled system will be reduced by two for every coordinate that does not appear in the Hamiltonian.[22]

---

[22]It is not always possible to choose a set of generalized coordinates in which all symmetries are simultaneously manifest. For these systems, the reduction of the phase space is more complicated. We have already encountered such a problem: the motion of a free rigid body. The system is invariant under rotation about any axis, yet no single coordinate system can reflect this symmetry. Nevertheless, we have already found that the dynamics is described by a system of lower dimension than the full phase space: the Euler equations.

Motion in a central potential

Consider the motion of a particle of mass m in a central potential. A natural choice for generalized coordinates that reflects the symmetry is polar coordinates. A Lagrangian is (equation 1.69):

$$L(t; r, \varphi; \dot{r}, \dot{\varphi}) = \tfrac{1}{2}m(\dot{r}^2 + r^2\dot{\varphi}^2) - V(r). \qquad (3.99)$$

The momenta are $p_r = m\dot{r}$ and $p_\varphi = mr^2\dot{\varphi}$. The kinetic energy is a homogeneous quadratic form in the velocities, so the Hamiltonian is $T + V$ with the velocities rewritten in terms of the momenta:

$$H(t; r, \varphi; p_r, p_\varphi) = \frac{p_r^2}{2m} + \frac{p_\varphi^2}{2mr^2} + V(r). \qquad (3.100)$$

Hamilton's equations are

$$Dr(t) = \frac{p_r(t)}{m}$$

$$D\varphi(t) = \frac{p_\varphi(t)}{m(r(t))^2}$$

$$Dp_r(t) = \frac{(p_\varphi(t))^2}{m(r(t))^3} - DV(r(t))$$

$$Dp_\varphi(t) = 0. \qquad (3.101)$$

The potential energy depends on the distance from the origin, r, as does the kinetic energy in polar coordinates, but neither the potential energy nor the kinetic energy depends on the polar angle φ. The angle φ does not appear in the Lagrangian so we know that p_φ, the momentum conjugate to φ, is conserved along realizable trajectories. The fact that p_φ is constant along realizable paths is expressed by one of Hamilton's equations. That p_φ has a constant value is immediately made use of in the other Hamilton's equations: the remaining equations are a self-contained subsystem with constant p_φ. To make a lower-dimensional subsystem in the Lagrangian formulation we have to use each conserved momentum to eliminate one of the other state variables, as we did for the axisymmetric top (see section 2.10).

We can check our derivations with the computer. A procedure implementing the Lagrangian has already been introduced (below equation 1.69). We can use this to get the Hamiltonian:

```
(show-expression
 ((Lagrangian->Hamiltonian
    (L-central-polar 'm (literal-function 'V)))
  (up 't (up 'r 'phi) (down 'p_r 'p-phi))))
```

$$V\left(r\right) + \frac{\frac{1}{2}p_\varphi^2}{mr^2} + \frac{\frac{1}{2}p_r^2}{m}$$

and to develop Hamilton's equations:

```
(show-expression
 (((Hamilton-equations
     (Lagrangian->Hamiltonian
       (L-central-polar 'm (literal-function 'V))))
   (up (literal-function 'r)
       (literal-function 'phi))
   (down (literal-function 'p_r)
         (literal-function 'p-phi)))
  't))
```

$$\begin{pmatrix} 0 \\ \begin{pmatrix} Dr\left(t\right) - \dfrac{p_r\left(t\right)}{m} \\ D\varphi\left(t\right) - \dfrac{p_\varphi\left(t\right)}{m\left(r\left(t\right)\right)^2} \end{pmatrix} \\ \begin{bmatrix} Dp_r\left(t\right) + DV\left(r\left(t\right)\right) - \dfrac{\left(p_\varphi\left(t\right)\right)^2}{m\left(r\left(t\right)\right)^3} \\ Dp_\varphi\left(t\right) \end{bmatrix} \end{pmatrix}$$

Axisymmetric top

We reconsider the axisymmetric top (see section 2.10) from the Hamiltonian point of view. Recall that a top is a rotating rigid body, one point of which is fixed in space. The center of mass is not at the fixed point, and there is a uniform gravitational field. An axisymmetric top is a top with an axis of symmetry. We consider here an axisymmetric top with the fixed point on the symmetry axis.

The axisymmetric top has two continuous symmetries we would like to exploit. It has the symmetry that neither the kinetic nor potential energy is sensitive to the orientation of the top about the symmetry axis. The kinetic and potential energy are also insensitive to a rotation of the physical system about the vertical axis, because the gravitational field is uniform. We take advantage of these symmetries by choosing coordinates that naturally express them. We already have an appropriate coordinate system that does the job—the Euler angles. We choose the reference orientation of the top so that the symmetry axis is vertical. The first Euler angle, ψ, expresses a rotation about the symmetry axis. The next Euler angle, θ, is the tilt of the symmetry axis of the top from the vertical. The third Euler angle, φ, expresses a rotation of the top about the fixed $\hat{z}$ axis. The symmetries of the problem imply that the first and third Euler angles do not appear in the Hamiltonian. As a consequence the momenta conjugate to these angles are conserved quantities. The problem of determining the motion of the axisymmetric top is reduced to the problem of determining the evolution of θ and p_θ. Let's work out the details.

In terms of Euler angles, a Lagrangian for the axisymmetric top is (see section 2.10):

```
(define ((L-axisymmetric-top A C gMR) local)
  (let ((q (coordinate local))
        (qdot (velocity local)))
    (let ((theta (ref q 0))
          (thetadot (ref qdot 0))
          (phidot (ref qdot 1))
          (psidot (ref qdot 2)))
      (+ (* 1/2 A
            (+ (square thetadot)
               (square (* phidot (sin theta)))))
         (* 1/2 C
            (square (+ psidot (* phidot (cos theta)))))
         (* -1 gMR (cos theta))))))
```

where gMR is the product of the gravitational acceleration, the mass of the top, and the distance from the point of support to the center of mass. The Hamiltonian is nicer than we have a right to expect:

```
(show-expression
 ((Lagrangian->Hamiltonian (L-axisymmetric-top 'A 'C 'gMR))
  (up 't
      (up 'theta 'phi 'psi)
      (down 'p_theta 'p_phi 'p_psi))))
```

$$\frac{\frac{1}{2}p_\psi^2}{C} + \frac{\frac{1}{2}p_\psi^2\left(\cos\left(\theta\right)\right)^2}{A\left(\sin\left(\theta\right)\right)^2} + \frac{\frac{1}{2}p_\theta^2}{A} - \frac{p_\varphi p_\psi\cos\left(\theta\right)}{A\left(\sin\left(\theta\right)\right)^2} + \frac{\frac{1}{2}p_\varphi^2}{A\left(\sin\left(\theta\right)\right)^2}$$
$$+ gMR\cos\left(\theta\right)$$

Note that the angles φ and ψ do not appear in the Hamiltonian, as expected. Thus the momenta p_φ and p_ψ are constants of the motion.

For given values of p_φ and p_ψ we must determine the evolution of θ and p_θ. The effective Hamiltonian for θ and p_θ has one degree of freedom, and does not involve the time. Thus the value of the Hamiltonian is conserved along realizable trajectories. So the trajectories of θ and p_θ trace contours of the effective Hamiltonian. This gives us a big picture of the possible types of motion and their relationship, for given values of p_φ and p_ψ.

If the top is standing vertically then $p_\varphi = p_\psi$. Let's concentrate on the case that $p_\varphi = p_\psi$, and define $p = p_\psi = p_\varphi$. The effective Hamiltonian becomes (after a little trigonometric simplification)

$$H_p(t,\theta,p_\theta) = \frac{p_\theta^2}{2A} + \frac{p^2}{2C} + \frac{p^2}{2A}\tan^2\frac{\theta}{2} + gMR\cos\theta. \tag{3.102}$$

Defining the effective potential energy

$$V_p(\theta) = \frac{p^2}{2C} + \frac{p^2}{2A}\tan^2\frac{\theta}{2} + gMR\cos\theta, \tag{3.103}$$

which parametrically depends on p, the effective Hamiltonian is

$$H_p(t,\theta,p_\theta) = \frac{p_\theta^2}{2A} + V_p(\theta). \tag{3.104}$$

If p is large, V_p has a single minimum at $\theta = 0$, as seen in figure 3.5 (top curve). For small p (bottom curve) there is a minimum for finite positive θ and a symmetrical minimum for negative θ; there is a local maximum at $\theta = 0$. There is a critical value of p at which $\theta = 0$ changes from a minimum to a local maximum. Denote the critical value by p_c. A simple calculation shows $p_c = \sqrt{4gMRA}$. For $\theta = 0$ we have $p = C\omega$, where ω is the rotation rate. Thus to p_c there corresponds a critical rotation rate

$$\omega_c = \sqrt{4gMRA}/C. \tag{3.105}$$

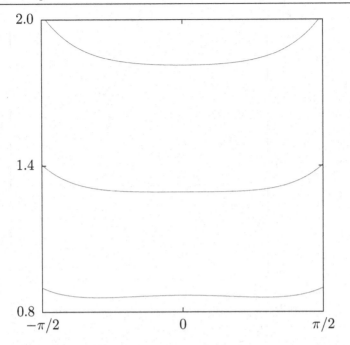

Figure 3.5   The effective potential energy V_p of the axisymmetric top as a function of the angle θ. The top curve is for an axial angular momentum $p > p_c$. For this value the top is stable standing vertically. The bottom curve is for $p < p_c$. Here the top is not stable standing vertically. The middle curve is for p at the critical angular momentum. We see the bifurcation of the stable equilibrium of the sleeping top into three equilibrium points, one of them unstable.

For $\omega > \omega_c$ the top can stand vertically; for $\omega < \omega_c$ the top falls if slightly displaced from the vertical. A top that stands vertically is called a "sleeping" top. For a more realistic top, friction gradually slows the rotation; the rotation rate eventually falls below the critical rotation rate and the top "wakes up."

We get additional insight into the sleeping top and the awake top by looking at the trajectories in the θ, p_θ phase plane. The trajectories in this plane are simply contours of the Hamiltonian, because the Hamiltonian is conserved. Figure 3.6 shows a phase portrait for $\omega > \omega_c$. All of the trajectories are loops around the vertical ($\theta = 0$). Displacing the top slightly from the vertical simply places the top on a nearby loop, so the top stays nearly vertical. Figure 3.7 shows the phase portrait for $\omega < \omega_c$. Here the vertical position is an unstable equilibrium. The trajectories that approach the vertical are asymptotic—they take an infinite amount of time to reach it, just as a pendulum with just the right

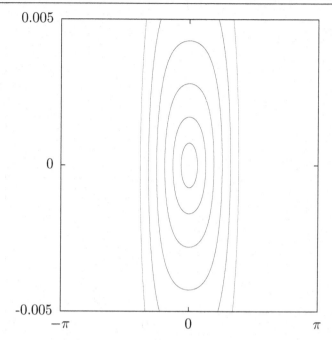

Figure 3.6   Trajectories of the axisymmetric top plotted on the (θ, p_θ) phase plane with $p_\varphi = p_\psi$ and $\omega = 145\,\mathrm{rad\,s^{-1}}$. The parameters are $A = 0.000696\,\mathrm{kg\,m^2}$, $C = 0.000132\,\mathrm{kg\,m^2}$, $gMR = 0.112\,\mathrm{kg\,m^2\,s^{-2}}$. For these parameters the critical frequency ω_c is about $133.8\,\mathrm{rad\,s^{-1}}$.

initial conditions can approach the vertical but never reach it. If the top is displaced slightly from the vertical then the trajectories loop around another center with nonzero θ. A top started at the center point of the loop stays there, and one started near this equilibrium point loops stably around it. Thus we see that when the top "wakes up" the vertical is unstable, but the top does not fall to the ground. Rather, it oscillates around a new equilibrium.

It is also interesting to consider the axisymmetric top when $p_\varphi \neq p_\psi$. Consider the case $p_\varphi > p_\psi$. Some trajectories in the θ, p_θ plane are shown in figure 3.8. Note that in this case trajectories do not go through $\theta = 0$. The phase portrait for $p_\varphi < p_\psi$ is similar and is not shown.

We have reduced the motion of the axisymmetric top to quadratures by choosing coordinates that express the symmetries. It turns out that the resulting integrals can be expressed in terms of elliptic functions. Thus, the axisymmetric top can be solved

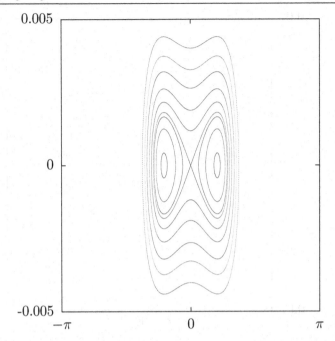

0.005

0

-0.005

$-\pi$ 0 π

Figure 3.7 Trajectories of the axisymmetric top plotted on the (θ, p_θ) phase plane with $p_\varphi = p_\psi$ and $\omega = 120\,\mathrm{rad\,s}^{-1}$. The other parameters are as before.

analytically. We do not dwell on this solution because it is not very illuminating: since most problems cannot be solved analytically, there is little profit in dwelling on the analytic solution of one of the rare problems that is analytically solvable. Rather, we have focused on the geometry of the solutions in the phase space and the use of conserved quantities to reduce the dimension of the problem. With the phase-space portrait we have found some interesting qualitative features of the motion of the top.

Exercise 3.8: Sleeping top

Verify that the critical angular velocity above which an axisymmetric top can sleep is given by equation (3.105).

3.4.1 Lagrangian Reduction

Suppose there are cyclic coordinates. In the Hamiltonian formulation, the equations of motion for the coordinates and momenta for the other degrees of freedom form a self-contained subsystem in

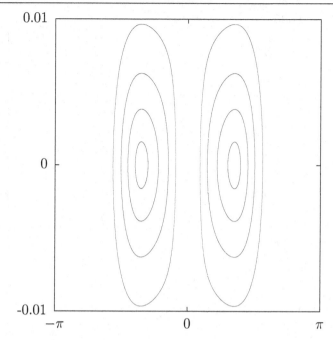

Figure 3.8   Trajectories of the axisymmetric top plotted on the (θ, p_θ) phase plane with $p_\varphi > p_\psi$. Most of the parameters are as in figure 3.6, but here $p_\varphi = 0.0145\,\mathrm{kg\,m^2\,s^{-1}}$ and $p_\psi = 0.0119\,\mathrm{kg\,m^2\,s^{-1}}$.

which the momenta conjugate to the cyclic coordinates are parameters. We can form a Lagrangian for this subsystem by performing a Legendre transform of the reduced Hamiltonian. Alternatively, we can start with the full Lagrangian and perform a Legendre transform for only those coordinates that are cyclic. The equations of motion are Hamilton's equations for those variables that are transformed and Lagrange's equations for the others. The momenta conjugate to the cyclic coordinates are conserved and can be treated as parameters in the Lagrangian for the remaining coordinates.

Divide the tuple q of coordinates into two subtuples $q = (x, y)$. Assume $L(t; x, y; v_x, v_y)$ is a Lagrangian for the system. Define the *Routhian* R as the Legendre transform of L with respect to the v_y slot:

$$p_y = \partial_{2,1}L(t; x, y; v_x, v_y) \tag{3.106}$$

$$p_y v_y = R(t; x, y; v_x, p_y) + L(t; x, y; v_x, v_y) \tag{3.107}$$

$$v_y = \partial_{2,1}R(t; x, y; v_x, p_y) \tag{3.108}$$

$$0 = \partial_0 R(t; x, y; v_x, p_y) + \partial_0 L(t; x, y; v_x, v_y) \tag{3.109}$$

$$0 = \partial_1 R(t; x, y; v_x, p_y) + \partial_1 L(t; x, y; v_x, v_y) \tag{3.110}$$

$$0 = \partial_{2,0}R(t; x, y; v_x, p_y) + \partial_{2,0}L(t; x, y; v_x, v_y). \tag{3.111}$$

To define the function R we must solve equation (3.106) for v_y in terms of the other variables, and substitute this into equation (3.107).

Define the state path Ξ:

$$\Xi(t) = (t; x(t), y(t); Dx(t), p_y(t)), \tag{3.112}$$

where

$$p_y(t) = \partial_{2,1}L(t; x(t), y(t); Dx(t), Dy(t)). \tag{3.113}$$

Realizable paths satisfy the equations of motion (see exercise 3.9)

$$D(\partial_{2,0}R \circ \Xi)(t) = \partial_{1,0}R \circ \Xi(t) \tag{3.114}$$

$$Dy(t) = \partial_{2,1}R \circ \Xi(t) \tag{3.115}$$

$$Dp_y(t) = -\partial_{1,1}R \circ \Xi(t), \tag{3.116}$$

which are Lagrange's equations for x and Hamilton's equations for y and p_y.

Now suppose that the Lagrangian is cyclic in y. Then $\partial_{1,1}L = \partial_{1,1}R = 0$, and $p_y(t)$ is a constant c on any realizable path. Equation (3.114) does not depend on y, by assumption, and we can replace p_y by its constant value c. So equation (3.114) forms a closed subsystem for the path x. The Lagrangian L_c

$$L_c(t, x, v_x) = -R(t; x, \bullet; v_x, c) \tag{3.117}$$

describes the motion of the subsystem (the minus sign is introduced for convenience, and $\bullet$ indicates that the function's value is independent of this argument). The path y can be found by integrating equation (3.115) using the independently determined path x.

Define the action

$$S'_c[x](t_1, t_2) = \int_{t_1}^{t_2} L_c \circ \Gamma[x]. \tag{3.118}$$

The realizable paths x satisfy the Lagrange equations with the Lagrangian L_c, so the action S'_c is stationary with respect to variations ξ of x that are zero at the end times:

$$\delta_\xi S'_c(t_1, t_2) = 0. \tag{3.119}$$

For realizable paths q the action $S[q](t_1, t_2)$ is stationary with respect to variations η of q that are zero at the end times. Along these paths the momentum $p_y(t)$ has the constant value c. For these same paths the action $S'_c[x](t_1, t_2)$ is stationary with respect to variations ξ of x that are zero at the end times. The dimension of ξ is smaller than the dimension of η.

The values of the actions $S'_c[x](t_1, t_2)$ and $S[q](t_1, t_2)$ are related:

$$\begin{aligned} S[q](t_1, t_2) &= S'_c[x] - \int_{t_1}^{t_2} cv_y \\ &= S'_c[x] - c(y(t_2) - y(t_1)). \end{aligned} \tag{3.120}$$

Exercise 3.9: Routhian equations of motion

Verify that the equations of motion are given by equations (3.114–3.116).

3.5   Phase Space Evolution

Most problems do not have enough symmetries to be reducible to quadrature. It is natural to turn to numerical integration to learn more about the evolution of such systems. The evolution in phase space may be found by numerical integration of Hamilton's equations.

As an illustration, consider again the periodically driven pendulum (see page 74). The Hamiltonian is

```
(show-expression
 ((Lagrangian->Hamiltonian
    (L-periodically-driven-pendulum 'm 'l 'g 'a 'omega))
  (up 't 'theta 'p_theta)))
```

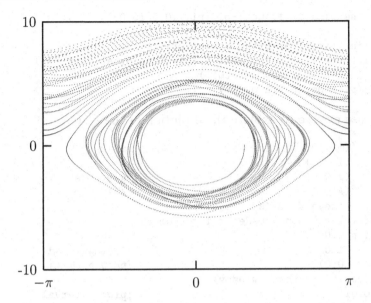

Figure 3.9   A trajectory of the periodically driven pendulum on the (θ, p_θ) phase plane. The trajectory starts in the oscillation region at $(1, 0)$. It oscillates for a while, but then escapes into circulation, only later to be recaptured into oscillation.

$$-\frac{1}{2}a^2 m\omega^2 \left(\cos\left(\theta\right)\right)^2 \left(\sin\left(\omega t\right)\right)^2 + agm\cos\left(\omega t\right)$$

$$+\frac{a\omega p_\theta \sin\left(\theta\right)\sin\left(\omega t\right)}{l} - glm\cos\left(\theta\right) + \frac{\frac{1}{2}p_\theta^2}{l^2 m}$$

Hamilton's equations for the periodically driven pendulum are unrevealing, so we will not show them. We build a system derivative from the Hamiltonian:

```
(define (H-pend-sysder m l g a omega)
  (Hamiltonian->state-derivative
    (Lagrangian->Hamiltonian
      (L-periodically-driven-pendulum m l g a omega))))
```

Now we integrate this system, with the same initial conditions as in section 1.7 (see figure 1.7), but display the trajectory in phase space (figure 3.9), using a monitor procedure:

```
(define window (frame :-pi :pi -10.0 10.0))

(define ((monitor-p-theta win) state)
  (let ((q ((principal-value :pi) (coordinate state)))
        (p (momentum state)))
    (plot-point win q p)))
```

We use evolve to explore the evolution of the system:

```
(let ((m 1.0)                        ;m=1kg
      (l 1.0)                        ;l=1m
      (g 9.8)                        ;g=9.8m/s²
      (A 0.1)                        ;A=1/10 m
      (omega (* 2 (sqrt 9.8)))))
  ((evolve H-pend-sysder m l g A omega)
   (up 0.0                           ;t₀=0
       1.0                           ;theta₀=1 rad
       0.0)                          ;p₀=0 kg m²/s
   (monitor-p-theta window)
   0.01                              ;plot interval
   100.0                             ;final time
   1.0e-12))
```

The trajectory sometimes oscillates and sometimes circulates. The patterns in the phase plane are reminiscent of the trajectories in the phase plane of the undriven pendulum shown in figure 3.4 on page 225.

3.5.1   Phase-Space Description Is Not Unique

We are familiar with the fact that a given motion of a system is expressed differently in different coordinate systems: the functions that express a motion in rectangular coordinates are different from the functions that express the same motion in polar coordinates. However, in a given coordinate system the evolution of the local state tuple for particular initial conditions is unique. The generalized velocity path function is the derivative of the generalized coordinate path function. On the other hand, the coordinate system alone does not uniquely specify the phase-space description. The relationship of the momentum to the coordinates and the velocities depends on the Lagrangian, and many different Lagrangians may be used to describe the behavior of the same physical system. When two Lagrangians for the same physical system are different, the phase-space descriptions of a dynamical state are different.

We have already seen two different Lagrangians for the driven pendulum (see section 1.6.4): one was found using $L = T - V$ and the other was found by inspection of the equations of motion. The two Lagrangians differ by a total time derivative. The momentum p_θ conjugate to θ depends on which Lagrangian we choose to work with, and the description of the evolution in the corresponding phase space also depends on the choice of Lagrangian, even though the behavior of the system is independent of the method used to describe it. The momentum conjugate to θ, using the $L = T - V$ Lagrangian for the periodically driven pendulum, is

$$p_\theta = ml^2\dot{\theta} - alm\omega \sin\theta \sin\omega t, \tag{3.121}$$

but with the alternative Lagrangian, it is

$$p_\theta = ml^2\dot{\theta}. \tag{3.122}$$

The two momenta differ by an additive distortion that varies periodically in time and depends on θ. That the phase-space descriptions are different is illustrated in figure 3.10. The evolution of the system is the same for each.

3.6   Surfaces of Section

Computing the evolution of mechanical systems is just the beginning of understanding the dynamics. Typically, we want to know much more than the phase space evolution of some particular trajectory. We want to obtain a qualitative understanding of the motion. We want to know what sorts of motion are possible, and how one type relates to others. We want to abstract the essential dynamics from the myriad particular evolutions that we can calculate. Paradoxically, it turns out that by throwing away most of the calculated information about a trajectory we gain essential new information about the character of the trajectory and its relation to other trajectories.

A remarkable tool that extracts the essence by throwing away information is a technique called the *surface of section* or *Poincaré section*.[23] A surface of section is generated by looking at successive

---

[23]The surface of section technique was introduced by Poincaré in his *Méthodes Nouvelles de la Mécanique Céleste* [35]. Poincaré proved remarkable results

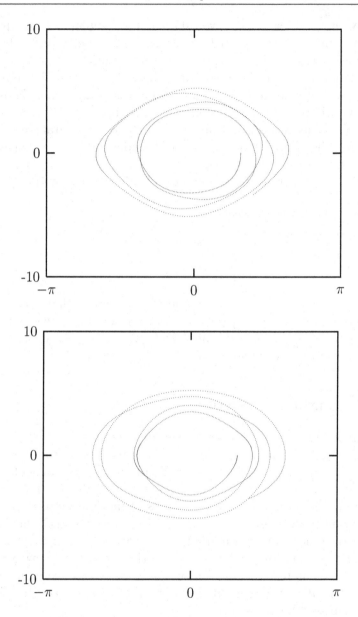

Figure 3.10   A trajectory of the periodically driven pendulum on the (θ, p_θ) phase plane. In the upper plot the trajectory is derived using the Lagrangian $L = T - V$ (see equation 1.88 on page 51. In the lower plot the trajectory is derived using the alternative Lagrangian of equation (1.120) on page 66. The evolution is the same, but the phase-space representations are not.

intersections of a trajectory or a set of trajectories with a plane in the phase space. Typically, the plane is spanned by a coordinate axis and the canonically conjugate momentum axis. We will see that surfaces of section made in this way have nice properties.

The surface of section technique was put to spectacular use in the 1964 landmark paper [22] by astronomers Michel Hénon and Carl Heiles. In their numerical investigations they found that some trajectories are chaotic, whereas other trajectories are regular. An essential characteristic of the chaotic motions is that initially nearby trajectories separate exponentially with time; the separation of regular trajectories is linear.[24] They found that these two types of trajectories are typically clustered in the phase space into regions of regular motion and regions of chaotic motion.

3.6.1   Periodically Driven Systems

For a periodically driven system the surface of section is a stroboscopic view of the evolution; we consider only the state of the system at the strobe times, with the period of the strobe equal to the drive period. We generate a surface of section for a periodically driven system by computing a number of trajectories and accumulating the phase-space coordinates of each trajectory whenever the drive passes through some particular phase. Let T be the period of the drive; then, for each trajectory, the surface of section accumulates the phase-space points $(q(t), p(t))$, $(q(t+T), p(t+T))$, $(q(t+2T), p(t+2T))$, and so on (see figure 3.11). For a system with a single degree of freedom we can plot the sequence of phase-space points on a q, p surface.

In the case of the stroboscopic section for the periodically driven system, the phase of the drive is the same for all section points;

---

about dynamical systems using the surface of section technique, and we shall return to some of these later. The surface of section technique is a key tool in the modern study of dynamical systems, for both analytical and numerical investigations.

[24]That solutions of ordinary differential equations can show exponential sensitivity to initial conditions was independently discovered by Edward Lorenz [31] in the context of a simplified model of convection in the Earth's atmosphere. Lorenz coined the picturesque term "butterfly effect" to describe this sensitivity: his weather system model is so sensitive to initial conditions that "the flapping of a butterfly's wings in Brazil can change the course of a typhoon in Japan."

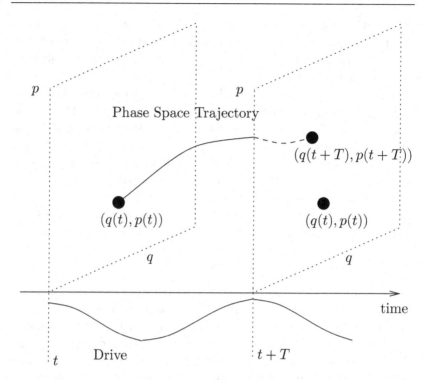

Figure 3.11 Stroboscopic surface of section for a periodically driven system. For each trajectory the surface of section accumulates the set of phase-space points after each full cycle of the drive.

thus each phase-space point in the section, with the known phase of the drive, may be considered as an initial condition for the rest of the trajectory. The absolute time of the particular section point does not affect the subsequent evolution; all that matters is that the phase of the drive have the value specified for the section. Thus we can think of the dynamical evolution as generating a map that takes a point in the phase space and generates a new point in the phase space after evolving the system for one drive period. This map of the phase space onto itself is called the *Poincaré map*.

Figure 3.12 shows an example Poincaré section for the driven pendulum. We plot the section points for a number of different initial conditions. We are immediately presented with a new facet of dynamical systems. For some initial conditions, the subsequent section points appear to fill out a set of curves in the section. For other initial conditions this is not the case: rather, the set

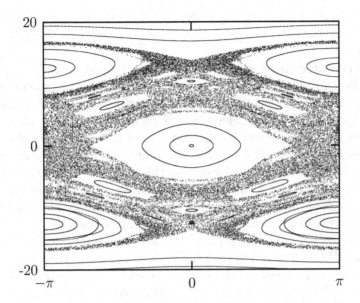

Figure 3.12   A surface of section for the driven pendulum on the (θ, p_θ) phase plane. The samples are taken at the peaks of the drive. For this section the parameters are: $m = 1\,\mathrm{kg}$, $l = 1\,\mathrm{m}$, $g = 9.8\,\mathrm{m\,s^{-2}}$, $A = 0.05\,\mathrm{m}$, and $\omega = 4.2\,\omega_0$, with $\omega_0 = \sqrt{g/l}$.

of section points appear scattered over a region of the section. In fact, *all* of the scattered points in figure 3.12 were generated from a single initial condition. The surface of section suggests that there are qualitatively different classes of trajectories distinguished by the dimension of the subspace of the section that they explore.

Trajectories that fill out curves on the surface of section are called *regular* or *quasiperiodic* trajectories. The curves that are filled out by the regular trajectories are *invariant curves*. They are invariant in that if any section point for a trajectory falls on an invariant curve, all subsequent points fall on the same invariant curve. Otherwise stated, the Poincaré map maps every point on an invariant curve onto the invariant curve.

The trajectories that appear to fill areas are called *chaotic* trajectories. For these points the distance in phase space between initially nearby points grows, on average, exponentially with time.[25]

---

[25]We saw an example of this extreme sensitivity to initial conditions in figure 1.7 (section 1.7) and also in the double-pendulum project (exercise 1.44).

In contrast, for the regular trajectories, the distance in phase space between initially nearby points grows, on average, linearly with time.

The phase space seems to be grossly clumped into different regions. Initial conditions in some regions appear to predominantly yield regular trajectories, and other regions appear to predominantly yield chaotic trajectories. This gross division of the phase space into qualitatively different types of trajectories is called the *divided phase space*. We will see later that there is much more structure here than is apparent at this scale, and that upon magnification there is a complicated interweaving of chaotic and regular regions on finer and finer scales. Indeed, we shall see that many trajectories that appear to generate curves on the surface of section are, upon magnification, actually chaotic and fill a tiny area. We shall also find that there are trajectories that lie on one-dimensional curves on the surface of section, but only explore a subset of this curve formed by cutting out an infinite number of holes.[26]

The features seen on the surface of section of the driven pendulum are quite general. The same phenomena are seen in most dynamical systems. In general, there are both regular and chaotic trajectories, and there is the clumping characteristic of the divided phase space. The specific details depend upon the system, but the basic phenomena are generic. Of course, we are interested in both aspects: the phenomena that are common to all systems, and the specific details for particular systems of interest.

The surface of section for the periodically driven pendulum has specific features that give us qualitative information about how this system behaves. The central island in figure 3.12 is the remnant of the oscillation region for the unforced pendulum (see figure 3.4 in section 3.3). There is a sizable region of regular trajectories here that are, in a sense, similar to the trajectories of the unforced pendulum. In this region, the pendulum oscillates back and forth, much as the undriven pendulum does, but the drive makes it wiggle as it does so. The section points are all collected at the same phase of the drive so we do not see these wiggles on the section.

---

[26] One-dimensional invariant sets with an infinite number of holes were discovered by John Mather. They are sometimes called *cantori* (singular *cantorus*), by analogy to the Cantor sets, but it really doesn't Mather.

The central island is surrounded by a large chaotic zone. Thus the region of phase space with regular trajectories similar to the unforced trajectories has finite extent. On the section, the boundary of this "stable" region is apparently rather well defined—there is a sudden transition from smooth regular invariant curves to chaotic motion that can take the system far from this region of regular motion.

There are two other sizeable regions of regular behavior with finite angular extent. The trajectories in these regions are resonant with the drive, on average executing one full rotation per cycle of the drive. The two islands differ in the direction of the rotation. In these regions the pendulum is making complete rotations, but the rotation is locked to the drive so that points on the section appear only in the islands. The fact that points for particular trajectories loop around the islands means that the pendulum sometimes completes a cycle faster than the drive and sometimes slower than the drive, but never loses lock.

Each regular region has finite extent. So from the surface of section we can see directly the range of initial conditions that remain in resonance with the drive. Outside of the regular region initial conditions lead to chaotic trajectories that evolve far from the resonant regions.

Various higher-order resonance islands are also visible, as are nonresonant regular circulating orbits. So, the surface of section has provided us with an overview of the main types of motion that are possible and their interrelationship.

Changing the parameters shows other interesting phenomena. Figure 3.13 shows the surface of section when the drive frequency is twice the natural small-amplitude oscillation frequency of the undriven pendulum. The section has a large chaotic zone, with an interesting set of islands. The central equilibrium has undergone an instability and instead of a central island we find two off-center islands. These islands are alternately visited one after the other. As the support goes up and down the pendulum alternately tips to one side and then the other. It takes two periods of the drive before the pendulum visits the same island. Thus, the system has "period-doubled." An island has been replaced by a period-doubled pair of islands. Note that other islands still exist. The islands in the top and bottom of the chaotic zone are the resonant islands, in which the pendulum loops on average a full turn for

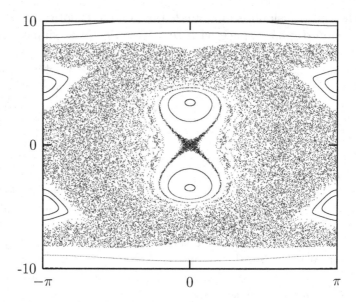

Figure 3.13   Another surface of section for the driven pendulum on the (θ, p_θ) phase plane. Here we see a period-doubled central island. For this section the frequency of the drive is resonant with the frequency of small-amplitude oscillations of the undriven pendulum. The angular momentum scale is -10 to $10\,\mathrm{kg\,m^2\,s^{-1}}$. For this section the parameters are: $m = 1\,\mathrm{kg}$, $l = 1\,\mathrm{m}$, $g = 9.8\,\mathrm{m\,s^{-2}}$, $A = 0.1\,\mathrm{m}$, $\omega = 2\omega_0$.

every cycle of the drive. Note that, as before, if the pendulum is rapidly circulating, the motion is regular.

It is a surprising fact that if we shake the support of a pendulum fast enough then the pendulum can stand upright. This phenomenon can be visualized with the surface of section. Figure 3.14 shows a surface of section when the drive frequency is large compared to the natural frequency. That the pendulum can stand upright is indicated by the existence of a regular island at the inverted equilibrium. The surface of section shows that the pendulum can remain upright for a range of initial displacements from the vertical.

3.6.2   Computing Stroboscopic Surfaces of Section

We already have the system derivative for the driven pendulum, and we can use it to make a parametric map for constructing Poincaré sections:

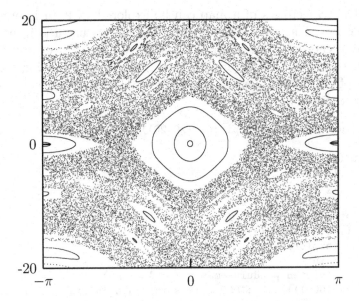

Figure 3.14   A surface of section for the rapidly driven pendulum on the (θ, p_θ) phase plane. Here we see the emergence of a vertical equilibrium. The angular momentum scale is -20 to $20 \, \mathrm{kg \, m^2 \, s^{-1}}$. For this section the parameters are: $m = 1 \, \mathrm{kg}$, $l = 1 \, \mathrm{m}$, $g = 9.8 \, \mathrm{m \, s^{-2}}$, $A = 0.2 \, \mathrm{m}$, $\omega = 10.1 \omega_0$.

```
(define (driven-pendulum-map m l g A omega)
  (let ((advance (state-advancer H-pend-sysder m l g A omega))
        (map-period (/ :2pi omega)))
    (lambda (theta ptheta return fail)
      (let ((ns (advance
                  (up 0 theta ptheta)   ; initial state
                  map-period)))         ; integration interval
        (return ((principal-value :pi) (coordinate ns))
                (momentum ns))))))
```

A map procedure takes the two section coordinates (here `theta` and `ptheta`) and two "continuation" procedures. If the section coordinates given are in the domain of the map, it produces two new section coordinates and passes them to the `return` continuation, otherwise the map procedure calls the `fail` continuation procedure with no arguments.[27]

---

[27]In the particular case of the driven pendulum there is no reason to call `fail`. This contingency is reserved for systems where orbits escape or cease to satisfy some constraint.

The trajectories of a map can be explored with an interactive interface. The procedure `explore-map` lets us use a pointing device to choose initial conditions for trajectories. For example, the surface of section in figure 3.12 was generated by plotting a number of trajectories, using a pointer to choose initial conditions, with the following program:

```
(define win (frame :-pi :pi -20 20))

(let ((m 1.0)                        ;m=1kg
      (l 1.0)                        ;l=1m
      (g 9.8)                        ;g=9.8m/s²
      (A 0.05))                      ;A=1/20m
  (let ((omega0 (sqrt (/ g l))))
    (let ((omega (* 4.2 omega0)))
      (explore-map
       win
       (driven-pendulum-map m l g A omega)
       1000))))    ;1000 points for each initial condition
```

Exercise 3.10: Fun with phase portraits

Choose some one-degree-of-freedom dynamical system that you are curious about and that can be driven with a periodic drive. Construct a map of the sort we made for the driven pendulum and do some exploring. Are there chaotic regions? Are all of the chaotic regions connected together?

3.6.3   Autonomous Systems

We illustrated the use of Poincaré sections to visualize qualitative features of the phase space for a one-degree-of-freedom system with periodic drive, but the idea is more general. Here we show how Hénon and Heiles [22] used the surface of section to elucidate the properties of an autonomous system.

Hénon–Heiles background

In the early '60s astronomers were up against a wall. Careful measurements of the motion of nearby stars in the galaxy had allowed particular statistical averages of the observed motions to be determined, and the averages were not at all what was expected. In particular, what was calculated was the velocity dispersion: the root mean square deviation of the velocity from the average. We use angle brackets to denote an average over nearby stars: $< w >$ is the average value of some quantity w for the stars in the ensemble. The average velocity is $< \dot{\vec{x}} >$. The components of the velocity dispersion are

$$\sigma_x = < (\dot{x} - < \dot{x} >)^2 >^{1/2} \qquad (3.123)$$

$$\sigma_y = < (\dot{y} - < \dot{y} >)^2 >^{1/2} \qquad (3.124)$$

$$\sigma_z = < (\dot{z} - < \dot{z} >)^2 >^{1/2} . \qquad (3.125)$$

If we use cylindrical polar coordinates (r, θ, z) and align the axes with the galaxy so that z is perpendicular to the galactic plane and r increases with the distance to the center of the galaxy, then two particular components of the velocity dispersion are

$$\sigma_z = < (\dot{z} - < \dot{z} >)^2 >^{1/2} \qquad (3.126)$$

$$\sigma_r = < (\dot{r} - < \dot{r} >)^2 >^{1/2} . \qquad (3.127)$$

It was expected at the time that these two components of the velocity dispersion should be equal. In fact they were found to differ by about a factor of 2: $\sigma_r \approx 2\sigma_z$. What was the problem? In the literature at the time there was considerable discussion of what could be wrong. Was the problem some observational selection effect? Were the velocities measured incorrectly? Were the assumptions used in the derivation of the expected ratio not adequately satisfied? For example, the derivation assumed that the galaxy was approximately axisymmetric. Perhaps non-axisymmetric components of the galactic potential were at fault. It turned out that the problem was much deeper. The understanding of motion was wrong.

Let's review the derivation of the expected relation among the components of the velocity dispersion. We wish to give a statistical description of the distribution of stars in the galaxy. We introduce the phase-space distribution function $f(\vec{x}, \vec{p})$, which gives the probability density of finding a star at position $\vec{x}$ with momentum $\vec{p}$.[28] Integrating this density over some finite volume of phase space gives the probability of finding a star in that phase-space volume (in that region of space within a specified region of

---

[28] We will see that it is convenient to look at distribution functions in the phase-space coordinates because the consequences of conserved momenta are more apparent, and also because volume in phase space is conserved by evolution (see section 3.8).

momenta). We assume the probability density is normalized so that the integral over all of phase space gives unit probability; the star is somewhere and has some momentum with certainty. In terms of f, the statistical average of any dynamical quantity w over some volume of phase space V is just

$$< w >_V = \int_V fw \tag{3.128}$$

where the integral extends over the phase-space volume V. In computing the velocity dispersion at some point $\vec{x}$, we would compute the averages by integrating over all momenta.

Individual stars move in the gravitational potential of the rest of the galaxy. It is not unreasonable to assume that the overall distribution of stars in the galaxy does not change much with time, or changes only very slowly. The density of stars in the galaxy is actually very small and close encounters of stars are very rare. Thus, we can model the gravitational potential of the galaxy as a fixed external potential in which individual stars move. The galaxy is approximately axisymmetric. We assume that the deviation from exact axisymmetry is not a significant effect and thus we take the model potential to be exactly axisymmetric.

Consider the motion of a point mass (a star) in an axisymmetric potential (of the galaxy). In cylindrical polar coordinates the Hamiltonian is

$$T + V = \frac{1}{2m} \left[p_r^2 + \frac{p_\theta^2}{r^2} + p_z^2 \right] + V(r, z), \tag{3.129}$$

where V does not depend on θ. Since θ does not appear, we know that the conjugate momentum p_θ is constant. For the motion of any particular star we can treat p_θ as a parameter. Thus the effective Hamiltonian has two degrees of freedom:

$$\frac{1}{2m} \left[p_r^2 + p_z^2 \right] + U(r, z) \tag{3.130}$$

where

$$U(r, z) = V(r, z) + \frac{p_\theta^2}{2mr^2}. \tag{3.131}$$

The value E of the Hamiltonian is constant since there is no explicit time dependence in the Hamiltonian. Thus, we have constants of the motion E and p_θ.

Jeans's "theorem" asserts that the distribution function f depends only on the values of the conserved quantities, also known as *integrals of motion*. That is, we can introduce a different distribution function f' that represents the same physical distribution:

$$f'(E, p_\theta) = f(\vec{x}, \vec{p}). \tag{3.132}$$

At the time, there was good reason to believe that this might be correct. First, it is clear that the distribution function surely depends at least on E and p_θ. The problem is, "Given an energy E and angular momentum p_θ, what motion is allowed?" The conserved quantities clearly confine the evolution. Does the evolution carry the system everywhere in the phase space subject to these known constraints? In the early part of the 20th century this appeared plausible. Statistical mechanics was successful, and statistical mechanics made exactly this assumption. Perhaps there are other conserved quantities of the motion that exist, but that we have not yet discovered?

Poincaré proved an important theorem with regard to conserved quantities. Poincaré proved that most of the conserved quantities of a dynamical system typically do not persist upon perturbation of the system. That is, if a small perturbation is added to a problem, then most of the conserved quantities of the original problem do not have analogs in the perturbed problem. The conserved quantities are destroyed. However, conserved quantities that result from symmetries of the problem continue to be preserved if the perturbed system has the same symmetries. Thus angular momentum continues to be preserved upon application of any axisymmetric perturbation. Poincaré's theorem is correct, but what came next was not.

As a corollary to Poincaré's theorem, in 1920 Fermi published a proof of a theorem stating that typically the motion of perturbed problems is ergodic[29] subject to the constraints imposed by the conserved quantities resulting from symmetries. Loosely speaking,

---

[29] A system is ergodic if time averages along trajectories are the same as phase-space averages over the region explored by the trajectories.

this means that trajectories go everywhere they are allowed to go by the conservation constraints. Fermi's theorem was later shown to be incorrect, but on the basis of this theorem we could expect that typically systems fully explore the phase space, subject only to the constraints imposed by the conserved quantities resulting from symmetries. Suppose then that the evolution of stars in the galactic potential is subject only to the constraints of conserving E and p_θ. We shall see that this is not true, but if it were we could then conclude that the distribution function for stars in the galaxy can also depend only on E and p_θ.

Given this form of the distribution function, we can deduce the stated ratios of the velocity dispersions. We note that p_z and p_r appear in the same way in the energy. Thus the average of any function of p_z computed with the distribution function must equal the average of the same function of p_r. In particular, the velocity dispersions in the z and r directions must be equal:

$$\sigma_z = \sigma_r. \tag{3.133}$$

But this is not what was observed, which was

$$\sigma_r \approx 2\sigma_z. \tag{3.134}$$

Hénon and Heiles [22] approached this problem differently from others at the time. Rather than improving the models for the motion of stars in the galaxy, they concentrated on what turned out to be the central issue: What is the qualitative nature of motion? The problem had nothing to do with galactic dynamics in particular, but with the problem of motion. They abstracted the dynamical problem from the particulars of galactic dynamics.

The system of Hénon and Heiles

We have seen that the study of the motion of a point with mass m and an axisymmetric potential energy reduces to the study of a reduced two-degree-of-freedom problem in r and z with potential energy $U(r, z)$. Hénon and Heiles chose to study the motion in a two-degree-of-freedom system with a particularly simple potential energy so that the dynamics would be clear and the calculation uncluttered. The Hénon–Heiles Hamiltonian is

$$H(t; x, y; p_x, p_y) = \frac{1}{2}\left(p_x^2 + p_y^2\right) + V(x, y) \tag{3.135}$$

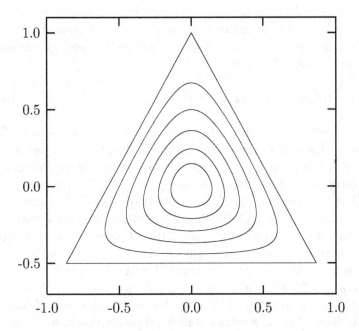

Figure 3.15 Contours of the Hénon–Heiles potential energy on the (x, y) plane. The contours shown, from the inside out, are for potential energies 1/100, 1/40, 1/20, 1/12, 1/8, and 1/6.

with potential energy

$$V(x, y) = \frac{1}{2} \left(x^2 + y^2 \right) + x^2 y - \frac{1}{3} y^3. \tag{3.136}$$

The potential energy is shaped like a distorted bowl. It has triangular symmetry, as is evident when it is rewritten in polar coordinates:

$$\frac{1}{2} r^2 + \frac{1}{3} r^3 \sin 3\theta. \tag{3.137}$$

Contours of the potential energy are shown in figure 3.15. At small values of the potential energy the contours are approximately circular; as the value of the potential energy approaches 1/6 the contours become triangular, and at larger potential energies the contours open to infinity.

The Hamiltonian is independent of time, so energy is conserved. In this case this is the only known conserved quantity. We first determine the restrictions that conservation of energy imposes on the evolution. We have

$$E = \frac{1}{2}\left(p_x^2 + p_y^2\right) + V(x,y) \geq V(x,y), \tag{3.138}$$

so the motion is confined to the region inside the contour $V = E$ because the sum of the squares of the momenta cannot be negative.

Let's compute some sample trajectories. For definiteness, we investigate trajectories with energy $E = 1/8$. There is a large variety of trajectories. There are trajectories that circulate in a regular way around the bowl, and there are trajectories that oscillate back and forth (figure 3.16). There are also trajectories that appear more irregular (figure 3.17). There is no end to the trajectories that could be computed, but let's face it, surely there is more to life than looking at trajectories.

The problem facing Hénon and Heiles was the issue of conserved quantities. Are there other conserved quantities besides the obvious ones? They investigated this issue with the surface of section technique. The surface of section is generated by looking at successive passages of trajectories through a plane in phase space.

Specifically, the surface of section is generated by recording and plotting p_y versus y whenever $x = 0$, as shown in figure 3.18. Given the value of the energy E and a point (y, p_y) on the section $x = 0$, we can recover p_x, up to a sign. If we restrict attention to intersections with the section plane that cross with, say, positive p_x, then there is a one-to-one relation between section points and trajectories. A section point thus corresponds to a unique trajectory.

How does this address the issue of the number of conserved quantities? A priori, there appear to be two possibilities: either there are hidden conserved quantities or there are not. Suppose there is no other conserved quantity besides the energy. Then the expectation was that successive intersections of the trajectory with the section plane would eventually explore all of the section plane that is consistent with conservation of energy. On the other hand, if there is a hidden conserved quantity then the successive intersections would be constrained to fall on a curve.

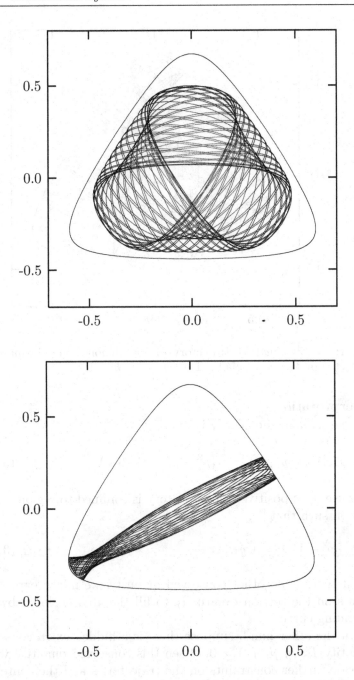

Figure 3.16 Two trajectories of the Hénon–Heiles Hamiltonian projected on the (x, y) plane. The energy is $E = 1/8$.

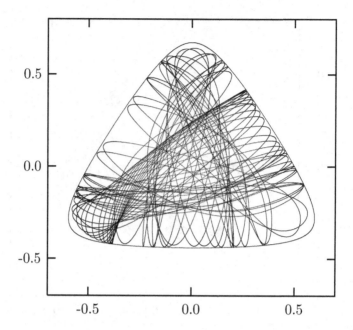

Figure 3.17   Another trajectory of the Hénon–Heiles Hamiltonian projected on the (x, y) plane. The energy is $E = 1/8$.

Interpretation

On the section, the energy is

$$E = H(t; 0, y; p_x, p_y) = \frac{1}{2} \left(p_x^2 + p_y^2 \right) + V(0, y). \qquad (3.139)$$

Because p_x^2 is positive, the trajectory is confined to regions of the section such that

$$E \geq \frac{1}{2} p_y^2 + V(x = 0, y). \qquad (3.140)$$

So, if there is no other conserved quantity, we might expect the points on the section eventually to fill the area enclosed by this bounding curve.

On the other hand, suppose there is a hidden extra conserved quantity $I(x, y; p_x, p_y) = 0$. Then this conserved quantity would provide further constraints on the trajectories and their intersections with the section plane. An extra conserved quantity I pro-

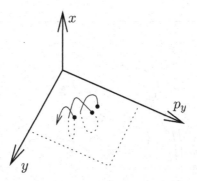

Figure 3.18   The surface of section for the Hénon–Heiles problem is
generated by recording and plotting the successive crossings of the $x = 0$
plane in the direction of increasing x.

vides a constraint among the four phase-space variables x, y, p_x,
and p_y. We can use E to solve for p_x, so for a given E, I gives
a relation among x, y, and p_y. Using the fact that on the section
$x = 0$, the I gives a relation between y and p_y on the section for a
given E. So we expect that if there is another conserved quantity
the successive intersections of a trajectory with the section plane
will fall on a curve.

  If there is no extra conserved quantity we expect the section
points to fill an area; if there is an extra conserved quantity we expect the section points to be restricted to a curve. What actually
happens? Figure 3.19 shows a surface of section for $E = 1/12$;
the section points for several representative trajectories are displayed. By and large, the points appear to be restricted to curves,
so there appears to be evidence for an extra conserved quantity.
Look closely though. Where the "curves" cross, the lines are a
little fuzzy. Hmmm.

  Let's try a little larger energy, $E = 1/8$. The appearance of the
section changes qualitatively (figure 3.20). For some trajectories
there still appear to be extra constraints on the motion. But other
trajectories appear to fill an area of the section plane, pretty much
as we expected of trajectories if there was no extra conserved
quantity. In particular, all of the scattered points on this section
were generated by a single trajectory. Thus, some trajectories
behave as if there is an extra conserved quantity, and others don't.
Wow!

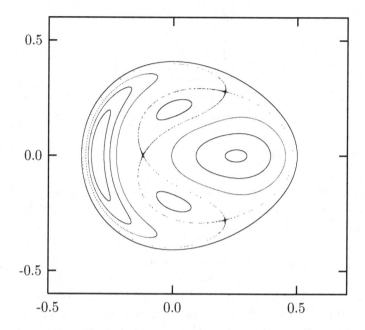

Figure 3.19   Surface of section for the Hénon–Heiles problem with energy $E = 1/12$.

Let's go on to a higher energy, $E = 1/6$, just at the escape energy. A section for this energy is shown in figure 3.21. Now, a single trajectory explores most of the region of the section plane allowed by energy conservation, but not entirely. There are still trajectories that appear to be subject to extra constraints.

We seem to have all possible worlds. At low energy, the system by and large behaves as if there is an extra conserved quantity, but not entirely. At intermediate energy, the phase space is divided: some trajectories explore areas whereas others are constrained. At high energy, trajectories explore most of the energy surface; few trajectories show extra constraints. We have just witnessed our first transition to chaos.

Two qualitatively different types of motion are revealed by this surface of section, just as we saw in the Poincaré sections for the driven pendulum. There are trajectories that seem to be constrained as if by an extra conserved quantity. And there are trajectories that explore an area on the section as though there were no extra conserved quantitiess. Regular trajectories appear to be constrained by an extra conserved quantity to a one-dimensional

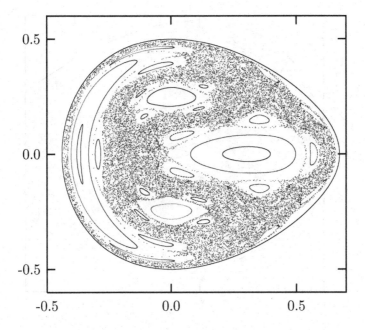

Figure 3.20 Surface of section for the Hénon–Heiles problem with energy $E = 1/8$.

set on the section; chaotic trajectories are not constrained in this way and explore an area.[30]

The surface of section not only reveals the existence of qualitatively different types of motion, but also provides an overview of the different types of trajectories. Take the surface of section for $E = 1/8$ (figure 3.20). There are four main islands, engulfed in a chaotic sea. The particular trajectories displayed above provide examples from different parts of the section. The trajectory that loops around the bowl (figure 3.16) belongs to the large island on the left side of the section. Similar trajectories that loop around the bowl in the other direction belong to the large island on the right side of the section. The trajectories that oscillate back and

---

[30] As before, upon close examination we may find that trajectories that appear to be confined to a curve on the section are chaotic trajectories that explore a highly confined region. It is known, however, that some trajectories really are confined to curves on the section. Trajectories that start on these curves remain on these curves forever, and they fill these curves densely. These invariant curves are preserved by the dynamical evolution. There are also invariant subsets of curves with an infinite number of holes.

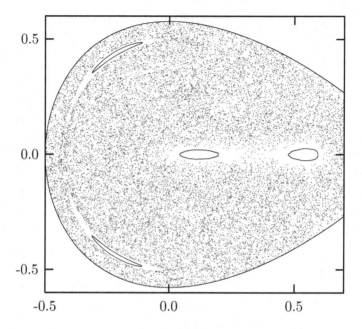

Figure 3.21 Surface of section for the Hénon–Heiles problem with energy $E = 1/6$. The section is clipped on the right.

forth across the bowl belong to the two islands above and below the center of the section. (By symmetry there should be three such islands. The third island is snugly wrapped against the boundary of the section.) Each of the main islands is surrounded by a chain of secondary islands. We will see that the types of orbits are inexhaustible, if we look closely enough. The chaotic trajectory (figure 3.17) lives in the chaotic sea. Thus the section provides a summary of the types of motion possible and how they are related to one another. It is much more useful than plots of a zillion trajectories.

The section for a particular energy summarizes the dynamics at that energy. A sequence of sections for various energies shows how the major features change with the energy. We have already noticed that at low energy the section is dominated by regular orbits, at intermediate energy the section is divided more or less equally into regular and chaotic regions, and at high energies the section is dominated by a single chaotic zone. We will see that such transitions from regular to chaotic behavior are quite common; similar phenomena occur in widely different systems, though the details depend on the system under study.

3.6.4   Computing Hénon–Heiles Surfaces of Section

The following procedures implement the Poincaré map for the Hénon–Heiles system:

```
(define (HHmap E dt sec-eps int-eps)
  (define ((make-advance advancer eps) s dt)
    (advancer s dt eps))
  (let ((adv
         (make-advance (state-advancer HHsysder) int-eps)))
    (lambda (y py cont fail)
      (let ((initial-state (section->state E y py)))
        (if (not initial-state)
            (fail)
            (find-next-crossing initial-state adv dt sec-eps
              (lambda (crossing-state running-state)
                (cont (ref (coordinate crossing-state)
                           1)
                      (ref (momentum crossing-state)
                           1)))))))))
```

Besides supplying the energy E of the section we must also supply a time step for the integrator to achieve, a tolerance for a point to be on the section sec-eps, and a local truncation error specification for the integrator int-eps.

For each initial point (y, p_y) on the surface of section, the map first finds the initial state that has the specified energy, if one exists. The procedure section->state handles this task:

```
(define (section->state E y py)
  (let ((d (- E (+ (HHpotential (up 0 (up 0 y)))
                   (* 1/2 (square py))))))
    (if (>= d 0.0)
        (let ((px (sqrt (* 2 d))))
          (up 0 (up 0 y) (down px py)))
        #f)))
```

The procedure section->state returns #f (false) if there is no state consistent with the specified energy.

The Hamiltonian procedure for the Hénon–Heiles problem is

```
(define (HHHam s)
  (+ (* 1/2 (square (momentum s)))
     (HHpotential s)))
```

with the potential energy

```
(define (HHpotential s)
  (let ((x (ref (coordinate s) 0))
        (y (ref (coordinate s) 1)))
    (+ (* 1/2 (+ (square x) (square y)))
       (- (* (square x) y) (* 1/3 (cube y))))))
```

The system derivative is computed directly from the Hamiltonian.

```
(define (HHsysder)
  (Hamiltonian->state-derivative HHHam))
```

The procedure `find-next-crossing` advances the initial state until successive states are on opposite sides of the section plane.

```
(define (find-next-crossing state advance dt sec-eps cont)
  (let lp ((s state))
    (let ((next-state (advance s dt)))
      (if (and (> (ref (coordinate next-state) 0) 0)
               (< (ref (coordinate s) 0) 0))
          (let ((crossing-state
                  (refine-crossing sec-eps advance s)))
            (cont crossing-state next-state))
          (lp next-state)))))
```

After finding states that straddle the section plane the crossing is refined by Newton's method, as implemented by the procedure `refine-crossing`. The procedure `find-next-crossing` returns both the crossing point and the next state produced by the integrator. The next state is not used in this problem but it is needed for other cases.

```
(define (refine-crossing sec-eps advance state)
  (let lp ((state state))
    (let ((x (ref (coordinate state) 0))
          (xd (ref (momentum state) 0)))
      (let ((zstate (advance state (- (/ x xd)))))
        (if (< (abs (ref (coordinate zstate) 0))
               sec-eps)
            zstate
            (lp zstate))))))
```

To explore the Hénon–Heiles map we use `explore-map` as before. The following exploration generated figure 3.20:

```
(define win (frame -0.5 0.7 -0.6 0.6))
(explore-map win (HHmap 0.125 0.1 1.0e-10 1.0e-12) 500)
```

3.6.5 Non-Axisymmetric Top

We have seen that the motion of an axisymmetric top can be essentially solved. A plot of the rate of change of the tilt angle versus the tilt angle is a simple closed curve. The evolution of the other angles describing the configuration can be obtained by quadrature once the tilting motion has been solved. Now let's consider a non-axisymmetric top. A non-axisymmetric top is a top with three unequal moments of inertia. The pivot is not at the center of mass, so uniform gravity exerts a torque. We assume the line between the pivot and the center of mass is one of the principal axes, which we take to be $\hat{c}$. There are no torques about the vertical axis, so the vertical component of the angular momentum is conserved. If we write the Hamiltonian in terms of the Euler angles, the angle φ, which corresponds to rotation about the vertical, does not appear. Thus the momentum conjugate to this angle is conserved. The nontrivial degrees of freedom are θ and ψ, with their conjugate momenta.

We can make a surface of section (see figure 3.22) for this problem by displaying p_θ versus θ when $\psi = 0$. There are in general two values of p_ψ possible for given values of energy and p_φ. We plot points only if the value of p_ψ at the crossing is the larger of the two possibilities. This makes the points of the section correspond uniquely to a trajectory.

In this section there is a large quasiperiodic island surrounding a fixed point that corresponds to the tilted equilibrium point of the awake axisymmetric top (see figure 3.7 in section 3.4). Surrounding this is a large chaotic zone that extends from $\theta = 0$ to angles near π. If this top is placed initially near the vertical, it exhibits chaotic motion that carries it to large tilt angles. If the top is started within the quasiperiodic island, the tilt is stable.

3.7 Exponential Divergence

Hénon and Heiles discovered that the chaotic trajectories had remarkable sensitivity to small changes in initial conditions—initially nearby chaotic trajectories separate roughly exponentially with time. On the other hand, regular trajectories do not exhibit this sensitivity—initially nearby regular trajectories separate roughly linearly with time.

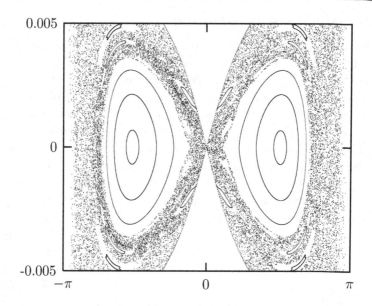

Figure 3.22   A surface of section for the non-axisymmetric top. The parameters are $A = 0.0003\,\mathrm{kg\,m^2}$, $B = 0.00025\,\mathrm{kg\,m^2}$, $C = 0.0001\,\mathrm{kg\,m^2}$, $gMR = 0.0456\,\mathrm{kg\,m^2\,s^{-2}}$. The energy and p_φ are those of the top initially standing vertically with rotation frequency $30\,\mathrm{rad\,s^{-1}}$. The angle θ is on the abscissa, and the momentum p_θ is on the ordinate.

Consider the evolution of two initially nearby trajectories for the Hénon–Heiles problem, with energy $E = 1/8$. Let $d(t)$ be the usual Euclidean distance in the x, y, p_x, p_y space between the two trajectories at time t. Figure 3.23 shows the common logarithm of $d(t)/d(0)$ as a function of time t. We see that the divergence is well described as exponential.

On the other hand, the distance between two initially nearby regular trajectories grows much more slowly. Figure 3.24 shows the distance between two regular trajectories as a function of time. The distance grows linearly with time.

It is remarkable that Hamiltonian systems have such radically different types of trajectories. On the surface of section the chaotic and regular trajectories differ in the dimension of the space that they explore. It is interesting that along with this dimensional difference there is a drastic difference in the way chaotic and regular trajectories separate. For higher-dimensional systems the surface of section technique is not as useful, but trajectories are still distinguished by the way neighboring trajectories diverge: some diverge

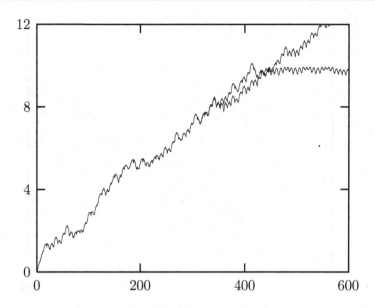

Figure 3.23   The common logarithm of the phase-space distance between two chaotic trajectories divided by the initial phase-space distance as a function of time. The initial distance was 10^{-10}. The logarithm of the distance grows approximately linearly; the distance grows exponentially. The two-trajectory method saturates when the distance between trajectories becomes comparable to that allowed by conservation of energy. Also displayed is the distance between trajectories calculated by integrating the linearized variational equations. This method does not saturate.

exponentially whereas others diverge approximately linearly. Exponential divergence is the hallmark of chaotic behavior.

The rate of exponential divergence is quantified by the slope of the graph of $\log(d(t)/d(0))$. We can estimate the rate of exponential divergence of trajectories from a particular phase-space trajectory σ by choosing a nearby trajectory σ' and computing

$$\gamma(t) = \frac{\log(d(t)/d(t_0))}{t - t_0}, \tag{3.141}$$

where $d(t) = \|\sigma'(t) - \sigma(t)\|$. A problem with this "two-trajectory" method is illustrated in figure 3.23. For strongly chaotic trajectories two initially nearby trajectories soon find themselves as far apart as they can get. Once this happens the distance no longer grows. The estimate of the rate of divergence of trajectories is limited by this *saturation*.

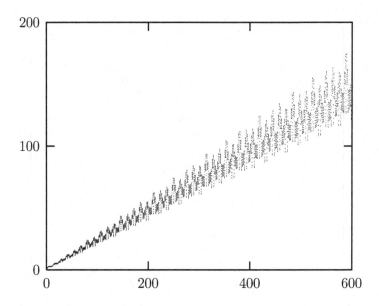

Figure 3.24   The phase-space distance between two regular trajectories divided by the initial phase-space distance as a function of time. The initial distance was 10^{-10}. The distance grows linearly.

We can improve on this method by studying a variational system of equations. Let

$$Dz(t) = F(t, z(t)) \tag{3.142}$$

be the system of equations governing the evolution of the system. A nearby trajectory z' satisfies

$$Dz'(t) = F(t, z'(t)). \tag{3.143}$$

The difference $\zeta = z' - z$ between these trajectories satisfies

$$\begin{aligned} D\zeta(t) &= F(t, z'(t)) - F(t, z(t)) \\ &= F(t, z(t) + \zeta(t)) - F(t, z(t)). \end{aligned} \tag{3.144}$$

If ζ is small we can approximate the right-hand side by a derivative

$$D\zeta(t) = \partial_1 F(t, z(t))\zeta(t). \tag{3.145}$$

This set of ordinary differential equations is called the *variational equations* for the system. It is linear in ζ and driven by z.

Let $d(t) = \|\zeta(t)\|$; then the rate of divergence can be estimated as before. The advantage of this "variational method" is that $\zeta(t)$ can become arbitrarily large and its growth still measures the divergence of nearby trajectories. We can see in figure 3.23 that the variational method gives nearly the same result as the two-trajectory method up to the point at which the two-trajectory method saturates.[31]

The *Lyapunov exponent* is defined to be the infinite time limit of $\gamma(t)$, defined by equation (3.141), in which the distance d is computed by the variational method. Actually, for each trajectory there are many Lyapunov exponents, depending on the initial direction of the variation ζ. For an N-dimensional system, there are N Lyapunov exponents. For a randomly chosen $\zeta(t_0)$, the subsequent growth of $\zeta(t)$ has components that grow with each of the Lyapunov exponents. In general, however, the growth of $\zeta(t)$ will be dominated by the largest exponent. The largest Lyapunov exponent thus can be interpreted as the typical rate of exponential divergence of nearby trajectories. The sum of the largest two Lyapunov exponents can be interpreted as the typical rate of growth of the area of two-dimensional elements. This interpretation can be extended to higher-dimensional elements: the rate of growth of volume elements is the sum of all the Lyapunov exponents.

In Hamiltonian systems, the Lyapunov exponents must satisfy the following constraints. Lyapunov exponents come in pairs; for every Lyapunov exponent λ, its negation $-\lambda$ is also an exponent. For every conserved quantity, one of the Lyapunov exponents is zero, as is its negation. So the Lyapunov exponents can be used to check for the existence of conserved quantities. The sum of the Lyapunov exponents for a Hamiltonian system is zero, so volume elements do not grow exponentially. We will see in the next section that phase-space volume is actually conserved for Hamiltonian systems.

---

[31]In strongly chaotic systems $\zeta(t)$ may become so large that the computer can no longer represent it. To prevent this we can replace ζ by ζ/c whenever $\zeta(t)$ becomes uncomfortably large. The equation governing ζ is linear, so except for the scale change, the evolution is unchanged. Of course we have to keep track of these scale changes when computing the average growth rate. This process is called "renormalization" to make it sound impressive.

3.8   Liouville's Theorem

If an ensemble of states occupies a particular volume of phase space at one moment, then the subsequent evolution of that volume by the flow described by Hamilton's equations may distort the ensemble but does not change the volume the ensemble occupies. The fact that phase-space volume is preserved by the phase flow is called *Liouville's theorem*.

We will first illustrate the preservation of phase-space volume with a simple example and then prove it in general.

The phase flow for the pendulum

Consider an undriven pendulum described by the Hamiltonian

$$H(t, \theta, p_\theta) = \frac{p_\theta^2}{2l^2m} + glm \cos \theta. \tag{3.146}$$

In figure 3.25 we see the evolution of an elliptic region around a point on the θ-axis, in the oscillation region of the pendulum. Three later positions of the region are shown. The region is stretched and sheared by the flow, but the area is preserved. After many cycles, the starting region will be stretched to be a thin layer distributed in the phase angle of the pendulum. Figure 3.26 shows a similar evolution (for smaller time intervals) of a region straddling the separatrix[32] near the unstable equilibrium point. The phase-space region rapidly stretches along the separatrix, while preserving the area. The initial conditions that start in the oscillation region (inside of the separatrix) will continue to spread into a thin ring-shaped region, while the initial conditions that start outside of the separatrix will spread into a thin region of rotation on the outside of the separatrix.

Proof of Liouville's theorem

Consider a set of ordinary differential equations of the form

$$Dz(t) = F(t, z(t)), \tag{3.147}$$

---

[32]The separatrix is the curve that separates the oscillating motion from the circulating motion. It is made up of several trajectories that are asymptotic to the unstable equilibrium.

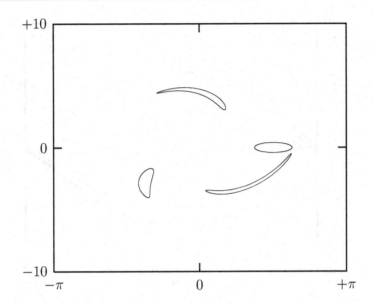

Figure 3.25   A swarm of initial points outlining an area in the phase space of the pendulum deforms as it evolves, but the area contained in the contour remains constant. The horizontal axis is the angle θ of the pendulum from the vertical; the vertical axis is the angular momentum p_θ. The initial contour is the "ellipse" on the horizontal axis. The pendulum has length 1 m in standard gravity ($9.8\,\mathrm{m\,s^{-2}}$), so its period is approximately 2 seconds. The flow proceeds clockwise and the deformed areas are shown at .9 seconds, 1.8 seconds, and 2.7 seconds. The successive positions exhibit "shearing" of the region because the pendulum is not isochronous.

where z is a tuple of N state variables. Let $R(t_1)$ be a region of the state space at time t_1. Each element of this region is an initial condition at time t_1 for the system, and evolves to an element at time t_2 according to the differential equations. The set of these elements at time t_2 is the region $R(t_2)$. Regions evolve to regions.

The evolution of the system for a time interval Δt defines a map $g_{t,\Delta t}$ from the state space to itself:

$$g_{t,\Delta t}(z(t)) = z(t + \Delta t). \tag{3.148}$$

Regions map to regions by mapping each element in the region:

$$g_{t,\Delta t}(R(t)) = R(t + \Delta t). \tag{3.149}$$

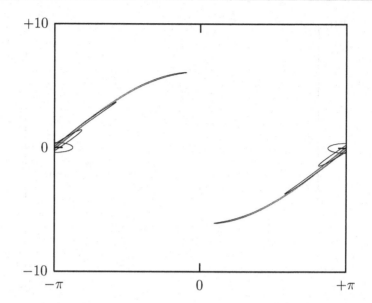

Figure 3.26   The pendulum here is the same as in the previous figure, but now the swarm of initial points surrounds the unstable equilibrium point for the pendulum in phase space, where $\theta = \pi$ and $p_\theta = 0$. The swarm is stretched out along the separatrix. The time interval between successively plotted contours is 0.3 seconds.

The volume $V(t)$ of a region $R(t)$ is $\int_{R(t)} \hat{1}$, where $\hat{1}$ is the function whose value is one for every input. The volume of the evolved region $R(t + \Delta t)$ is

$$
\begin{aligned}
V(t + \Delta t) &= \int_{R(t+\Delta t)} \hat{1} \\
&= \int_{g_{t,\Delta t}(R(t))} \hat{1} \\
&= \int_{R(t)} \mathrm{Jac}(g_{t,\Delta t}),
\end{aligned}
\tag{3.150}
$$

where $\mathrm{Jac}(g_{t,\Delta t})$ is the Jacobian of the mapping $g_{t,\Delta t}$. The Jacobian is the determinant of the derivative of the mapping.

For small Δt

$$
g_{t,\Delta t}(z(t)) = z(t) + \Delta t F(t, z(t)) + o(\Delta t^2),
\tag{3.151}
$$

and thus

$$Dg_{t,\Delta t}(z(t)) = DI(z(t)) + \Delta t \partial_1 F(t, z(t)) + o(\Delta t^2), \qquad (3.152)$$

where I is the identity function, so $DI(z(t))$ is a unit multiplier. We can use the fact that if $\mathbf{A}$ is an $N \times N$ square matrix then

$$\det(1 + \epsilon \mathbf{A}) = 1 + \epsilon \ \text{trace} \ \mathbf{A} + o(\epsilon^2) \qquad (3.153)$$

to show that

$$\text{Jac}(g_{t,\Delta t})(z) = 1 + \Delta t G_t(z) + o(\Delta t^2), \qquad (3.154)$$

where

$$G_t(z) = \text{trace}(\partial_1 F(t, z)). \qquad (3.155)$$

Thus

$$V(t + \Delta t) = \int_{R(t)} [\hat{1} + \Delta t G_t + o(\Delta t^2)]$$

$$= V(t) + \Delta t \int_{R(t)} G_t + o(\Delta t^2). \qquad (3.156)$$

So the rate of change of the volume at time t is

$$DV(t) = \int_{R(t)} G_t. \qquad (3.157)$$

Now we compute G_t for a system described by a Hamiltonian H. The components of z are the components of the coordinates and the momenta: $z^k = q^k$ and $z^{k+n} = p_k$ for $k = 0, \ldots, n-1$. The components of F are

$$F^k(t, z) = (\partial_2 H)^k(t, q, p)$$
$$F^{k+n}(t, z) = -(\partial_1 H)_k(t, q, p), \qquad (3.158)$$

for $k = 0, \ldots, n-1$. The diagonal components of the derivative $\partial_1 F$ are

$$(\partial_1)_k F^k(t, z) = (\partial_1)_k (\partial_2)^k H(t, q, p)$$
$$(\partial_1)_{k+n} F^{k+n}(t, z) = -(\partial_2)^k (\partial_1)_k H(t, q, p). \qquad (3.159)$$

The component partial derivatives commute, so the diagonal components with index k and index $k + n$ are equal and opposite. We see that the trace, which is the sum of these diagonal components, is zero. Thus the integral of G_t over the region $R(t)$ is zero, so the derivative of the volume at time t is zero. Because t is arbitrary, the volume does not change. This proves *Liouville's theorem*: the phase-space flow conserves phase-space volume.

Notice that the proof of Liouville's theorem does not depend upon whether or not the Hamiltonian has explicit time dependence. Liouville's theorem holds for systems with time-dependent Hamiltonians.

We may think of the ensemble of all possible states as a fluid flowing around under the control of the dynamics. Liouville's theorem says that this fluid is incompressible for Hamiltonian systems.

Exercise 3.11: Determinants and traces

Show that equation (3.153) is correct.

Area preservation of stroboscopic surfaces of section

Surfaces of section for periodically driven Hamiltonian systems are area preserving if the section coordinates are the phase-space coordinate and momentum. This is an important feature of surfaces of section. It is a consequence of Liouville's theorem for one-degree-of-freedom problems.

It is also the case that surfaces of section such as those we have used for the Hénon–Heiles problem are area preserving, but we are not ready to prove this yet!

Poincaré recurrence

The *Poincaré recurrence theorem* is a remarkable theorem that is a trivial consequence of Liouville's theorem. Loosely, the theorem states that almost all trajectories eventually return arbitrarily close to where they started. This is true regardless of whether the trajectories are chaotic or regular.

More precisely, consider a Hamiltonian dynamical system for which the phase space is a bounded domain D. We identify some initial point in the phase space, say z_0. Then, for any finite neighborhood U of z_0 we choose, there are trajectories that emanate from initial points in that neighborhood and eventually return to the neighborhood.

We can prove this by considering the successive images of U under the time evolution. For simplicity, we restrict consideration to time evolution for a time interval Δ. The map of the phase space onto itself generated by time evolution for an interval Δ we call C. Subsequent applications of the map generate a discrete time evolution. Sets of points in phase space transform by evolving all the points in the set; the image of the set U is denoted $C(U)$. Now consider the trajectory of the set U, that is, the sets $C^n(U)$ where C^n indicates the n-times composition of C. Now there are two possibilities: either the successive images $C^i(U)$ intersect or they do not. If they do not intersect, then with each iteration, a volume of D equal to the volume of U gets "used up" and cannot belong to the further image. But the volume of D is finite, so we cannot fit an infinite number of non-intersecting finite volumes into it. Therefore, after some number of iterations the images intersect. Suppose $C^i(U)$ intersects with $C^j(U)$, with $j < i$, for definiteness. Then the pre-image of each must also intersect, since the pre-image of a point in the intersection belongs to both sets. Thus $C^{i-1}(U)$ intersects $C^{j-1}(U)$. This can be continued until finally we have that $C^{i-j}(U)$ intersects U. So we have proven that after $i - j$ iterations of the map C there are a set of points initially in U that return to the neighborhood U.

So for every neighborhood of every point in the phase space there is a subneighborhood such that the trajectories emanating from all of the points in that subneighborhood return to that subneighborhood. Thus almost every trajectory returns arbitrarily close to where it started.

The gas in the corner of the room
Suppose we have a collection of N classical atoms in a perfectly sealed room. The phase-space dimension of this system is $6N$. A point in this phase space is denoted z. Suppose that initially all the atoms are, say, within one centimeter of one corner, with arbitrarily chosen finite velocities. This corresponds to some initial point z_0 in the phase space. The phase space of the system is limited in space by the room and in momentum by energy conservation; the phase space is bounded. The recurrence theorem then says that in the neighborhood of z_0 there is an initial condition of the system that returns to the neighborhood of z_0 after some time. For the individual atoms this means that after some time

all of the atoms will be found in the corner of the room again, and again, and again. Makes one wonder about the second law of thermodynamics, doesn't it?[33]

Nonexistence of attractors in Hamiltonian systems

Some systems have attractors. An *attractor* is a region of phase space that gobbles volumes of trajectories. For an attractor there is some larger region, the basin of attraction, such that sets of trajectories with nonzero volume eventually end up in the attractor and never leave it. The recurrence theorem shows that Hamiltonian systems with bounded phase space do not have attractors. Consider some candidate volume in the proposed basin of attraction. The recurrence theorem guarantees that some trajectories in the candidate volume return to the volume repeatedly. Therefore, the volume is not in a basin of attraction. Attractors do not exist in Hamiltonian systems with bounded phase space.

This does not mean that every trajectory always returns. A simple example is the pendulum. Suppose we take a blob of trajectories that spans the separatrix, the trajectory that asymptotically approaches the unstable equilibrium with the pendulum pointed up. Trajectories with more energy than the separatrix make a full loop around and return to their initial point; trajectories with lower energy than the separatrix oscillate once across and back to their initial position; but the separatrix trajectory itself leaves the initial region permanently, and continually approaches the unstable point.

Conservation of phase volume in a dissipative system

The definition of a dissipative system is not so clear. For some, "dissipative" implies that phase-space volume is not conserved, which is the same as saying the evolution of the system is not governed by Hamilton's equations. For others, "dissipative" implies that friction is present, representing loss of energy to unmodeled degrees of freedom. Here is a curious example. The damped harmonic oscillator is the paradigm of a dissipative system. Here we show that the damped harmonic oscillator can be described by Hamilton's equations and that phase-space volume is conserved.

---

[33]It is reported that when Boltzmann was confronted with this problem he responded, "You should wait that long!"

The damped harmonic oscillator is governed by the ordinary differential equation

$$mD^2x + \alpha Dx + kx = 0 \tag{3.160}$$

where α is a coefficient of damping. We can formulate this system with the Lagrangian[34]

$$L(t, x, \dot{x}) = (\frac{m}{2}\dot{x}^2 - \frac{k}{2}x^2)e^{\frac{\alpha}{m}t}. \tag{3.161}$$

The Lagrange equation for this Lagrangian is

$$(mD^2x(t) + \alpha Dx(t) + kx(t))e^{\frac{\alpha}{m}t} = 0. \tag{3.162}$$

Since the exponential is never zero this equation has the same trajectories as equation (3.160) above.

The momentum conjugate to x is

$$p = m\dot{x}e^{\frac{\alpha}{m}t}, \tag{3.163}$$

and the Hamiltonian is

$$H(t, x, p) = (\frac{1}{2m}p^2)e^{-\frac{\alpha}{m}t} + (\frac{k}{2}x^2)e^{\frac{\alpha}{m}t}. \tag{3.164}$$

For this system, the Hamiltonian is not the sum of the kinetic energy of the motion of the mass and the potential energy stored in the spring. The value of the Hamiltonian is not conserved ($\partial_0 H \neq 0$). Hamilton's equations are

$$Dx(t) = \frac{p(t)}{m}e^{-\frac{\alpha}{m}t}$$
$$Dp(t) = -kx(t)e^{\frac{\alpha}{m}t}. \tag{3.165}$$

Let's consider a numerical case. Let $m = 5$, $k = 1/4$, $\alpha = 3$. Here the characteristic roots of the linear constant-coefficient ordinary differential equation (3.160) are $s = -1/10, -1/2$. Thus the solutions are

$$\begin{pmatrix} x(t) \\ p(t) \end{pmatrix} = \begin{pmatrix} e^{-\frac{1}{10}t} & e^{-\frac{1}{2}t} \\ -\frac{1}{2}e^{+\frac{1}{2}t} & -\frac{5}{2}e^{+\frac{1}{10}t} \end{pmatrix} \begin{pmatrix} A_1 \\ A_2 \end{pmatrix}, \tag{3.166}$$

---

[34]This is just the product of the Lagrangian for the undamped harmonic oscillator with an increasing exponential of time.

for A_1 and A_2 determined by the initial conditions

$$\begin{pmatrix} x(0) \\ p(0) \end{pmatrix} = \begin{pmatrix} 1 & 1 \\ -\frac{1}{2} & -\frac{5}{2} \end{pmatrix} \begin{pmatrix} A_1 \\ A_2 \end{pmatrix}. \tag{3.167}$$

Thus we can form the transformation from the initial state to the final state:

$$\begin{pmatrix} x(t) \\ p(t) \end{pmatrix} = \begin{pmatrix} e^{-\frac{1}{10}t} & e^{-\frac{1}{2}t} \\ -\frac{1}{2}e^{+\frac{1}{2}t} & -\frac{5}{2}e^{+\frac{1}{10}t} \end{pmatrix} \begin{pmatrix} 1 & 1 \\ -\frac{1}{2} & -\frac{5}{2} \end{pmatrix}^{-1} \begin{pmatrix} x(0) \\ p(0) \end{pmatrix}. \tag{3.168}$$

The transformation is linear, so the area is transformed by the determinant, which is 1 in this case. Thus, contrary to intuition, the phase-space volume is conserved. So why is this not a contradiction with the statement that there are no attractors in Hamiltonian systems? The answer is that the Poincaré recurrence argument is true only for bounded phase spaces. Here, the momentum expands exponentially with time (as the coordinate contracts), so it is unbounded.

We shouldn't really be too surprised by the way the theory protects itself from an apparent paradox—that the phase volume is conserved even though all trajectories decay to zero velocity and coordinates. The proof of Liouville's theorem allows time-dependent Hamiltonians. In this case we are able to model the dissipation by just such a time-dependent Hamiltonian.

Exercise 3.12: Time-dependent systems

To make the fact that Liouville's theorem holds for time-dependent systems even more concrete, extend the results of section 3.8 to show how a swarm of initial points outlining an area in the phase space of the *driven* pendulum deforms as it evolves. Construct pictures analogous to figures 3.25 and 3.26 for one of the interesting cases where we have surfaces of section. Does the distortion look different in different parts of the phase space? How?

Distribution functions

We know the state of a system only approximately. It is reasonable to model our state of knowledge by a probability density function on the set of possible states. Given such incomplete knowledge, what are the probable consequences? As the system evolves, the density function also evolves. Liouville's theorem gives us a handle on this kind of problem.

Let $f(t, q, p)$ be a probability density function on the phase space at time t. For this to be a good probability density function we require that the integral of f over all coordinates and momenta be 1—it is certain that the system is somewhere.

There is a set of trajectories that pass through any particular region of phase space at a particular time. These trajectories are neither created nor destroyed, and they proceed as a bundle to another region of phase space at a later time. Liouville's theorem tells us that the volume of the source region is the same as the volume of the target region, so the density must remain constant. Thus $D(f \circ \sigma) = 0$. If we have a system described by the Hamiltonian H then

$$D(f \circ \sigma) = \partial_0 f \circ \sigma + \{f, H\} \circ \sigma, \tag{3.169}$$

so we may conclude that

$$\partial_0 f \circ \sigma + \{f, H\} \circ \sigma = 0, \tag{3.170}$$

or

$$(\partial_0 f + \{f, H\}) \circ \sigma = 0. \tag{3.171}$$

Since this must be true at each moment and since there is a solution trajectory that emanates from every point in phase space, we may abstract from solution paths and deduce a constraint on f:

$$\partial_0 f + \{f, H\} = 0. \tag{3.172}$$

This linear partial differential equation governs the evolution of the density function, and thus shows how our state of knowledge evolves.

3.9   Standard Map

We have seen that the surfaces of section for a number of different problems are qualitatively very similar. They all show two qualitatively different types of motion: regular motion and chaotic motion. They show that these types of orbits are clustered: there are regions of the surface of section that have mostly regular trajectories and other regions dominated by chaotic behavior. We have also seen a transition to large-scale chaotic behavior as some parameter is varied. Now we have learned that the map that takes points on a two-dimensional surface of section to new points on the surface of section is area preserving. The sole property that these maps of the section onto itself have in common (that we know of at this point) is that they preserve area. Otherwise they are quite distinct. Suppose we consider an abstract map of the section onto itself that is area preserving, without regard for whether the map is generated by some dynamical system. Do area-preserving maps typically show similar phenomena, or is the dynamical origin of the map crucial to the phenomena we have found?[35]

Consider a map of the phase plane onto itself defined in terms of the dynamical variables θ and its "conjugate momentum" I. The map is

$$I' = (I + K \sin \theta) \bmod 2\pi \tag{3.173}$$
$$\theta' = (\theta + I') \bmod 2\pi. \tag{3.174}$$

This map is known as the "standard map."[36] A curious feature of the standard map is that the momentum variable I is treated as an angular quantity. The derivative of the map has determinant one, implying the map is area preserving.

We can implement the standard map:

```
(define ((standard-map K) theta I return failure)
  (let ((nI (+ I (* K (sin theta)))))
    (return ((principal-value :2pi) (+ theta nI))
            ((principal-value :2pi) nI))))
```

We use the `explore-map` procedure introduced earlier to use a pointing device to interactively explore the surface of section. For example, to explore the surface of section for parameter $K = 0.6$ we use:

---

[35] This question was also addressed in the remarkable paper by Hénon and Heiles, but with a different map from what we use here.

[36] The standard map has been extensively studied. Early investigations were by Chirikov [12] and by Taylor [44], so the map is sometimes called the Chirikov–Taylor map. Chirikov coined the term "standard map," which we adopt.

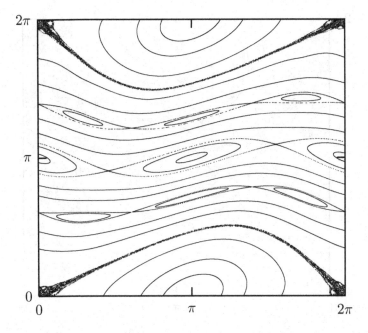

Figure 3.27 Surface of section for the standard map for $K = 0.6$. The section shows mostly regular trajectories, with a few dominant islands, but also a number of small chaotic zones.

```
(define window (frame 0.0 :2pi 0.0 :2pi))
(explore-map window (standard-map 0.6) 2000)
```

The resulting surface of section, for a variety of orbits chosen with the pointer, is shown in figure 3.27. The surface of section does indeed look qualitatively similar to the surfaces of section generated by dynamical systems.

The surface of section for $K = 1.4$ (as shown in figure 3.28) is dominated by a large chaotic zone. The standard map exhibits a transition to large-scale chaos near $K = 1$. So this abstract area-preserving map of the phase plane onto itself shows behavior that is similar to behavior in the sections generated by a Hamiltonian dynamical system. Evidently, the area-preservation property of the dynamics in the phase space plays a determining role for many interesting properties of trajectories of mechanical systems.

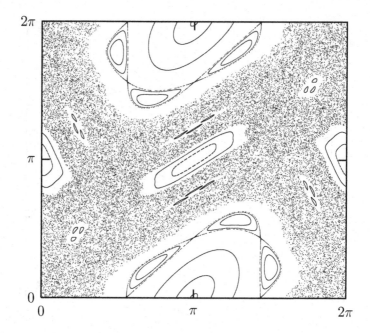

Figure 3.28   Surface of section for the standard map for $K = 1.4$. The dominant feature is a large chaotic zone. There are also some large islands of regular behavior. In this case there are also some interesting secondary islands—islands around islands.

Exercise 3.13: Fun with Hénon's quadratic map

Consider the map of the plane defined by the equations:

$$x' = x \cos \alpha - (y - x^2) \sin \alpha$$

$$y' = x \sin \alpha + (y - x^2) \cos \alpha$$

a. Show that the map preserves area.

b. Implement the map as a procedure. The interesting range of x and y is $(-1, 1)$. There will be orbits that escape. You should check for values of x and y that escape from this range and call the failure continuation when this occurs.

c. Explore the phase portrait of this map for a few values of the parameter α. The map is particularly interesting for $\alpha = 1.32$ and $\alpha = 1.2$. What happens in between?

3.10 Summary

Lagrange's equations are a system of n second-order ordinary differential equations in the time, the generalized coordinates, the generalized velocities, and the generalized accelerations. Trajectories are determined by the coordinates and the velocities at a moment.

Hamilton's equations specify the dynamics as a system of first-order ordinary differential equations in the time, the generalized coordinates, and the conjugate momenta. Phase-space trajectories are determined by an initial point in phase space at a moment.

The Hamiltonian formulation and the Lagrangian formulation are equivalent in that equivalent initial conditions produce the same configuration path.

If there is a symmetry of the problem that is naturally expressed as a cyclic coordinate, then the conjugate momentum is conserved. In the Hamiltonian formulation, such a symmetry naturally results in the reduction of the dimension of the phase space of the difficult part of the problem. If there are enough symmetries, then the problem of determining the time evolution may be reduced to evaluation of definite integrals (reduced to quadratures).

Systems without enough symmetries to be reducible to quadratures may be effectively studied with the surface of section technique. This is particularly advantageous in systems for which the reduced problem has two degrees of freedom or has one degree of freedom with explicit periodic time dependence.

Surfaces of section reveal tremendous structure in the phase space. There are chaotic zones and islands of regular behavior. There are interesting transitions as parameters are varied between mostly regular motion and mostly chaotic motion.

Chaotic trajectories exhibit sensitive dependence on initial conditions, separating exponentially from nearby trajectories. Regular trajectories do not show such sensitivity. Curiously, chaotic trajectories are distinguished both by the dimension of the space they explore and by their exponential divergence.

The time evolution of a $2n$-dimensional region in phase space preserves the volume. Hamiltonian flow is "incompressible" flow of the "phase fluid."

Surfaces of section for two-degree-of-freedom systems and for periodically driven one-degree-of-freedom systems are area preserving. Abstract area-preserving maps of a phase plane onto itself show the same division of the phase space into chaotic and regular regions as surfaces of section generated by dynamical systems. They also show transitions to large-scale chaos.

3.11   Projects

Exercise 3.14: Periodically driven pendulum

Explore the dynamics of the driven pendulum, using the surface of section method. We are interested in exploring the regions of parameter space over which various phenomena occur. Consider a pendulum of length 9.8 m, mass 1 kg, and acceleration of gravity $g = 9.8 \, \mathrm{m\,s^{-2}}$, giving $\omega_0 = 1 \, \mathrm{rad\,s^{-1}}$. Explore the parameter plane of the amplitude A and frequency ω of the periodic drive.

Examples of the phenomena to be investigated:

a. Inverted equilibrium. Show the region of parameter space (A, ω) in which the inverted equilibrium is stable. If the inverted equilibrium is stable there is some range of stability, i.e., there is a maximum angle of displacement from the equilibrium that stable oscillations reach. If you have enough time, plot contours in the parameter space for different amplitudes of the stable region.

b. Period doubling of the normal equilibrium. For this case, plot the angular momenta of the stable and unstable equilibria as functions of the frequency for some given amplitude.

c. Transition to large-scale chaos. Show the region of parameter space (A, ω) for which the chaotic zones around the three principal resonance islands are linked.

Exercise 3.15: Spin-orbit surfaces of section

Write a program to compute surfaces of section for the spin-orbit problem, with the section points being recorded at pericenter. Investigate the following:

a. Give a Hamiltonian formulation of the spin-orbit problem introduced in section 2.11.2.

b. For out-of-roundness parameter $\epsilon = 0.1$ and eccentricity $e = 0.1$, measure the widths, in momentum of the regular islands associated with the 1:1, 3:2, and 1:2 resonances.

c. Explore the surfaces of section for a range of ϵ for fixed $e = 0.1$. Estimate the critical value of ϵ above which the main chaotic zones around the 3:2 and the 1:1 resonance islands are merged.

d. For a fixed eccentricity $e = 0.1$ trace the location on the surface of section of the stable and unstable fixed points associated with the 1:1 resonance as a function of the out-of-roundness ϵ.

Exercise 3.16: Restricted three-body problem

Investigate the dynamics of the restricted three-body problem for the equal mass case where $M_0 = M_1$.

a. Derive the Hamiltonian for the restricted three-body problem, starting with Lagrangian (1.150).

b. The Jacobi constant, equation (1.151), is the sum of a positive definite quadratic term in the velocities and a potential energy term, equation (1.152), so the boundaries of the allowed motion are contours of the potential energy function. Write a program to display these boundaries for a given value of the Jacobi constant. Where is motion allowed relative to these contours? (Note that for some values of the Jacobi constant there is more than one allowed region of motion.)

c. Evolve some trajectories for a Jacobi constant of $\mathcal{E} = -1.75$ ($C_J = 3.5$). Display the trajectories on the same plot as the boundaries of allowed motion.

d. Write a program to compute surfaces of section for the restricted three-body problem. This program is similar to the Hénon-Heiles program starting on page 261. Plot section points when the trajectory crosses the $y_r = 0$ axis with $\dot{y}_r$ positive; plot $\dot{x}_r$ versus x_r. Note that $p_x = m\dot{x}_r - m\Omega y_r$, but on this section $y_r = 0$, so the velocity is proportional to the momentum, and thus the section is area preserving. Plot the boundaries of the allowed motion on the surface of section for the Jacobi constant suggested above. Explore the section and plot typical orbits for each major region in the section.

4

Phase Space Structure

> When we try to represent the figure formed by
> these two curves and their intersections in a finite
> number, each of which corresponds to a doubly
> asymptotic solution, these intersections form a
> type of trellis, tissue, or grid with infinitely
> serrated mesh. Neither of these two curves must
> ever cut across itself again, but it must bend back
> upon itself in a very complex manner in order to
> cut across all of the meshes in the grid an infinite
> number of times. The complexity of this figure will
> be striking, and I shall not even try to draw it.
>
> Henri Poincaré, *New Methods of Celestial
> Mechanics, volume III* [35], chapter XXXIII,
> section 397

We have seen rather complicated features appear as part of the
Poincaré sections of a variety of systems. We have seen fixed
points, invariant curves, resonance islands, and chaotic zones in
systems as diverse as the driven pendulum, the non-axisymmetric
top, the Hénon–Heiles system, and the spin-orbit coupling of a
satellite. Indeed, even in the standard map, where there is no
continuous process sampled by the surface of section, the phase
space shows similar features.

The motion of other systems is simpler. For some systems conserved quantities can be used to reduce the solution to the evaluation of definite integrals. Such a system is traditionally called
integrable. An example is the axisymmetric top. Two symmetries
imply the existence of two conserved momenta, and time independence of the Hamiltonian implies energy conservation. With
these conserved quantities, determining the motion is reduced to
the evaluation of definite integrals of the periodic motion of the
tilt angle as a function of time. Such systems do not exhibit
chaotic behavior; on a surface of section the conserved quantities
constrain the points to fall on curves. If points on a surface of
section do not apparently fall on curves then we may take this as

evidence that not enough conserved quantities exist to reduce the
solution to quadratures.

We have seen a number of instances in which the behavior of
a system changes qualitatively with the inclusion of additional
effects. The free rigid body can be reduced to quadratures, but
the addition of gravity-gradient torques in the spin-orbit system
yields the familiar mixture of regular and chaotic motions. The
motion of an axisymmetric top is also reducible to quadratures,
but if the top is made non-axisymmetric then the divided phase
space appears. The system studied by Hénon and Heiles, with
the classic divided phase space, can be thought of as a solvable
pair of harmonic oscillators with nonlinear coupling terms. The
pendulum is solvable, but the driven pendulum has the divided
phase space.

We observe that as additional effects are turned on, qualitative changes occur in the phase space. Resonance islands appear,
chaotic zones appear, some invariant curves disappear, but others persist. Why do resonance islands appear? How does chaotic
behavior arise? When do invariant curves persist? Can we draw
any general conclusions?

4.1   Emergence of the Divided Phase Space

We can get some insight into these qualitative changes of behavior
by considering systems in which the additional effects are turned
on by varying a parameter. For some value of the parameter
the system has enough conserved quantities to be reducible to
quadratures; as we vary the parameter away from this value we can
study how the divided phase space appears. The driven pendulum
offers an archetypal example of such a system. If the amplitude
of the drive is zero, then solutions of the driven pendulum are the
same as the solutions of the undriven pendulum. We have seen
surfaces of section for the strongly driven pendulum, illustrating
the divided phase space. Here we crank up the drive slowly and
study how the phase portrait changes.

The motion of the driven pendulum with zero-amplitude drive
is the same as that of an undriven pendulum, as described in section 3.3. Energy is conserved, so all orbits are level curves of the
Hamiltonian in the phase plane (see figure 4.1). There are three
regions of the phase plane that have qualitatively different types

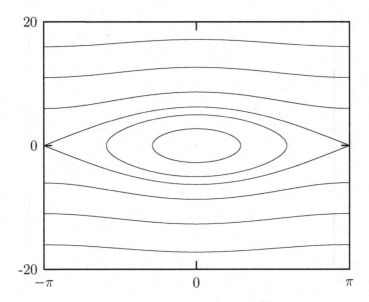

Figure 4.1 The phase plane of the undriven pendulum has three regions displaying two distinct kinds of behavior. Trajectories lie on the contours of the Hamiltonian. Trajectories may oscillate, making ovoid curves around the equilibrium point, or they may circulate, producing wavy tracks outside the eye-shaped region. The eye-shaped region is delimited by the separatrix. This pendulum has length 1 m and a bob of mass 1 kg, and the acceleration of gravity is 9.8 m s⁻².

of motion: the region in which the pendulum oscillates, the region in which the pendulum circulates in one direction, and the region of circulation in the other direction. In the center of the oscillation region there is a stable equilibrium, at which the pendulum is hanging motionless. At the boundaries between these regions the pendulum is asymptotic to the unstable equilibrium, at which the pendulum is standing upright. There are two asymptotic trajectories, corresponding to the two ways the equilibrium can be approached. Each of these is also asymptotic to the unstable equilibrium going backward in time.

Driven pendulum sections with zero-amplitude drive
Now consider the periodically driven pendulum, but with zero-amplitude drive. The state of the driven pendulum is specified by an angle coordinate, its conjugate momentum, and the phase of the periodic drive. With zero-amplitude drive the evolution of the "driven" pendulum is the same as the undriven pendulum. The

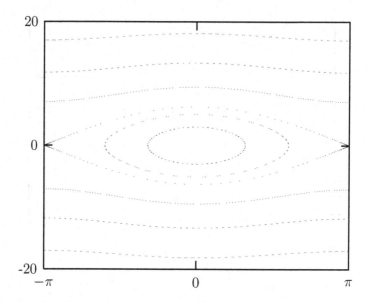

Figure 4.2   A surface of section for the driven pendulum, with zero-amplitude drive. The effect is to sample the trajectories of the undriven pendulum, which lie on the contours of the Hamiltonian. Only a small number of points are plotted for each trajectory to illustrate the fact that for zero-amplitude drive the surface of section samples the continuous trajectories of the undriven pendulum.

phase of the drive does not affect the evolution, but we consider the phase of the drive as part of the state so we can give a uniform description that allows us to include the zero-amplitude drive case with the nonzero-amplitude case.

For the driven pendulum we make stroboscopic surfaces of section by sampling the state at the drive period and plotting the angular momentum versus the angle (see figure 4.2). For zero-amplitude drive, the section points are confined to the curves traced by trajectories of the undriven pendulum. For each kind of orbit that we saw for the undriven pendulum, there are orbits of the driven pendulum that generate a corresponding pattern of points on the section.

The two stationary orbits at the equilibrium points of the pendulum appear as points on the surface of section. Equilibrium points are fixed points of the Poincaré map.

Section points for the oscillating orbits of the pendulum fall on the corresponding contour of the Hamiltonian. Section points for the circulating orbits of the pendulum are likewise confined to the corresponding contour of the Hamiltonian. We notice that the pattern of the points generated by orbits varies from contour to contour. Typically, if we collected more points on the surface of section the points would eventually fill in the contours. However, there are actually two possibilities. Remember that the period of the pendulum is different for different trajectories. If the period of the pendulum is commensurate with the period of the drive, then only a finite number of points will appear on the section. Two periods are commensurate if one is a rational multiple of the other. If the two periods are incommensurate then the section points never repeat. In fact, the points fill the contour densely, coming arbitrarily close to every point on the contour.

Section points for the asymptotic trajectories of the pendulum fall on the contour of the Hamiltonian containing the saddle point. Each asymptotic orbit generates a sequence of isolated points that accumulate near the fixed point. No individual orbit fills the separatrix on the section.

Driven pendulum sections for small drive

Now consider the surface of section for small-amplitude drive (see figure 4.3). The amplitude of the drive is $A = 0.001$ m; the drive frequency is $4.2\,\omega_0$, where $\omega_0 = \sqrt{g/l}$. The overall appearance of the surface of section is similar to the section with zero-amplitude drive. Many orbits appear to lie on invariant curves similar to the invariant curves of the zero-drive case. However, there are several new features.

There are now resonance regions that correspond to the pendulum rotating in lock with the drive. These features are found in the upper and lower circulating region of the surface of section. Each island has a fixed point for which the pendulum rotates exactly once per cycle of the drive. In general, fixed points on the surface of section correspond to periodic motions of the system in the full phase space. The fixed point is at $\pm\pi$, indicating that the pendulum is vertical at the section phase of the drive. For orbits in the resonance region away from the fixed point the points on the section apparently generate curves that surround the fixed point.[1] For these orbits the pendulum rotates on average once per

---

[1] Keep in mind that the abscissa is an angle.

drive, but the phase of the pendulum is sometimes ahead of the drive and sometimes behind it.

There are other islands that appear with nonzero-amplitude drive. In the central oscillation region there is a sixfold chain of secondary islands. For this orbit the pendulum is oscillating, and the period of the oscillation is commensurate with the drive. The six islands are all generated by a single orbit. In fact, the islands are visited successively in a clockwise direction. After six cycles of the drive the section point returns to the same island but falls at a different point on the island curve, accumulating the island curve after many iterations. The motion of the pendulum is not periodic, but is locked in a resonance so that on average it oscillates once for every six cycles of the drive.

Another feature that appears is a narrow chaotic region near where the separatrix was in the zero-amplitude drive pendulum. We find that chaotic behavior typically makes its most prominent appearance near separatrices. This is not surprising because the difference in velocities that distinguish whether the pendulum rotates or oscillates is small for orbits near the separatrix. As the pendulum approaches the top, whether it receives the extra nudge it needs to go over the top depends on the phase of the drive.

Actually, the apparent separatrices of the resonance islands for which the pendulum period is equal to the drive period are each generated by a chaotic orbit. To see that this orbit appears to occupy an area one would have to magnify the picture by about a factor of 10^4.

As the drive amplitude is increased the main qualitative changes are the appearance of resonance islands and chaotic zones. Some qualitative characteristics of the zero-amplitude case remain. For instance, many orbits appear to lie on invariant curves. This behavior is not peculiar to the driven pendulum; similar features quite generally arise as additional effects are added to problems that are reducible to quadratures. This chapter is devoted to understanding in greater detail how these generic features arise.

4.2   Linear Stability

Qualitative changes are associated with fixed points of the surface of section. As the drive is turned on, chaotic zones appear at fixed points on separatrices of the undriven system, and we observe the

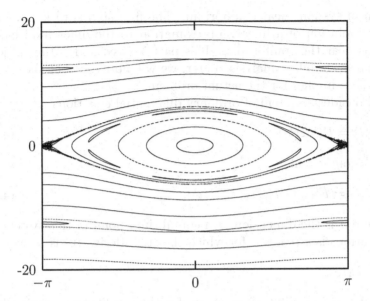

Figure 4.3 A surface of section for the driven pendulum, with nonzero drive amplitude $A = 0.001$ m and drive frequency $4.2\,\omega_0$. Many trajectories apparently generate invariant curves, as in the zero-amplitude drive case. Here, in addition, some orbits belong to island chains and others are chaotic. The most apparent chaotic orbit is near the separatrix of the undriven pendulum.

appearance of new fixed points and periodic points associated with resonance islands. Here we investigate the behavior of systems near fixed points. We can distinguish two types of fixed points on a surface of section: there are fixed points that correspond to equilibria of the system and there are fixed points that correspond to periodic orbits of the system. We first consider the stability of equilibria of systems governed by differential equations, then discuss the stability of fixed points of maps.

4.2.1 Equilibria of Differential Equations

Consider first the case of an equilibrium of a system of differential equations. If a system is initially at an equilibrium point, the system remains there. What can we say about the evolution of the system for points near such an equilibrium point? This is actually a very difficult question, which has not been completely answered. We can, however, understand quite a lot about the motion of systems near equilibrium. The first step is to investigate the evolution of a linear approximation to the differential equations near the equilibrium. This part is easy, and is the subject of linear stability analysis. Later, we will address what the linear analysis implies for the actual problem.

Consider a system of ordinary differential equations

$$Dz(t) = F(t, z(t)) \tag{4.1}$$

with components

$$Dz^i(t) = F^i(t, z^0(t), \ldots, z^{n-1}(t)), \tag{4.2}$$

where n is the dimension. An equilibrium point of this system of equations is a point z_e for which the state derivative is zero:

$$0 = F(t, z_e). \tag{4.3}$$

That this is zero at all moments for the equilibrium solution implies $\partial_0 F(t, z_e) = 0$.

Next consider a path z' that passes near the equilibrium point. The path displacement ζ is defined so that at time t

$$z'(t) = z_e + \zeta(t). \tag{4.4}$$

We have

$$D\zeta(t) = Dz'(t) = F(t, z_e + \zeta(t)). \tag{4.5}$$

If ζ is small we can write the right-hand side as a Taylor series in ζ:

$$D\zeta(t) = F(t, z_e) + \partial_1 F(t, z_e)\zeta(t) + \cdots, \tag{4.6}$$

but the first term is zero because z_e is an equilibrium point, so

$$D\zeta(t) = \partial_1 F(t, z_e)\zeta(t) + \cdots. \tag{4.7}$$

If ζ is small the evolution is approximated by the linear terms. Linear stability analysis investigates the evolution of the approximate equation

$$D\zeta(t) = \partial_1 F(t, z_e)\zeta(t). \tag{4.8}$$

These are the variational equations (3.145) with the equilibrium solution substituted for the reference trajectory. The relationship of the solutions of this linearized system to the full system is a difficult mathematical problem, which has not been fully resolved.

If we restrict attention to autonomous systems ($\partial_0 F = 0$), then the variational equations at an equilibrium are a linear system of ordinary differential equations with constant coefficients.[2] Such systems can be solved analytically. To simplify the notation, let $M = \partial_1 F(t, z_e)$, so

$$D\zeta(t) = M\zeta(t). \tag{4.9}$$

We seek a solution of the form

$$\zeta(t) = \alpha e^{\lambda t}, \tag{4.10}$$

where α is a structured constant with the same number of components as ζ, and λ is a complex number called a *characteristic exponent*. Substituting, we find

$$\lambda \alpha e^{\lambda t} = M \alpha e^{\lambda t}. \tag{4.11}$$

The exponential factor is not zero, so we find

$$M\alpha = \lambda\alpha, \tag{4.12}$$

which is an equation for the eigenvalue λ and (normalized) eigenvector α. In general, there are n eigenvalues and n eigenvectors, so we must add a subscript to both α and λ indicating the particular solution. The general solution is an arbitrary linear combination of these individual solutions. The eigenvalues are solutions of the characteristic equation

$$0 = \det(\mathbf{M} - \lambda\mathbf{I}) \tag{4.13}$$

where $\mathbf{M}$ is the matrix representation of M, and $\mathbf{I}$ is the identity matrix of the same dimension. The elements of $\mathbf{M}$ are real, so we know that the eigenvalues λ either are real or come in complex-conjugate pairs. We assume the eigenvalues are all distinct.[3]

---

[2] Actually, all we need is $\partial_0 \partial_1 F(t, z_e) = 0$.

[3] If the eigenvalues are not distinct then the form of the solution is modified.

If the eigenvalue is real then the solution is exponential, as assumed. If the eigenvalue $\lambda > 0$ then the solution expands exponentially along the direction α; if $\lambda < 0$ then the solution contracts exponentially along the direction α.

If the eigenvalue is complex we can form real solutions by combining the two solutions for the complex-conjugate pair of eigenvalues. Let $\lambda = a + ib$, with real a and b, be one such complex eigenvalue. Let $\alpha = u + iv$, where u and v are real, be the eigenvector corresponding to it. So there is a complex solution of the form

$$
\begin{aligned}
\zeta_c(t) &= (u + iv)e^{(a+ib)t} \\
&= (u + iv)e^{at}(\cos bt + i \sin bt) \\
&= e^{at}(u \cos bt - v \sin bt) + ie^{at}(u \sin bt + v \cos bt).
\end{aligned} \tag{4.14}
$$

The complex conjugate of this solution is also a solution, because the ordinary differential equation is linear with real linear coefficients. This complex-conjugate solution is associated with the eigenvalue that is the complex conjugate of the original complex eigenvalue. So the real and imaginary parts of ζ_c are real solutions:

$$
\begin{aligned}
\zeta_a(t) &= e^{at}(u \cos bt - v \sin bt) \\
\zeta_b(t) &= e^{at}(u \sin bt + v \cos bt).
\end{aligned} \tag{4.15}
$$

These two solutions reside in the plane containing the vectors u and v. If a is positive both solutions spiral outwards exponentially, and if a is negative they both spiral inwards. If a is zero, both solutions trace the same ellipse, but with different phases.

Again, the general solution is an arbitrary linear combination of the particular real solutions corresponding to the various eigenvalues. So if we denote the kth real eigensolution $\zeta_k(t)$, then the general solution is

$$
\zeta(t) = \sum_k A_k \zeta_k(t), \tag{4.16}
$$

where A_k may be determined by the initial conditions (the state at a given time).

Exercise 4.1: Pendulum

Carry out the details of finding the eigensolutions for the two equilibria of the pendulum ($\theta = 0$ and $\theta = \pi$, both with $p_\theta = 0$). How is the small-amplitude oscillation frequency related to the eigenvalues? How are the eigendirections related to the contours of the Hamiltonian?

4.2.2 Fixed Points of Maps

Fixed points on a surface of section correspond either to equilibrium points of the system or to a periodic motion of the system. Linear stability analysis of fixed points of maps is similar to the linear stability analysis for equilibrium points of systems governed by differential equations.

Let T be a map of the state space onto itself, as might be generated by a surface of section. A trajectory sequence is generated by successive iteration of the map T. Let $x(n)$ be the nth point of the sequence. The map carries one point of the trajectory sequence to the next: $x(n + 1) = T(x(n))$. We can represent successive iterations of the map by a superscript, so that T^i indicates T composed i times. For example, $T^2(x) = T(T(x))$. Thus $x(n) = T^n(x(0))$.[4]

A *fixed point* x_0 of the map T satisfies

$$x_0 = T(x_0). \tag{4.17}$$

A *periodic point* of the map T is a point that is visited every k iterations of T. Thus it is a fixed point of the map T^k. So the behavior near a periodic point can be ascertained by looking at the behavior near an associated fixed point of T^k.

Let x be some trajectory initially near the fixed point x_0 of T, and ξ be the deviation from x_0: $x(n) = x_0 + \xi(n)$. The trajectory satisfies

$$x_0 + \xi(n + 1) = T(x_0 + \xi(n)). \tag{4.18}$$

Expanding the right-hand side as a Taylor series, we obtain

$$x_0 + \xi(n + 1) = T(x_0) + DT(x_0)\xi(n) + \cdots, \tag{4.19}$$

but $x_0 = T(x_0)$ so

$$\xi(n + 1) = DT(x_0)\xi(n) + \cdots. \tag{4.20}$$

---

[4]The map T is being used as an operator: multiplication is interpreted as composition.

Linear stability analysis considers the evolution of the system truncated to the linear terms

$$\xi(n+1) = DT(x_0)\xi(n). \tag{4.21}$$

This is a system of linear difference equations, with constant coefficients $DT(x_0)$.

We assume there are solutions of the form

$$\xi(n) = \rho^n \alpha, \tag{4.22}$$

where ρ is some (complex) number, called a *characteristic multiplier*.[5] Substituting this solution into the linearized evolution equation, we find

$$\rho\alpha = DT(x_0)\alpha, \tag{4.23}$$

or

$$(DT(x_0) - \rho I)\alpha = 0, \tag{4.24}$$

where I is the identity multiplier. We see that ρ is an eigenvalue of the linear transformation $DT(x_0)$ and α is the associated (normalized) eigenvector. Let $M = DT(x_0)$, and $\mathbf{M}$ be its matrix representation. The eigenvalues are determined by

$$\det(\mathbf{M} - \rho\mathbf{I}) = 0. \tag{4.25}$$

The elements of $\mathbf{M}$ are real, so the eigenvalues ρ are either real or come in complex-conjugate pairs.[6]

For the real eigenvalues the solutions are just exponential expansion or contraction along the associated eigenvector α:

$$\xi(n) = \rho^n \alpha. \tag{4.26}$$

The solution is expanding if $|\rho| > 1$ and contracting if $|\rho| < 1$.

If the eigenvalues are complex, then the solution is complex, but the complex solutions corresponding to the complex-conjugate pair of eigenvalues can be combined to form two real solutions, as was done for the equilibrium solutions. Let $\rho = e^{A+iB}$ with real

---

[5] A characteristic multiplier is also sometimes called a Floquet multiplier.

[6] We assume for now that the eigenvalues are distinct.

A and B, and $\alpha = u + iv$. A calculation similar to that for the equilibrium case shows that there are two real solutions

$$\xi_a(n) = e^{An}\left(u \cos Bn - v \sin Bn\right)$$
$$\xi_b(n) = e^{An}\left(u \sin Bn + v \cos Bn\right). \tag{4.27}$$

We see that if $A > 0$ then the solution exponentially expands, and if $A < 0$ the solution exponentially contracts. Exponential expansion, $A > 0$, corresponds to $|\rho| > 1$; exponential contraction, $A < 0$, corresponds to $|\rho| < 1$. If $A = 0$ then the two real solutions trace an ellipse and any linear combination of them traces an ellipse.

The general solution is an arbitrary linear combination of the eigensolutions. Let ξ_k be the kth real eigensolution. The general solution is

$$\xi(n) = \sum_k A_k \xi_k(n), \tag{4.28}$$

where A_k may be determined by the initial conditions.

Exercise 4.2: Elliptical oscillation

Show that the arbitrary linear combination of ξ_a and ξ_b traces an ellipse for $A = 0$.

Exercise 4.3: Standard map

The standard map (see section 3.9) has fixed points at $I = 0$ for $\theta = 0$ and $\theta = \pi$. Find the full eigensolutions for these two fixed points. For what ranges of the parameter K are the fixed points linearly stable or unstable?

4.2.3 Relations Among Exponents

For maps that are generated by stroboscopic sampling of the evolution of a system of autonomous differential equations, equilibrium points are fixed points of the map. The eigensolutions of the equilibrium of the flow and the eigensolutions of the map at the fixed point are then related. Let τ be the sampling period. Then $\rho_i = e^{\lambda_i \tau}$.

The Lyapunov exponent is a measure of the rate of exponential divergence of nearby trajectories from a reference trajectory. If the reference trajectory is an equilibrium of a flow, then the Lyapunov exponents are the real parts of the linearized characteristic exponents λ_i. If the reference trajectory is a fixed point of a map generated by a flow (either a periodic orbit or an equilibrium), then the Lyapunov exponents are real parts of the logarithm of the characteristic multipliers, divided by the period of the map. So if the characteristic multiplier is $\rho = e^{A+iB}$ and the period of the map is τ, then the Lyapunov exponent is A/τ. A positive Lyapunov exponent of a fixed point indicates linear instability of the fixed point.

The Lyapunov exponent has less information than the characteristic multipliers or exponents because the imaginary part is lost. However, the Lyapunov exponent is more generally applicable in that it is well defined even for reference trajectories that are not periodic.

In the linear analysis of the fixed point, each characteristic multiplier corresponds to a subspace of possible linear solutions. For instance, for a real characteristic multiplier there is a corresponding eigendirection, and for any initial displacement along this direction successive iterates are also along this direction. Complex-conjugate pairs of multipliers correspond to a plane of solutions. For a displacement initially on this plane, successive iterates are also on this plane.

It turns out that something like this is also the case for the linearized solutions near a reference trajectory that is not at a fixed point. For each nonzero Lyapunov exponent there is a twisting subspace, so that for an initial displacement in this subspace successive iterates also belong to the subspace. At different points along the reference trajectory the unit displacement vector that characterizes the direction of this subspace is different.

Hamiltonian specialization

For Hamiltonian systems there are additional constraints among the eigenvalues.

Consider first the case of two-dimensional surfaces of section. We have seen that Hamiltonian surfaces of section preserve area. As we saw in the proof of Liouville's theorem, area preservation implies that the determinant of the derivative of the transformation is 1. At a fixed point x_0 the linearized map is $\xi(n+1) = DT(x_0)\xi(n)$. So $M = DT(x_0)$ has unit determinant. The determinant is the product of the eigenvalues, so for a fixed point on a Hamiltonian surface of section the two eigenvalues must be inverses of each other. We also have the constraint that if an eigen-

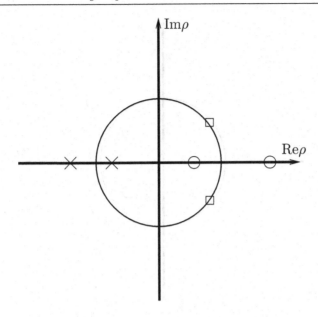

Figure 4.4 The eigenvalues for fixed points of a two-dimensional Hamiltonian map. The eigenvalues either are real or are complex-conjugate pairs that lie on the unit circle. For each eigenvalue the inverse is also an eigenvalue.

value is complex then the complex conjugate of the eigenvalue is also an eigenvalue. These two conditions imply that the eigenvalues must either be real and inverses, or be complex-conjugate pairs on the unit circle (see figure 4.4).

Fixed points for which the characteristic multipliers all lie on the unit circle are called *elliptic* fixed points. The solutions of the linearized variational equations trace ellipses around the fixed point. Elliptic fixed points are linearly stable.

Fixed points with positive real characteristic multipliers are called *hyperbolic* fixed points. For two-dimensional maps, there is an exponentially expanding subspace and an exponentially contracting subspace. The general solution is a linear combination of these. Fixed points for which the characteristic multipliers are negative are called *hyperbolic with reflection.*

The edge case of a double root of the characteristic equation is called *parabolic.* In this case the general solution grows linearly. This happens at points of bifurcation where elliptic points become hyperbolic points or vice versa.

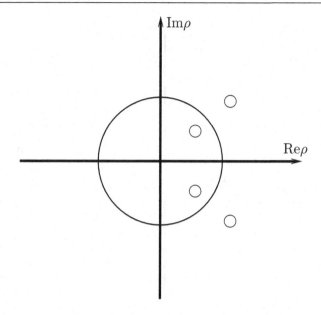

Figure 4.5   If there is more than one degree of freedom the eigenvalues for fixed points of a Hamiltonian map may lie in a quartet, with two complex-conjugate pairs. The magnitudes of the pairs must be inverses. This enforces the constraint that the expansion produced by the roots with magnitude greater than one is counterbalanced by the contraction produced by the roots with magnitude smaller than one.

For two-dimensional Hamiltonian maps these are the only possibilities. For higher-dimensional Hamiltonian maps, we can get combinations of these: some characteristic multipliers can be real and others complex-conjugate pairs. We might imagine that in addition there would be many other types of fixed points that occur in higher dimensions. In fact, there is only one additional type, shown in figure 4.5. For Hamiltonian systems of arbitrary dimensions it is still the case that for each eigenvalue the complex conjugate and the inverse are also eigenvalues. We can prove this starting from a result in chapter 5. Consider the map T_β of the phase space onto itself that is generated by time evolution of a Hamiltonian system by time increment β. Let $z = (q, p)$; then the map T_β satisfies $z(t + \beta) = T_\beta(z(t))$ for solutions z of Hamilton's equations. We will show in chapter 5 that the derivative of the map T_β is symplectic, whether or not the starting point is at a fixed point. A $2n \times 2n$ matrix $\mathbf{M}$ is *symplectic* if it satisfies

$$\mathbf{MJM}^\mathsf{T} = \mathbf{J}, \tag{4.29}$$

where $\mathbf{J}$ is the $2n$-dimensional symplectic unit:

$$\mathbf{J} = \begin{pmatrix} \mathbf{0}_{n \times n} & \mathbf{1}_{n \times n} \\ -\mathbf{1}_{n \times n} & \mathbf{0}_{n \times n} \end{pmatrix}, \tag{4.30}$$

with the $n \times n$ unit matrix $\mathbf{1}_{n \times n}$ and the $n \times n$ zero matrix $\mathbf{0}_{n \times n}$.

Using the symplectic property, we can show that in general for each eigenvalue its inverse is also an eigenvalue. Assume ρ is an eigenvalue, so that ρ satisfies $\det(\mathbf{M} - \rho\mathbf{I}) = 0$. This equation is unchanged if $\mathbf{M}$ is replaced by its transpose, so ρ is also an eigenvalue of $\mathbf{M}^\mathsf{T}$:

$$\mathbf{M}^\mathsf{T}\boldsymbol{\alpha}' = \rho\boldsymbol{\alpha}'. \tag{4.31}$$

From this we can see that

$$\frac{1}{\rho}\boldsymbol{\alpha}' = (\mathbf{M}^\mathsf{T})^{-1}\boldsymbol{\alpha}'. \tag{4.32}$$

Now, from the symplectic property we have

$$\mathbf{MJ} = \mathbf{J}(\mathbf{M}^\mathsf{T})^{-1}. \tag{4.33}$$

So

$$\mathbf{MJ}\boldsymbol{\alpha}' = \mathbf{J}(\mathbf{M}^\mathsf{T})^{-1}\boldsymbol{\alpha}' = \frac{1}{\rho}\mathbf{J}\boldsymbol{\alpha}', \tag{4.34}$$

and we can conclude that $1/\rho$ is an eigenvalue of $\mathbf{M}$ with the eigenvector $\mathbf{J}\boldsymbol{\alpha}'$. From the fact that for every eigenvalue its inverse is also an eigenvalue we deduce that the determinant of the transformation $\mathbf{M}$, which is the product of the eigenvalues, is one.

Thus the constraints that the eigenvalues must be associated with inverses and complex conjugates yields exactly one new pattern of eigenvalues in higher dimensions. Figure 4.5 shows the only new pattern that is possible.

We have seen that the Lyapunov exponents for fixed points are related to the characteristic multipliers for the fixed points, so the Hamiltonian constraints on the multipliers correspond to Hamiltonian constraints for Lyapunov exponents at fixed points. For each characteristic multiplier, the inverse is also a characteristic multiplier. This means that at fixed points, for each positive Lyapunov exponent there is a corresponding negative Lyapunov exponent with the same magnitude. It turns out that this is also true if the reference trajectory is not at a fixed point. For Hamiltonian systems, for each positive Lyapunov exponent there is a corresponding negative exponent of equal magnitude.

Exercise 4.4: Quartet

Describe (perhaps by drawing cross sections) the orbits that are possible with quartets.

Linear and nonlinear stability

A fixed point that is linearly unstable indicates that the full system is unstable at that point. This means that trajectories starting near the fixed point diverge from the fixed point. On the other hand, linear stability of a fixed point does not generally guarantee that the full system is stable at that point. For a two-degree-of-freedom Hamiltonian system, the Kolmogorov–Arnold–Moser theorem proves under certain conditions that linear stability implies nonlinear stability. In higher dimensions, though, it is not known whether linear stability implies nonlinear stability.

4.3   Homoclinic Tangle

For the driven pendulum we observe that as the amplitude of the drive is increased the separatrix of the undriven pendulum is where the most prominent chaotic zone appears. Here we examine in great detail the motion in the vicinity of the separatrix. What emerges is a remarkably complicated picture, first discovered by Henri Poincaré. Indeed, Poincaré stated (see the epigraph to this chapter) that the picture that emerged was so complicated that he was not even going to attempt to draw it. We will review the argument leading to the picture, and compute enough of it to convince ourselves of its reality.

The separatrix of the undriven pendulum is made up of two trajectories that are asymptotic to the unstable equilibrium. In the driven pendulum with zero drive, an infinite number of distinct

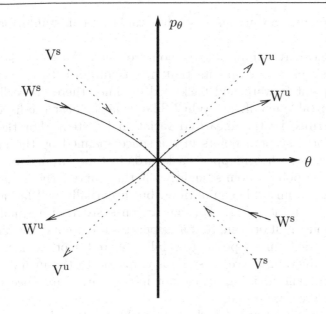

Figure 4.6 The stable and unstable manifolds of the unstable fixed point for the pendulum are compared to the stable and unstable manifolds of the linearized variational system in the vicinity of the fixed point. The axes are centered at the fixed point $(\pm\pi, 0)$. The linear stable and unstable manifolds are labeled by V^s and V^u respectively; the nonlinear stable and unstable manifolds are labeled by W^s and W^u.

orbits lie on the separatrix; they are distinguished by the phase of the drive. These orbits are asymptotic to the unstable fixed point both forward and backward in time.

Notice that close to the unstable fixed point the sets of points that are asymptotic to the unstable equilibrium must be tangent to the linear variational eigenvectors at the fixed point. (See figure 4.6.) In a sense, the sets of orbits that are asymptotic to the fixed point are extensions to the nonlinear problem of the sets of orbits that are asymptotic to the fixed point in the linearized problem.

The set of points that are asymptotic to an unstable fixed point forward in time is called the *stable manifold* of the fixed point. The set of points that are asymptotic to an unstable fixed point backward in time is called the *unstable manifold*. For the driven pendulum with zero-amplitude drive, all points on the separatrix are asymptotic both forward and backward in time to the unstable fixed point. So in this case the stable and unstable manifolds coincide.

If the drive amplitude is nonzero then there are still one-dimensional sets of points that are asymptotic to the unstable fixed point forward and backward in time: there are still stable and unstable manifolds. Why? The behavior near the fixed point is described by the linearized variational system. For the linear variational system, points in the space spanned by the unstable eigenvector, when mapped backwards in time, are asymptotic to the fixed point. Points slightly off this curve may initially approach the unstable equilibrium, but eventually will fall away to one side or the other. For the driven system with small drive, there must still be a curve that separates the points that fall away to one side from the points that fall away to the other side. Points on the dividing curve must be asymptotic to the unstable equilibrium. The dividing set cannot have positive area because the map is area preserving.

For the zero-amplitude drive case, the stable and unstable manifolds are contours of the conserved Hamiltonian. For nonzero amplitude the Hamiltonian is no longer conserved, and the stable manifolds and unstable manifolds no longer coincide. This is generally true for non-integrable systems: stable and unstable manifolds do not coincide.

If the stable and unstable manifolds no longer coincide, where do they go? A stable manifold cannot cross another stable manifold, and an unstable manifold cannot cross another unstable manifold, because the crossing point would be asymptotic to two different fixed points. A stable manifold or unstable manifold may not cross itself, as shown below. However, a stable and an unstable manifold may cross one another.

Actually, the stable and unstable manifolds must cross at some point. The only other possibilities are that they run off to infinity or spiral around. We will see that in general there are barriers to running away. Area preservation excludes the existence of attractors, and this can be used to exclude the spiraling case. A finite region of initial conditions between two successive arms of the spiral will eventually run out of area as the spiral progresses.

So the only possibility is that the stable and unstable manifolds cross, as is illustrated in figure 4.7. The point of crossing of a stable and unstable manifold is called a *homoclinic intersection* if the stable and unstable manifolds belong to the same unstable

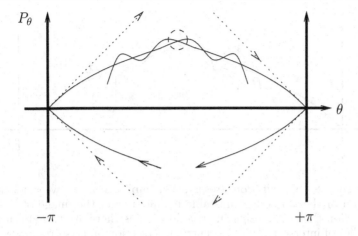

Figure 4.7 For nonzero drive the stable and unstable manifolds no longer coincide and in general cross. The dashed circle indicates the central intersection. Forward and backward images of this intersection are themselves intersections. Because the orbits are asymptotic to the fixed point there is an infinity of such intersections.

fixed point. It is called a *heteroclinic intersection* if the stable and unstable manifolds belong to different fixed points.

If the stable and unstable manifolds cross once then there is an infinite number of other crossings. The intersection point belongs to both the stable and unstable manifolds. That it is on the unstable manifold means that all images forward and backward in time also belong to the unstable manifold, and likewise for points on the stable manifold. Thus all images of the intersection belong to both the stable and unstable manifolds. So these images must be additional crossings of the two manifolds.

We can deduce that there are still more intersections of the stable and unstable manifolds. The maps we are considering not only preserve area but also orientation. In the proof of Liouville's theorem we showed that the determinant of the transformation is one, not just magnitude one. If we consider little segments of the stable and unstable manifolds near the intersection point, then these segments must map near the image of the intersection point. That the map preserves orientation implies that the manifolds are crossing one another in the same sense as at the previous intersection. Therefore there must have been at least one more crossing of the stable and unstable manifolds in between these two. This

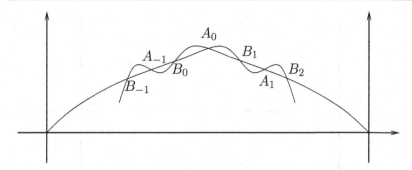

Figure 4.8 Orientation preservation implies that between an intersection of the stable and unstable manifolds and the image of this intersection there is another intersection. Thus there are two alternating families of intersections. The central intersection and its pre-images and post-images are labeled A_i. Another family is labeled B_i.

is illustrated in figure 4.8. Of course, all forward and backward images of these intermediate intersections are also intersections.

As the picture gets more complicated, keep in mind that the stable manifold cannot cross itself and the unstable manifold cannot cross itself. Suppose one did, say by making a little loop. The image of this loop under the map must also be a loop. So if there were a loop there would have to be an infinite number of loops. That would be okay, but what happens as the loop gets close to the fixed point? There would still have to be loops, but then the stable and unstable manifolds would not have the right behavior: the stable and unstable manifolds of the linearized map do not have loops. Therefore, the stable and unstable manifolds cannot cross themselves.[7]

We are not done yet! The lobes that are defined by successive crossings of the stable and unstable manifolds enclose a certain area. The map is area preserving so all images of these lobes must have the same area. As the lobes approach the fixed point, we get an infinite number of lobes with a base of exponentially shrinking length. The stable and unstable manifolds cannot cross

---

[7]Sometimes it is argued that the stable and unstable manifolds cannot cross themselves on the basis of the uniqueness of solutions of differential equations. This argument is incorrect. The stable and unstable manifolds are not themselves solutions of a differential equation, they are sets of points whose solutions are asymptotic to the unstable fixed points.

themselves, so to pack these lobes together on the plane the lobes must stretch out to preserve area. We see that the length of the lobe must grow roughly exponentially (it may not be uniform in width so it need not be exactly exponential). This exponential lengthening of the lobes no doubt bears some responsibility for the exponential divergence of nearby trajectories of chaotic orbits, but does not prove it. It does, however, suggest a connection between the fact that chaotic orbits appear to occupy an area on the section and the fact that nearby chaotic orbits diverge exponentially.

Actually, the situation is even more complicated. As the lobes stretch, they form tendrils that wrap around the separatrix region. The tendrils of the unstable manifold can cross the tendrils of the stable manifold. Each point of crossing is a new homoclinic intersection, and so each pre- and post-image of this point belongs to both the stable and unstable manifolds, indicating another crossing of these curves. We could go on and on. No wonder Poincaré refused to draw this mess.

Exercise 4.5: Homoclinic paradox

How do we fit an infinite number of copies of a finite area in a finite region, without allowing the stable and unstable manifolds to cross themselves? Resolve this apparent paradox.

4.3.1 Computation of Stable and Unstable Manifolds

The homoclinic tangle is not just a bad dream. We can actually compute it.

Very close to an unstable fixed point the stable and unstable manifolds become indistinguishable from the rays along the eigenvectors of the linearized system. So one way to compute the unstable manifold is to take a line of initial conditions close to the fixed point along the unstable manifold of the linearized system and evolve them forward in time. Similarly, the stable manifold can be constructed by taking a line of initial conditions along the stable manifold of the linearized system and evolving them backward in time.

We can do better than this by choosing a parameter (like arc length) along the manifold and for each value of the parameter deciding how many iterations of the map would be required to take the point back to within some small region of the fixed point. We then choose an initial condition along the linearized eigenvectors and iterate the point back with the map. This idea is implemented in the following program:[8]

```
(define ((unstable-manifold T xe ye dx dy rho eps) param)
  (let ((n (floor->exact (/ (log (/ param eps)) (log rho)))))
    ((iterated-map T n) (+ xe (* dx (/ param (expt rho n))))
                        (+ ye (* dy (/ param (expt rho n))))
                        make-point
                        (lambda () (error "Failed")))))
```

where T is the map, xe and ye are the coordinates of the fixed point, dx and dy are components of the linearized eigenvector, rho is the characteristic multiplier, eps is a scale within which the linearized map is a good enough approximation to T, and param is a continuous parameter along the manifold. The procedure make-point, supplied as the success continuation for the iterated map, packages two numbers. They can be split with abscissa and ordinate.

The program assumes that there is a basic exponential divergence along the manifold—that is why we take the logarithm of param to get initial conditions in the linear regime. This assumption is not exactly true, but it is good enough for now.

The curve is generated by a call to plot-parametric-fill, which recursively subdivides intervals of the parameter until there are enough points to get a smooth curve.

```
(define (plot-parametric-fill win f a b near?)
  (let loop ((a a) (xa (f a)) (b b) (xb (f b)))
    (if (not (close-enuf? a b (* 10 *machine-epsilon*)))
        (let ((m (/ (+ a b) 2)))
          (let ((xm (f m)))
            (plot-point win (abscissa xm) (ordinate xm))
            (if (not (near? xa xm))
                (loop a xa m xm))
            (if (not (near? xb xm))
                (loop m xm b xb)))))))
```

The near? argument is a test for whether two points are within a given distance of each other in the graph. Because some coordinates are angle variables, this may involve a principal value comparison. For example, for the driven pendulum section, the

---

[8]The procedure iterated-map takes a map and an integer n. It returns a new map that is the result of iterating the given map n times.

horizontal axis is an angle but the vertical axis is not, so the picture is on a cylinder:

```
(define (cylinder-near? eps)
  (let ((eps2 (square eps)))
    (lambda (point1 point2)
      (< (+ (square ((principal-value pi)
                     (- (abscissa point1) (abscissa point2))))
            (square (- (ordinate point1) (ordinate point2))))
         eps2))))
```

Figure 4.9 shows a computation of the homoclinic tangle for the driven pendulum. The parameters are $m = 1\,\text{kg}$, $g = 9.8\,\text{m}\,\text{s}^{-2}$, $l = 1\,\text{m}$, $\omega = 4.2\sqrt{g/l}$, and amplitude $A = 0.05\,\text{m}$. For reference, figure 4.9 shows a surface of section for these parameters on the same scale.

Exercise 4.6: Computing homoclinic tangles

a. Compute stable and unstable manifolds for the standard map.

b. Identify the features on the homoclinic tangle that entered the argument about its existence, such as the central crossing of the stable and unstable manifolds, etc.

c. Investigate the errors in the process. Are the computed manifolds really correct or a figment of wishful thinking? One could imagine that the errors are exponential and the computed manifolds have nothing to do with the actual manifolds.

d. How much actual space is taken up by the homoclinic tangle? Consider a value of the coupling constant $K = 0.8$. Does the homoclinic tangle actually fill out the apparent chaotic zone?

4.4 Integrable Systems

Islands appear near commensurabilities, and commensurabilities are present even in integrable systems.[9] In integrable systems an infinite number of periodic orbits are associated with each commensurability, but upon perturbation only a finite number of periodic orbits survive. How does this happen? First we have to learn more about integrable systems.

---

[9] A commensurability occurs when the frequencies involved are not linearly independent over the integers. We will define this carefully on page 312.

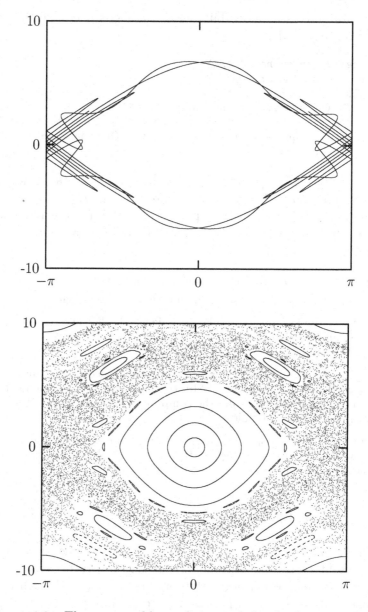

Figure 4.9   The computed homoclinic tangle for the driven pendulum exhibits the features described in the text. Notice how the excursions of the stable and unstable manifolds become longer and thinner as they approach the unstable fixed point. A surface of section with the same parameters is also shown.

If an n-degree-of-freedom system has n independent conserved quantities then the solution of the problem can be reduced to quadratures. Such a system is called *integrable*. Typically, the phase space of integrable systems is divided into regions of qualitatively different behavior. For example, the motion of a pendulum is reducible to quadratures and has three distinct types of solutions: the oscillating solutions and the clockwise and counterclockwise circulating solutions. The different regions of the pendulum phase space are separated by the trajectories that are asymptotic to the unstable equilibrium. It turns out that for any system that is reducible to quadratures, a set of phase-space coordinates can be chosen for each region of the phase space so that the Hamiltonian describing the motion in that region depends only on the momenta. Furthermore, if the phase space is bounded then the generalized coordinates can be chosen to be angles (that are 2π-periodic). The configuration space described by n angles is an n-torus. The momenta conjugate to these angles are called *actions*. Such phase-space coordinates are called *action-angle* coordinates. We will see later how to reformulate systems in this way. Here we explore the consequences of such a formulation; this formulation is especially useful for finding out what happens as additional effects are added to integrable problems.

Orbit types in integrable systems

Suppose we have a time-independent n-degree-of-freedom system that is reducible to quadratures. For each region of phase space there is a local formulation of the system so that the evolution of the system is described by a time-independent Hamiltonian that depends only on the momenta. Suppose further that the coordinates are all angles. Let θ be the tuple of angles and J be the tuple of conjugate momenta. The Hamiltonian is

$$H(t, \theta, J) = f(J). \tag{4.35}$$

Hamilton's equations are simply

$$DJ(t) = -\partial_1 H(t, \theta(t), J(t)) = 0$$
$$D\theta(t) = \;\;\partial_2 H(t, \theta(t), J(t)) = \omega(J(t)), \tag{4.36}$$

where $\omega(J) = Df(J)$ is a tuple of frequencies with a component for each degree of freedom. The momenta are all constant because the Hamiltonian does not depend on any of the coordinates. The motion of the coordinate angles is uniform; the rates of change of the angles are the frequencies ω, which depend only on the constant momenta. Given initial values $\theta(t_0)$ and $J(t_0)$ at time t_0, the solutions are simple:

$$J(t) = J(t_0)$$
$$\theta(t) = \omega(J(t_0))(t - t_0) + \theta(t_0). \tag{4.37}$$

Though the solutions are simple, there are two distinct orbit types: periodic orbits and quasiperiodic orbits, depending on the frequency ratios.

A solution is *periodic* if all the coordinates (and momenta) of the system return to their initial values at some later time. Each coordinate θ^i with nonzero frequency $\omega^i(J(t_0))$ is periodic with a period $\tau_i = 2\pi/\omega^i(J(t_0))$. The period of the system must therefore be an integer multiple k_i of each of the individual coordinate periods τ_i. If the system is periodic with some set of integer multiples, then it is also periodic with any common factors divided out. Thus the period of the system is $\tau = (k_i/d)\tau_i$ where d is the greatest common divisor of the integers k_i.

For a system with two degrees of freedom, a solution is periodic if there exists a pair of relatively prime integers k and j such that $k\omega^0(J(t_0)) = j\omega^1(J(t_0))$. The period of the system is $\tau = 2\pi j/\omega^0(J(t_0)) = 2\pi k/\omega^1(J(t_0))$; the frequency is $\omega^0(J(t_0))/j = \omega^1(J(t_0))/k$. A periodic motion on the 2-torus is illustrated in figure 4.10.

If the frequencies $\omega^i(J(t_0))$ satisfy an integer-coefficient relation $\sum_i n_i\omega^i(J(t_0)) = 0$, we say that the frequencies satisfy a *commensurability*. If there is no commensurability for any nonzero integer coefficients, we say that the frequencies are linearly independent (with respect to the integers) and the solution is *quasiperiodic*. One can prove that for n incommensurate frequencies all solutions come arbitrarily close to every point in the configuration space.[10]

For a system with two degrees of freedom the solutions in a region described by a particular set of action-angle variables are

---

[10]Motion with n incommensurate frequencies is dense on the n-torus. Furthermore, such motion is *ergodic* on the n-torus. This means that time averages of time-independent phase-space functions computed along trajectories are equal to the phase-space average of the same function over the torus.

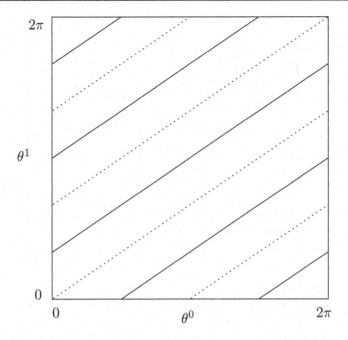

Figure 4.10   The solid and dotted lines show two periodic trajectories on the configuration coordinate plane. For commensurate frequencies the configuration motion is periodic, independent of the initial angles. In this illustration the frequencies satisfy $2\omega^0(J(t_0)) = 3\omega^1(J(t_0))$. The orbits close after three cycles of θ^0 and two cycles of θ^1, for any initial θ^0 and θ^1.

either periodic or quasiperiodic.[11]   For systems with more than two degrees of freedom there are trajectories that are neither periodic nor quasiperiodic with n frequencies. These are quasiperiodic with fewer frequencies and dense over a corresponding lower-dimensional torus.

Surfaces of section for integrable systems

As we have seen, in action-angle coordinates the angles move with constant angular frequencies, and the momenta are constant. Thus surfaces of section in action-angle coordinates are particularly simple. We can make surfaces of section for time-independent two-degree-of-freedom systems or one-degree-of-freedom systems

---

[11]For time-independent systems with two degrees of freedom the boundary between regions described by different action-angle coordinates has asymptotic solutions and unstable periodic orbits or equilibrium points. The solutions on the boundary are not described by the action-angle Hamiltonian.

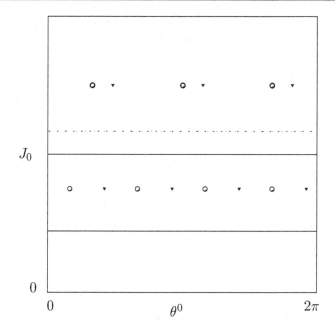

Figure 4.11   On surfaces of section for systems in action-angle coordinates a trajectory generates points on a horizontal line. Trajectories with frequencies that are commensurate with the sampling frequency produce a finite number of points, independent of the initial angle. Here we use different symbols to indicate section points for distinct trajectories with the same momentum J_0. Trajectories with frequencies that are incommensurate with the sampling frequency fill out a horizontal line densely.

with periodic drive. In the latter case, one of the angles in the action-angle system is the phase of the drive. We make surfaces of section by accumulating points in one pair of canonical coordinates as the other coordinate goes through some particular value, such as zero. If we plot the section points with the angle coordinate on the abscissa and the conjugate momentum on the ordinate then the section points for all trajectories lie on horizontal lines, as illustrated in figure 4.11.

For definiteness, let the plane of the surface of section be the (θ^0, J_0) plane, and the section condition be $\theta^1 = 0$. The other momentum J_1 is chosen so that all the trajectories have the same energy. The momenta are all constant, so for a given trajectory all points that are generated are constrained to a line of constant J_0.

The time between section points is the period of θ^1: $\Delta t = 2\pi/\omega^1(J(t_0))$ because a section point is generated for every cycle of θ^1. The angle between successive points on the section is $\omega^0(J(t_0))\Delta t = \omega^0(J(t_0))2\pi/\omega^1(J(t_0)) = 2\pi\nu(J(t_0))$, where $\nu(J) = \omega^0(J)/\omega^1(J)$ is called the *rotation number* of the trajectory. Let $\hat{\theta}(i)$ and $\hat{J}(i)$ be the ith point (i is an integer) in a sequence of points on the surface of section generated by a solution trajectory:

$$\hat{\theta}(i) = \theta^0(i\Delta t + t_0)$$
$$\hat{J}(i) = J_0(i\Delta t + t_0), \tag{4.38}$$

where the system is assumed to be on the section at $t = t_0$. Along a trajectory, the map from one section point $\left(\hat{\theta}(i), \hat{J}(i)\right)$ to the next $\left(\hat{\theta}(i+1), \hat{J}(i+1)\right)$ is of the form:[12]

$$\begin{pmatrix} \hat{\theta}(i+1) \\ \hat{J}(i+1) \end{pmatrix} = T\begin{pmatrix} \hat{\theta}(i) \\ \hat{J}(i) \end{pmatrix} = \begin{pmatrix} \hat{\theta}(i) + 2\pi\hat{\nu}(\hat{J}(i)) \\ \hat{J}(i) \end{pmatrix}. \tag{4.39}$$

As a function of the action on the section, the rotation number is $\hat{\nu}(\hat{J}(0)) = \nu(\hat{J}(0), J_1(t_0))$, where $J_1(t_0)$ has the value required to be on the section, as for example by giving the correct energy. If the rotation number function $\hat{\nu}$ is strictly monotonic in the action coordinate on the section then the map is called a *twist map*.[13]

On a surface of section the different types of orbits generate different patterns. If the two frequencies are commensurate, then the trajectory is periodic and only a finite number of points are generated on the surface of section. Each of the periodic solutions illustrated in figure 4.10 generates two points on the surface of section defined by $\theta^1 = 0$. If the frequencies are commensurate they satisfy a relation of the form $k\omega^0(J(t_0)) = j\omega^1(J(t_0))$, where $J(t_0) = (\hat{J}(0), J_1(t_0))$ is the initial and constant value of the momentum tuple. The motion is periodic with frequency $\omega^0(J(t_0))/j = \omega^1(J(t_0))/k$, so the period is $2\pi j/\omega^0(J(t_0)) = 2\pi k/\omega^1(J(t_0))$. Thus this periodic orbit generates k points on this

---

[12]The coordinate $\hat{\theta}(i)$ is an angle. It can be brought to a standard interval such as 0 to 2π.

[13]For a map to be a twist map we require that there is a positive number K such that $|D\nu(J)| > K > 0$ over some interval of J.

surface of section. For trajectories with commensurate frequencies the rotation number is rational: $\hat{\nu}(\hat{J}(0)) = \nu(\hat{J}(0), J_1(t_0)) = j/k$. The coordinate θ^1 makes k cycles while the coordinate θ^0 makes j cycles (figure 4.10 shows a system with a rotation number of $3/2$). The frequencies depend on the momenta but not on the coordinates, so the motion is periodic with the same period and rotation number for all initial angles given these momenta. Thus there is a continuous family of periodic orbits with different initial angles.

If the two frequencies are incommensurate, then the 2-torus is filled densely. Thus the line on which the section points are generated is filled densely. Again, this is the case for any initial coordinates, because the frequencies depend only on the momenta. There are infinitely many such orbits that are distinct for a given set of frequencies.[14]

4.5   Poincaré–Birkhoff Theorem

How does this picture change if we add additional effects?

One peculiar feature of the orbits in integrable systems is that there are continuous families of periodic orbits. The initial angles do not matter; the frequencies depend only on the actions. Contrast this with our earlier experience with surfaces of section in which periodic points are isolated, and associated with island chains. Henri Poincaré and George Birkhoff investigated periodic orbits of near-integrable systems, and found that typically for each rational rotation number there are a finite number of periodic points, half of which are linearly stable and half linearly unstable. Here we show how to construct the Poincaré–Birkhoff periodic points.

Consider an integrable system described in action-angle coordinates by the Hamiltonian $H_0(t, \theta, J) = f(J)$. We add some small additional effect described by the term H_1 in the Hamiltonian

$$H = H_0 + \epsilon H_1. \tag{4.40}$$

An example of such a system is the periodically driven pendulum with small-amplitude drive. For zero-amplitude drive the driven

---

[14]The section points for any particular orbit are countable and dense, but they have zero measure on the line.

pendulum is integrable, but not for small drive. Unfortunately, we do not yet have the tools to develop action-angle coordinates for the pendulum. A simpler problem that is already in action-angle form is the driven rotor, which is just the driven pendulum with gravity turned off. We can implement this by turning our driven pendulum on its side, making the plane of the pendulum horizontal. A Hamiltonian for the driven rotor is

$$H(t, \theta, p_\theta) = \frac{p_\theta^2}{2ml^2} + mlA\omega^2 \cos\omega t \cos\theta, \tag{4.41}$$

where A is the amplitude of the drive with frequency ω, m is the mass of the bob, and l is the length of the rotor. For zero amplitude, the Hamiltonian is already in action-angle form in that it depends only on the momentum p_θ and the coordinate is an angle.

For an integrable system, the map generated on the surface of section is of the twist map form (4.39). With the addition of a small perturbation to the Hamiltonian, small corrections are added to the map

$$\begin{pmatrix} \hat\theta(i+1) \\ \hat J(i+1) \end{pmatrix} = T_\epsilon \begin{pmatrix} \hat\theta(i) \\ \hat J(i) \end{pmatrix}$$
$$= \begin{pmatrix} \hat\theta(i) + 2\pi\hat\nu(\hat J(i)) + \epsilon f(\hat\theta(i), \hat J(i)) \\ \hat J(i) + \epsilon g(\hat\theta(i), \hat J(i)) \end{pmatrix}. \tag{4.42}$$

Both the map T and the perturbed map T_ϵ are area preserving because the maps are generated as surfaces of section for Hamiltonian systems.

Suppose we are interested in determining whether periodic orbits of a particular rational rotation number $\hat\nu(\hat J(0)) = j/k$ exist in some interval of the action $\alpha < \hat J(0) < \beta$. If the rotation number is strictly monotonic in this interval and orbits with the rotation number $\hat\nu(\hat J(0))$ occur in this interval for the unperturbed map T, then by a simple construction we can show that periodic orbits with this rotation number also exist for T_ϵ for sufficiently small ϵ.

If a point is periodic for rational rotation number $\hat\nu(\hat J(0)) = j/k$, with relatively prime j and k, we expect k distinct images of the point to appear on the section. So if we consider the kth iterate of the map then the point is a fixed point of the map. For rational rotation number j/k the map T^k has a fixed point for every initial angle.

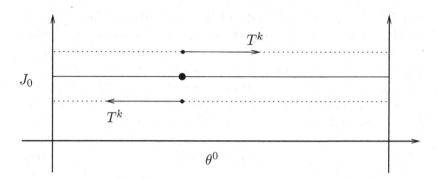

Figure 4.12 The map T^k has a line of fixed points if the rotation number is the rational j/k. Points above this line map to larger θ^0; points below this line map to smaller θ^0.

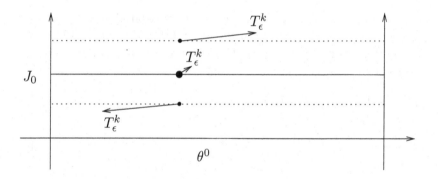

Figure 4.13 The map T_ϵ^k is slightly different from T^k, but above the central region points still map to larger θ^0 and below the central region they map to smaller θ^0. By continuity there are points between for which θ^0 does not change.

The rotation number of the map T is strictly monotonic. Suppose for definiteness we assume the rotation number $\hat{\nu}(\hat{J}(0))$ increases with $\hat{J}(0)$. For some $\hat{J}^*$ such that $\alpha < \hat{J}^* < \beta$ the rotation number is j/k, and $(\hat{\theta}^*, \hat{J}^*)$ is a fixed point of T^k for any initial $\hat{\theta}^*$. For $\hat{J}^*$ the rotation number of T^k is zero. The rotation number of the map T is monotonically increasing so for $\hat{J}(0) > \hat{J}^*$ the rotation number of T^k is positive, and for $\hat{J}(0) < \hat{J}^*$ the rotation number of T^k is negative, as long as $\hat{J}(0)$ is not too far from $\hat{J}^*$. See figure 4.12.

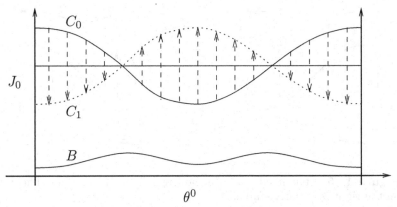

Figure 4.14   The solid curve C_0 consists of points that map to the same θ^0 under T_ϵ^k. The image C_1 of C_0 under T_ϵ^k is the dotted curve. Area preservation implies that these curves cross.

Now consider the map T_ϵ^k. In general, for small ϵ, points map to slightly different points under T_ϵ than they do under T, but not too different. So we can expect that there is still some interval in $\hat{J}(0)$ near $\hat{J}^*$ such that for $\hat{J}(0)$ in the upper end of the interval, T_ϵ^k maps points to larger θ^0, and for points in the lower end of the interval, T_ϵ^k maps points to smaller θ^0. If this is the case then for every $\hat{\theta}(0)$ there is a point somewhere in the interval, some $\hat{J}^+(\hat{\theta}(0))$, for which θ^0 does not change, by continuity. These are not fixed points because the momentum J_0 generally changes. See figure 4.13.

The map is continuous, so we can expect that $\hat{J}^+$ is a continuous function of the θ^0. The twist-map condition (see footnote 13) ensures that $\hat{J}^+$ is periodic, so $\hat{J}^+(0) = \hat{J}^+(2\pi)$. The twist-map condition also guarantees that for sufficiently small perturbations there cannot be more than one radially-mapping point for any angle. So the set of points that do not change θ^0 under T_ϵ^k form some periodic function of θ^0. Call this curve C_0. See figure 4.14.

The map T_ϵ^k takes the curve C_0 to another curve C_1 that, like C_0, is continuous and periodic. The two curves C_0 and C_1 must cross each other, as a consequence of area preservation. How do we see this? Typically, there is a lower boundary or upper boundary in J_0 for the evolution. In some situations, we have such a lower boundary because J_0 cannot be negative. For example, in action-angle variables for motion near an elliptic fixed point we will see

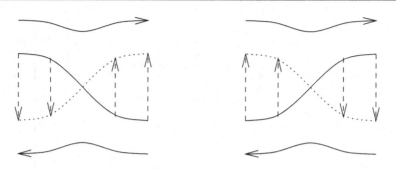

Figure 4.15   The fixed point on the left is linearly unstable. The one on the right is linearly stable.

that the action is the area enclosed on the phase plane, which cannot be negative. For other situations, we might use the fact that there are invariant curves for large positive or negative J_0. In any case, suppose there is such a barrier B. Then the area of the region between the barrier and C_0 must be equal to the area of the image of this region, which is the region between the barrier and C_1. So if C_0 and C_1 are not the same curve they must cross to contain the same area. In fact, they must cross an even number of times: they are both periodic, so if they cross once they must cross again to get back to the same side they started on. The points at which the curves C_0 and C_1 cross are fixed points because the angle does not change (that is what it means to be on C_0) and the action does not change (that is what it means for C_0 and C_1 to be the same at this point). So we have deduced that there must be an even number of fixed points of T_ϵ^k. For each fixed point of T_ϵ^k there are k images of this fixed point generated on the surface of section for the map T_ϵ. Each of these image points is a periodic point of the map T_ϵ.

We can deduce the stability of these fixed points of T_ϵ^k just from the construction. The fixed points come in two types, elliptic and hyperbolic. An elliptic (stable) fixed point appears where the steps from C_0 to C_1 join with the flow of the background twist map to encircle the fixed point. A hyperbolic (unstable) fixed point appears where the steps from C_0 to C_1 join with the flow of the background twist map to move away from the fixed point. So just from the way the arrows connect we can determine the character of the fixed point. See figure 4.15.

As we develop a Poincaré section, we find that some orbits leave traces that circulate around the stable fixed points, resulting in the

Poincaré–Birkhoff islands. If we look at a particular island we see that orbits in the island circulate around the fixed point at a rate that is monotonically dependent upon the distance from the fixed point. In the vicinity of the fixed point the evolution is governed by a twist map. So the entire Poincaré–Birkhoff construction can be carried out again. We expect that there will be concentric families of stable periodic points surrounded by islands and separated by separatrices emanating from unstable periodic points. Around each of these stable periodic orbits, the construction is repeated. So the Poincaré–Birkhoff construction is recursive, leading to the development of an infinite hierarchy of structure.

4.5.1 Computing the Poincaré–Birkhoff Construction

There are so many conditions in our construction of the fixed points that one might be suspicious. We can make the construction more convincing by actually computing the various pieces for a specific problem. Consider the periodically driven rotor, with Hamiltonian (4.41). We set $m = 1\,\text{kg}$, $l = 1\,\text{m}$, $A = 0.1\,\text{m}$, $\omega = 4.2\sqrt{9.8}\,\text{rad s}^{-1}$.

We call points that map to the same angle "radially mapping points." We find them with a simple bisection search:

```
(define (radially-mapping-points Tmap Jmin Jmax phi eps)
  (bisect
    (lambda (J)
      ((principal-value pi)
       (Tmap phi J
             (lambda (phip Jp) (- phi phip))
             (lambda () (error "should not get here")))))
    Jmin Jmax eps))
```

The procedure `Tmap` implements some map, which may be an iterate of some more primitive map. We give the procedure an angle `phi` to study, a range of actions `Jmin` to `Jmax` to search, and a tolerance `eps` for the solution.

In figure 4.16 we show the Poincaré–Birkhoff construction of the fixed points for the driven rotor. These particular curves are constructed for the two 1:1 commensurabilities between the rotation and the drive. One set of fixed points is constructed for each sense of rotation. The corresponding section is in figure 4.17. We see that the section shows the existence of fixed points exactly where the Poincaré–Birkhoff construction shows the crossing of the curves C_0 and C_1. Indeed, the nature of the fixed point is clearly reflected in the relative configuration of the C_0 and C_1 curves.

In figure 4.18 we show the result for a rotation number of 1/3. The curves are the radially mapping points for the third iterate of the section map (solid) and the images of these points (dotted). These curves are distorted by their proximity to the 1:1 islands shown in figure 4.17. The corresponding section is shown in figure 4.19.

Exercise 4.7: Computing the Poincaré–Birkhoff construction

Consider figure 3.27. Find the fixed points for the three major island chains, using the Poincaré–Birkhoff construction.

4.6　Invariant Curves

We started with an integrable system, where there are invariant curves. Do any invariant curves survive if a perturbation is added?

The Poincaré–Birkhoff construction for twist maps shows that invariant curves with rational rotation number typically do not survive perturbation. Upon perturbation the invariant curves with rational rotation numbers are replaced by an alternating sequence of stable and unstable periodic orbits. So if there are invariant curves that survive perturbation they must have irrational rotation numbers.

The perturbed system has chains of alternating stable and unstable fixed points for every rational rotation number. Each stable fixed point is surrounded by an island that occupies some region of the section. Each irrational is arbitrarily close to a rational, so it is not obvious that any invariant curve can survive an arbitrarily small perturbation.

Nevertheless, the Kolmogorov–Arnold–Moser (KAM) theorem proves that invariant curves do exist if the perturbation is small enough that the perturbed problem is "close enough" to an integrable problem, and if the rotation number is "irrational enough." We will not prove this theorem here. Instead we will develop methods for finding particular invariant curves.

Stable periodic orbits have a stable island surrounding them on the surface of section. The largest islands are associated with rationals with small denominators. In general, the size of the island

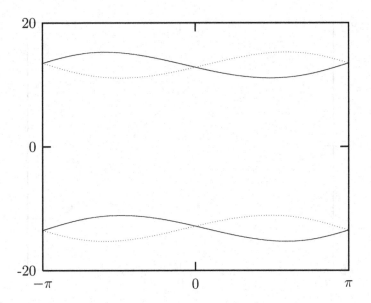

Figure 4.16   The curves C_0 (solid) and C_1 (dotted) for the 1:1 commensurability.

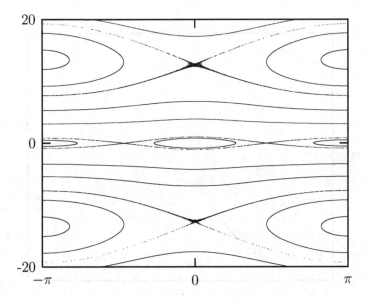

Figure 4.17   A surface of section displaying the 1:1 commensurability.

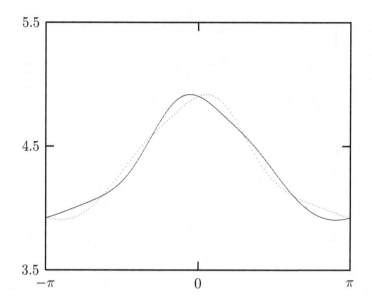

Figure 4.18   The curves C_0 (solid) and C_1 (dotted) for the 1:3 commensurability. The angle runs from $-\pi$ to π. The momentum runs from 3.5 to 4.5 in appropriate units.

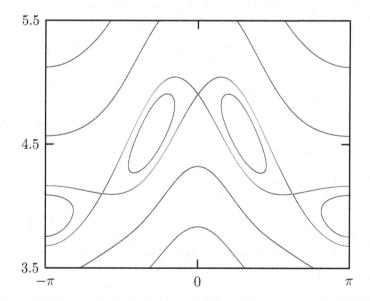

Figure 4.19   A surface of section displaying the 1:3 commensurability. The angle runs from $-\pi$ to π. The momentum runs from 3.5 to 4.5 in appropriate units.

is limited to a size that decreases as the denominator increases. These islands are a local indication of the effect of the perturbation. Similarly, the chaotic zones appear near unstable periodic orbits and their homoclinic tangles. The homoclinic tangle is a continuous curve so it cannot cross an invariant curve, which is also continuous. If we are looking for invariant curves that persist upon perturbation, we would be wise to avoid regions of phase space where the islands or homoclinic tangles are major features.

The Poincaré–Birkhoff islands are ordered by rotation number. Because of the twist condition, the rotation number is monotonic in the momentum of the unperturbed problem. If there is an invariant curve with a given rotation number, it is sandwiched between island chains associated with rational rotation numbers. The rotation number of the invariant curve must be between the rotation numbers of the island chains on either side of it.

The fact that the size of the islands decreases with the size of the denominator suggests that invariant curves with rotation numbers for which nearby rationals require large denominators are the most likely to exist. So we will begin our search for invariant curves by examining rotation numbers that are not near rationals with small denominators.

Any irrational can be approximated by a sequence of rationals, and for each of these rationals we expect there to be stable and unstable periodic orbits with stable islands and homoclinic tangles. An invariant curve for a given rotation number has the best chance of surviving if the size of the islands associated with each rational approximation to the rotation number is smaller than the separation of the islands from that invariant curve.

For any particular size denominator, the best rational approximation to an irrational number is given by an initial segment of a simple continued fraction. If the approximating continued fraction converges slowly to the irrational number, then that number is not near rationals with small denominators. Thus, we will look for invariant curves with rotation numbers that have slowly converging continued-fraction approximations. The continued fractions that converge most slowly have tails that are all one. Such a number is called a *golden number*. For example, the golden ratio,

$$\phi = \frac{1 + \sqrt{5}}{2} = 1 + \cfrac{1}{1 + \cfrac{1}{1 + \cfrac{1}{1 + \cdots}}}, \tag{4.43}$$

is just such a number.

4.6.1   Finding Invariant Curves

Invariant curves, if there are any, are characterized by a particular rotation number. Points on the invariant curve map to points on the invariant curve. Neighboring points map to neighboring points, preserving the order.

On the section for the unperturbed integrable system, the angle between successive section points is constant: $\Delta\theta = 2\pi\nu(J)$ for rotation number $\nu(J)$. This map of the circle onto itself with constant angular step we call a *uniform circle map*.

For a given rotation number, points on the section are laid down in a particular order characteristic of the rotation number only. As a perturbation is turned on, the invariant curve with a particular rotation number will be distorted and the angle between successive points will no longer be constant. All that is required to have a particular rotation number is that the average change in angle be $\Delta\theta$. Nevertheless, the ordering of the points on the surface of section is preserved, and is characteristic of the rotation number.

The fact that the sequence of points on the surface of section for an invariant curve with a given rotation number must have a particular order can be used to find the invariant curve. At a specified angle we perform a bisection search for the momentum that lies on the invariant curve. We can tell whether the initial point is on the desired invariant curve or which side of the invariant curve it is on by evolving a candidate initial point with both the perturbed map and the uniform circle map and comparing the ordering of the sequences of points that are generated.

A program to implement this plan of attack is[15]

```
(define (find-invariant-curve the-map rn theta0 Jmin Jmax eps)
  (bisect (lambda (J) (which-way? rn theta0 J the-map))
          Jmin Jmax eps))
```

Since ordering inconsistencies are found near the initial angle we do not need to keep the whole list of angles. Instead, we can keep track of a small list of angles near the initial angle. In fact, keeping

---

[15]This method depends on the assumptions that Jmin and Jmax bracket the actual momentum, and that the rotation number is sufficiently continuous in momentum in that region.

track of the nearest angle on either side of the initial angle works well.

The procedure which-way? is implemented as a simple loop with state variables for the two orbits and the endpoints of the intervals. The z variables keep track of the angle of the uniform circle map; the x variables keep track of the angle of the map under study. The y variable is the momentum for the map under study. On each iteration we determine if the angle of the uniform circle map is in the interval of interest below or above the initial angle. If it is in neither interval then the map is further iterated. However, if it is in the region of interest then we check to see if the angle of the other map is in the corresponding interval. If so, the intervals for the uniform circle map and the other map are narrowed and the iteration proceeds. If the angle is not in the required interval, a discrepancy is noted and the sign of the discrepancy is reported. For this process to make sense the differences between the angles for successive iterations of both maps must be less than π.

```
(define (which-way? rotation-number x0 y0 the-map)
  (let ((pv (principal-value (+ x0 pi))))
    (let lp ((z x0) (zmin (- x0 :2pi)) (zmax (+ x0 :2pi))
             (x x0) (xmin (- x0 :2pi)) (xmax (+ x0 :2pi))
             (y y0))
      (let ((nz (pv (+ z (* :2pi rotation-number)))))
        (the-map x y
                 (lambda (nx ny)
                   (let ((nx (pv nx)))
                     (cond ((< x0 z zmax)
                            (if (< x0 x xmax)
                                (lp nz zmin z nx xmin x ny)
                                (if (> x xmax) 1 -1)))
                           ((< zmin z x0)
                            (if (< xmin x x0)
                                (lp nz z zmax nx x xmax ny)
                                (if (< x xmin) -1 1)))
                           (else
                            (lp nz zmin zmax nx xmin xmax ny)))))
                 (lambda ()
                   (error "Map failed" x y)))))))
```

With this method of comparing rotation numbers we can find the initial momentum (for a given initial angle) for an invariant curve with a given rotation number to high precision.

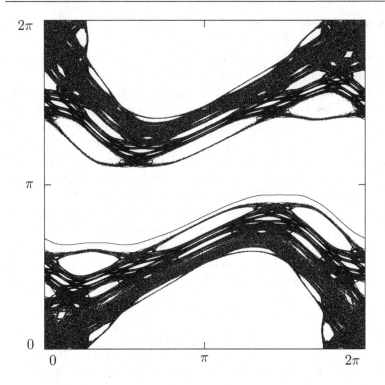

Figure 4.20   A surface of section displaying the invariant curve at rotation number $1 - 1/\phi$ for the standard map with $K = .95$. The invariant curve is in context: there is a chaotic region that almost eats the curve. The angle and momentum run from 0 to 2π.

We search the standard map for an invariant curve with a golden rotation number:[16]

```
(find-invariant-curve (standard-map 0.95)
                      (- 1 (/ 1 golden-ratio))
                      0.0
                      2.0
                      2.2
                      1e-16)
;Value: 2.1144605494391726
```

Using initial conditions computed in this way, we can produce the invariant curve (see figure 4.20). If we expand the putative

---

[16]There is no invariant curve in the standard map that has rotation number $\phi = 1.618....$ However, $1 - 1/\phi$ has the same continued-fraction tail as ϕ and this rotation number appears in the standard map.

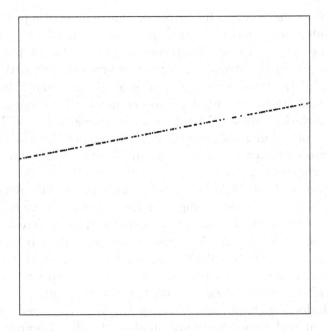

Figure 4.21 Here is a small portion of the invariant curve shown in figure 4.20, magnified by $2\pi \times 10^7$. We see that even at this magnification the points appear to lie on a line. We also see that the visitation frequency of points is highly nonuniform.

invariant curve it should remain a curve for all magnifications—it should show no sign of chaotic fuzziness (see figure 4.21).

Exercise 4.8: Invariant curves in the standard map

Find an invariant curve of the standard map with a different golden rotation number. Expand it to show that it retains the features of a curve at high magnification.

4.6.2 Dissolution of Invariant Curves

As can be seen in figure 4.21, the points on an invariant curve are not uniformly visited, unlike the picture we would get plotting the angles for the uniform circle map. This is because an interval may be expanded or compressed when mapped. We can compute the relative probability density for visitation of each angle on the invariant curve. A crude way to obtain this result is to count the number of points that fall into equal incremental angle bins. It is more effective to use the linear variational map constructed from the map being investigated to compute the change in incremental angle from one point to its successor. Since all of the points in a small interval around the source point are mapped to points (in the same order) in a small interval around the target point, the relative probability density at a point is inversely proportional to the size of the incremental interval around that point. In order to get this started we need a good estimate of the initial slope for the invariant curve. We can estimate the slope by a difference quotient of the momentum and angle increments for the interval that we used to refine the momentum of the invariant curve with a given rotation number.

Figures 4.22 and 4.23 show the relative probability density of visitation as a function of angle for the invariant curve of golden rotation number in the standard map for three different values of the parameter K. As K increases, certain angles become less likely. Near $K = 0.971635406$ some angles are never visited. But the invariant curve must be continuous. Thus it appears that for larger K the invariant curve with this rotation number will not exist. Indeed, if the invariant set persists with the given rotation number it will have an infinite number of holes (because it has an irrational rotation number). Such a set is sometimes called a *cantorus* (plural *cantori*).

4.7 Summary

Surfaces of section of a typical Hamiltonian system exhibit a menagerie of features including fixed points, invariant curves, resonance islands, and chaotic zones. Integrable systems have much simpler surfaces of section. By adding small effects to integrable systems we get insight into how this complicated behavior emerges.

Surfaces of section for integrable systems display only certain characteristic orbit types. There are fixed points, which correspond to equilibria or periodic orbits. A fixed point may be stable or unstable, depending on the stability of the corresponding equilibrium or orbit. There are sets of points on the section that are asymptotic forward and backward in time to the unstable fixed point. And there are sets of trajectories that fall on invariant curves. If the rotation number of the invariant curve is irrational, each of these trajectories densely covers the invariant curve; if the rotation number is rational, then each trajectory visits only a finite number of points on the invariant curve.

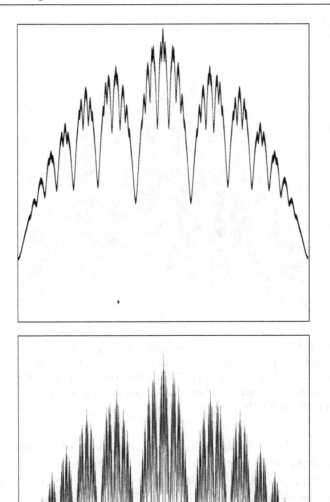

Figure 4.22 The relative probability density of visitation as a function of angle for the invariant curve of golden rotation number in the standard map with $K = 0.95$ (above) and $K = 0.97$ (below). As K increases, the function becomes more complex and certain angles become less likely to be visited.

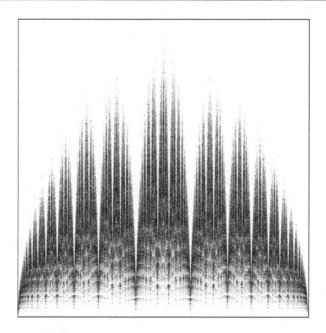

Figure 4.23   The relative probability density of visitation as a function of angle for the invariant curve of golden rotation number in the standard map with $K = 0.971635406$. Here the function is very complex and appears self-similar. The valleys appear to reach to zero, so there are discrete angles that are never visited.

Linear stability analysis addresses the nature of the motion near the fixed points on the section. These points correspond to either equilibrium points or periodic orbits. There are characteristic frequencies of the motion, each with an associated characteristic direction. For Hamiltonian systems only certain patterns of characteristic frequencies are possible. On two-dimensional area-preserving surfaces of section, as generated by Hamiltonian systems, fixed points are linearly stable (elliptic fixed points) or linearly unstable (hyperbolic fixed points).

With the addition of small effects, the surface of section changes in certain typical ways. One characteristic change occurs near the unstable fixed points. The stable and unstable manifolds, those curves consisting of the sets of points that are asymptotic to the unstable fixed points forward and backward in time, no longer join smoothly, but instead cross. A first crossing implies that there are an infinite number of other crossings, and the stable and unstable manifolds develop an extremely complicated tangle.

The Poincaré–Birkhoff construction shows how the infinite number of periodic orbits on an invariant curve with rational rotation number that is characteristic of an integrable system degenerates into a finite number of alternating stable and unstable fixed points when the system becomes nonintegrable. This phenomenon is recursive, so we find that it develops an infinite hierarchy of structure: The region around every stable fixed point is itself filled with commensurabilities with alternating stable and unstable fixed points.

Some invariant curves survive the addition of small effects to an integrable system. The Kolmogorov–Arnold–Moser theorem proves that some invariant curves persist upon perturbation. We can find invariant curves of particular rotation numbers by comparing the pattern of points generated for a candidate initial point on the invariant curve to the expected pattern of points for the invariant curve being sought. As the additional effect is made stronger, the invariant curves that survive longest are those with the most irrational rotation number. At the point of breakup, the probability of visitation of various points on the invariant curve develops a self-similar appearance. For larger perturbations, the invariant curve disappears, leaving an invariant set with an infinite number of holes.

4.8   Projects

Exercise 4.9: Secondary islands

In figure 4.3 (section 4.1) we see a chain of six secondary islands in the oscillation region. Carry out the Poincaré–Birkhoff construction to obtain the alternating sequence of stable and unstable fixed points for this island chain.

Exercise 4.10: Invariant curves of the standard map

a. Make programs that reproduce figures 4.22 and 4.23. You will need to develop an effective method of estimating the probability of visitation. There is one suggestion of how to do that in the text, but you may find a better way.

b. As the parameter K is increased beyond the critical value, the golden invariant curve ceases to exist. Investigate how the method for finding invariant curves fails beyond the critical value of K.

5
Canonical Transformations

> We have done considerable mountain climbing.
> Now we are in the rarefied atmosphere of theories
> of excessive beauty and we are nearing a high
> plateau on which geometry, optics, mechanics, and
> wave mechanics meet on common ground. Only
> concentrated thinking, and a considerable amount
> of re–creation, will reveal the beauty of our subject
> in which the last word has not been spoken.
>
> Cornelius Lanczos, *The Variational Principles of
> Mechanics* [29], p. 229

One way to simplify the analysis of a problem is to express it in a form in which the solution has a simple representation. However, it may not be easy to formulate the problem in such a way initially. It is often useful to start by formulating the problem in one way, and then transform it. For example, the formulation of the problem of the motion of a number of gravitating bodies is simple in rectangular coordinates, but it is easier to understand aspects of the motion in terms of orbital elements, such as the semimajor axes, eccentricities, and inclinations of the orbits. The semimajor axis and eccentricity of an orbit depend on both the configuration and the velocity of the body. Such transformations are more general than those that express changes in configuration coordinates. Here we investigate transformations of phase-space coordinates that involve both the generalized coordinates and the generalized momenta.

Suppose we have two different Hamiltonian systems, and suppose the trajectories of the two systems are in one-to-one correspondence. In this case both Hamiltonian systems can be mathematical models of the same physical system. Some questions about the physical system may be easier to answer by reference to one model and others may be easier to answer in the other model. For example, it may be easier to formulate the physical system in one model and to discover a conserved quantity in the other. Canonical transformations are maps between Hamiltonian systems that preserve the dynamics.

A *canonical transformation* is a phase-space coordinate transformation and an associated transformation of the Hamiltonian such that the dynamics given by Hamilton's equations in the two representations describe the same evolution of the system.

5.1   Point Transformations

A *point transformation* is a canonical transformation that extends a possibly time-dependent transformation of the configuration coordinates to a phase-space transformation. For example, one might want to reexpress motion in terms of polar coordinates, given a description in terms of rectangular coordinates. In order to extend a transformation of the configuration coordinates to a phase-space transformation we must specify how the momenta and Hamiltonian are transformed.

We have already seen how coordinate transformations can be carried out in the Lagrangian formulation (see section 1.6.1). In that case, we found that if the Lagrangian transforms by composition with the coordinate transformation, then the Lagrange equations are equivalent.

Lagrangians that differ by the addition of a total time derivative have the same Lagrange equations, but may have different momenta conjugate to the generalized coordinates. So there is more than one way to make a canonical extension of a coordinate transformation.

Here, we find the particular canonical extension of a coordinate transformation for which the Lagrangians transform by composition with the transformation, with no extra total time derivative terms added to the Lagrangian.

Let L be a Lagrangian for a system. Consider the coordinate transformation $q = F(t, q')$. The velocities transform by

$$v = \partial_0 F(t, q') + \partial_1 F(t, q')v'. \tag{5.1}$$

We obtain a Lagrangian L' in the transformed coordinates by composition of L with the coordinate transformation. We require that $L'(t, q', v') = L(t, q, v)$, so:

$$L'(t, q', v') = L(t, F(t, q'), \partial_0 F(t, q') + \partial_1 F(t, q')v'). \tag{5.2}$$

The momentum conjugate to q' is

$$
\begin{aligned}
p' &= \partial_2 L'(t, q', v') \\
&= \partial_2 L(t, F(t, q'), \partial_0 F(t, q') + \partial_1 F(t, q')v') \, \partial_1 F(t, q') \\
&= p\partial_1 F(t, q'),
\end{aligned}
\tag{5.3}
$$

where we have used

$$
\begin{aligned}
p &= \partial_2 L(t, q, v) \\
&= \partial_2 L(t, F(t, q'), \partial_0 F(t, q') + \partial_1 F(t, q')v').
\end{aligned}
\tag{5.4}
$$

So, from equation (5.3),[1]

$$
p = p'(\partial_1 F(t, q'))^{-1}.
\tag{5.5}
$$

We can collect these results to define a canonical phase-space transformation C_H:[2]

$$
\begin{aligned}
(t, q, p) &= C_H(t, q', p') \\
&= (t, F(t, q'), p'(\partial_1 F(t, q'))^{-1}).
\end{aligned}
\tag{5.6}
$$

The Hamiltonian is obtained by the Legendre transform

$$
\begin{aligned}
H'&(t, q', p') \\
&= p'v' - L'(t, q', v') \\
&= (p\partial_1 F(t, q')) \left((\partial_1 F(t, q')^{-1}(v - \partial_0 F(t, q'))) \right) - L(t, q, v) \\
&= pv - L(t, q, v) - p\partial_0 F(t, q') \\
&= H(t, q, p) - p\partial_0 F(t, q'),
\end{aligned}
\tag{5.7}
$$

using relations (5.1) and (5.3) in the second step. Fully expressed in terms of the transformed coordinates and momenta, the trans-

---

[1] Solving for p in terms of p' involves multiplying equation (5.3) on the right by $(\partial_1 F(t, q'))^{-1}$. This inverse is the structure that when multiplying $\partial_1 F(t, q')$ on the right gives an identity structure. Structures representing linear transformations may be represented in terms of matrices. In this case, the matrix representation of the inverse structure is the inverse of the matrix representing the given structure.

[2] In chapter 1 the transformation C takes a local tuple in one coordinate system and gives a local tuple in another coordinate system. In this chapter C_H is a phase-space transformation.

formed Hamiltonian is

$$H'(t, q', p') = H(t, F(t, q'), p'(\partial_1 F(t, q'))^{-1})$$
$$- (p'(\partial_1 F(t, q'))^{-1})\partial_0 F(t, q'). \tag{5.8}$$

The Hamiltonians H' and H are equivalent because L and L' have the same value for a given dynamical state and so have the same paths of stationary action. In general H and H' do not have the same values for a given dynamical state, but differ by a term that depends on the coordinate transformation.

For time-independent transformations, $\partial_0 F = 0$, there are a number of simplifications. The relationship of the velocities (5.1) becomes

$$v = \partial_1 F(t, q')v'. \tag{5.9}$$

Comparing this to the relation (5.5) between the momenta, we see that in this case the momenta transform "oppositely" to the velocities[3]

$$pv = p'(\partial_1 F(t, q'))^{-1}\partial_1 F(t, q')v' = p'v', \tag{5.10}$$

so the product of the momenta and the velocities is not changed by the transformation. This, combined with the fact that by construction $L(t, q, v) = L'(t, q', v')$, shows that

$$H(t, q, p) = pv - L(t, q, v)$$
$$= p'v' - L'(t, q', v')$$
$$= H'(t, q', p'). \tag{5.11}$$

For time-independent coordinate transformations the Hamiltonian transforms by composition with the associated phase-space transformation. We can also see this from the general relationship (5.7) between the Hamiltonians.

---

[3]The velocities and the momenta are dual geometric objects with respect to time-independent point transformations. The velocities are coordinates of a vector field on the configuration manifold, and the momenta are coordinates of a covector field on the configuration manifold. The invariance of the inner product pv under time-independent point transformations provides a motivation for our use of superscripts for velocity components and subscripts for momentum components.

Implementing point transformations

The procedure F->CH takes a procedure F implementing a transformation of configuration coordinates and returns a procedure
implementing a transformation of phase-space coordinates:[4]

```
(define ((F->CH F) state)
  (up (time state)
      (F state)
      (solve-linear-right (momentum state)
                          (((partial 1) F) state))))
```

Consider a particle moving in a central field. In rectangular
coordinates a Hamiltonian is

```
(define ((H-central m V) state)
  (let ((x (coordinate state))
        (p (momentum state)))
    (+ (/ (square p) (* 2 m))
       (V (sqrt (square x))))))
```

Let's look at this Hamiltonian in polar coordinates. The phase-
space transformation is obtained by applying F->CH to the procedure p->r that takes a time and a polar tuple and returns a
tuple of rectangular coordinates (see section 1.6.1). The transformation is time independent so the Hamiltonian transforms by
composition. In polar coordinates the Hamiltonian is

```
(show-expression
 ((compose (H-central 'm (literal-function 'V))
           (F->CH p->r))
  (up 't (up 'r 'phi) (down 'p_r 'p_phi))))
```

$$
V\left(r\right) + \frac{\frac{1}{2}p_r^2}{m} + \frac{\frac{1}{2}p_\varphi^2}{mr^2}
$$

There are three terms. There is the potential energy, which depends on the radius, there is the kinetic energy due to radial motion, and there is the kinetic energy due to tangential motion. As
expected, the angle φ does not appear and thus the angular mo-

---

[4]The procedure solve-linear-right multiplies its first argument by the inverse of its second argument on the right. So, if $u = vM$ then $v = uM^{-1}$;
(solve-linear-right u M) produces v.

mentum is a conserved quantity. By going to polar coordinates we have decoupled one of the two degrees of freedom in the problem.

If the transformation is time varying the Hamiltonian must be adjusted by adding a correction to the composition of the Hamiltonian and the transformation (see equation 5.8):

$$H' = H \circ C_{\mathrm{H}} + K \qquad\qquad (5.12)$$

The correction is computed by

```
(define ((F->K F) state)
  (- (* (solve-linear-right (momentum state)
                            (((partial 1) F) state))
        (((partial 0) F) state))))
```

For example, consider a transformation to coordinates translating with velocity v:

```
(define ((translating v) state)
  (+ (coordinates state) (* v (time state)))))
```

We compute the additive adjustment required for the Hamiltonian:

```
((F->K (translating (up 'v^x 'v^y 'v^z)))
 (up 't (up 'x 'y 'z) (down 'p_x 'p_y 'p_z)))
(+ (* -1 p_x v^x) (* -1 p_y v^y) (* -1 p_z v^z))
```

Notice that this is the negation of the inner product of the momentum and the velocity of the coordinate system.

Let's see how a simple free-particle Hamiltonian is transformed:

```
(define ((H-free m) s)
  (/ (square (momentum s)) (* 2 m)))
```

The transformed Hamiltonian is:

```
(define H-prime
  (+ (compose (H-free 'm)
              (F->CH (translating (up 'v^x 'v^y 'v^z))))
     (F->K (translating (up 'v^x 'v^y 'v^z)))))
```

```
(H-prime
 (up 't
     (up 'xprime 'yprime 'zprime)
     (down 'pprime_x 'pprime_y 'pprime_z)))
(+ (* -1 pprime_x v^x)
   (* -1 pprime_y v^y)
   (* -1 pprime_z v^z)
   (/ (* 1/2 (expt pprime_x 2)) m)
   (/ (* 1/2 (expt pprime_y 2)) m)
   (/ (* 1/2 (expt pprime_z 2)) m))
```

Exercise 5.1: Galilean invariance

Is this result what you expected? Let's investigate.

Recall that in exercise 1.29 we showed that if the kinetic energy is $\frac{1}{2}mv^2$ then the translation to a uniformly moving coordinate system introduces extra terms that can be identified as a total time derivative. Since these terms do not affect the Lagrange equations, we can take the kinetic energy in the transformed coordinates to also be $\frac{1}{2}m(v')^2$.

Let C_{H} be the phase space extension of the translation transformation, and C be the local tuple extension. The transformed Hamiltonian is $H' = H \circ C_{\mathrm{H}} + K$; the transformed Lagrangian is $L' = L \circ C$.

a. Derive the relationship between p and p' both from C_{H} and from the Lagrangians. Are they the same? Derive the relationship between v and v' by taking the derivative of the Hamiltonians with respect to the momenta (Hamilton's equation). Show that the Legendre transform of L' gives the same H'.

b. We have shown that L and L' differ by a total time derivative. So for any uniformly moving coordinate system we can write the Lagrangian as $\frac{1}{2}mv^2$. Similarly, we would expect to always be able to write the Hamiltonian as $p^2/(2m)$. Show that this differs from H' by a total time derivative in the corresponding Lagrangians.

Exercise 5.2: Rotations

Let q and q' be rectangular coordinates that are related by a rotation R: $q = Rq'$. The Lagrangian for the system is $L(t, q, v) = \frac{1}{2}mv^2 - V(q)$. Find the corresponding phase-space transformation C_{H}. Compare the transformation equations for the rectangular components of the momenta to those for the rectangular components of the velocities. Are you surprised, considering equation (5.10)?

5.2   General Canonical Transformations

Although we have shown how to extend any coordinate transformation of the configuration space to a canonical transformation, there are other ways to construct canonical transformations. How do we know if we have a canonical transformation? To test if a transformation is canonical we may use the fact that if the transformation is canonical, then Hamilton's equations of motion for the transformed system and the original system will be equivalent.

Consider a Hamiltonian H and a phase-space transformation C_H. Let D_s be the function that takes a Hamiltonian and gives the Hamiltonian state-space derivative:[5]

$$D_s H(t,q,p) = (1, \partial_2 H(t,q,p), -\partial_1 H(t,q,p)). \tag{5.13}$$

Hamilton's equations are

$$D\sigma = D_s H \circ \sigma, \tag{5.14}$$

for any realizable phase-space path σ.

The transformation C_H transforms the phase-space path $\sigma'(t) = (t, q'(t), p'(t))$ into $\sigma(t) = (t, q(t), p(t))$:

$$\sigma = C_H \circ \sigma'. \tag{5.15}$$

The rates of change of the phase-space coordinates are transformed by the derivative of the transformation

$$D\sigma = D(C_H \circ \sigma') = (DC_H \circ \sigma') \, D\sigma'. \tag{5.16}$$

The transformation is canonical if the equations of motion obtained from the new Hamiltonian are the same as those that could be obtained by transforming the equations of motion derived from the original Hamiltonian to the new coordinates:

$$D\sigma = (DC_H \circ \sigma') \, D\sigma' = (DC_H \circ \sigma') \, (D_s H' \circ \sigma'). \tag{5.17}$$

Using equation (5.14), we see that

$$D_s H \circ \sigma = (DC_H \circ \sigma') \, (D_s H' \circ \sigma'). \tag{5.18}$$

---

[5]D_s is not a derivative operator. It is not linear because the time component is a nonzero constant.

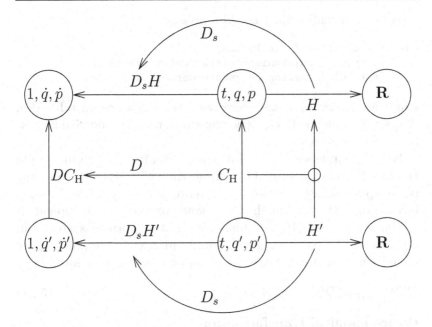

Figure 5.1   A canonical transformation C_H relates the descriptions of a dynamical system in two phase-space coordinate systems. The transformation shows how Hamilton's equations in one coordinate system may be derived from Hamilton's equations in the other coordinate system.

With $\sigma = C_H \circ \sigma'$, we find

$$D_s H \circ C_H \circ \sigma' = (DC_H \circ \sigma')\,(D_s H' \circ \sigma'). \tag{5.19}$$

This condition must hold for any realizable phase-space path σ'. Certainly this is true if the following condition holds for every phase-space point:[6]

$$D_s H \circ C_H = DC_H \cdot D_s H'. \tag{5.20}$$

Any transformation that satisfies equation (5.20) is a canonical transformation among phase-space representations of a dynamical system. In one phase-space representation the system's dynamics is characterized by the Hamiltonian H' and in the other by H. The idea behind this equation is illustrated in figure 5.1.

---

[6]Sometimes we use a center dot to indicate multiplication, to avoid the ambiguity of the use of juxtaposition to indicate both multiplication and function application. This is not to be interpreted as a vector dot product.

We can formalize this test as a program:

```
(define (canonical? C H Hprime)
  (- (compose (Hamiltonian->state-derivative H) C)
     (* (D C) (Hamiltonian->state-derivative Hprime))))
```

where `Hamiltonian->state-derivative`, which was introduced in chapter 3, implements D_s. The transformation is canonical if these residuals are zero.

For time-independent point transformations an appropriate Hamiltonian can be formed by composition with the corresponding phase-space transformation. For more general canonical transformations, we will see that if a transformation is independent of time, a suitable Hamiltonian for the transformed system can be obtained by composing the Hamiltonian with the phase-space transformation. In this case we obtain a more specific formula:

$$D_s H \circ C_H = DC_H \cdot D_s(H \circ C_H). \tag{5.21}$$

Polar-canonical transformation

The analysis of the harmonic oscillator illustrates the use of a general canonical transformation in the solution of a problem. The harmonic oscillator is a mathematical model of a simple spring-mass system. A Hamiltonian for a spring-mass system with mass m and spring constant k is

$$H(t, x, p_x) = \frac{p_x^2}{2m} + \frac{1}{2}kx^2. \tag{5.22}$$

Hamilton's equations of motion are

$$
\begin{aligned}
Dx &= p_x/m \\
Dp_x &= -kx,
\end{aligned}
\tag{5.23}
$$

giving the second-order system

$$mD^2x + kx = 0. \tag{5.24}$$

The solution is

$$x(t) = A\sin(\omega t + \varphi), \tag{5.25}$$

where

$$\omega = \sqrt{k/m} \tag{5.26}$$

and where A and φ are determined by initial conditions.

We use the polar-canonical transformation:

$$(t, x, p_x) = C_\alpha\,(t, \theta, I) \tag{5.27}$$

where

$$x = \sqrt{\frac{2I}{\alpha}}\,\sin\theta \tag{5.28}$$

$$p_x = \sqrt{2\alpha I}\,\cos\theta. \tag{5.29}$$

Here α is an arbitrary parameter. We define:

```
(define ((polar-canonical alpha) state)
  (let ((t (time state))
        (theta (coordinate state))
        (I (momentum state)))
    (let ((x (* (sqrt (/ (* 2 I) alpha)) (sin theta)))
          (p_x (* (sqrt (* 2 alpha I)) (cos theta))))
      (up t x p_x))))
```

And now we just run our test:

```
(define ((H-harmonic m k) s)
  (+ (/ (square (momentum s)) (* 2 m))
     (* 1/2 k (square (coordinate s)))))

((canonical? (polar-canonical 'alpha)
             (H-harmonic 'm 'k)
             (compose (H-harmonic 'm 'k)
                      (polar-canonical 'alpha)))
 (up 't 'theta 'I))
(up 0 0 0)
```

So the transformation is canonical for the harmonic oscillator.[7]

---

[7]Actually, for $I = 0$ the transform is not well defined and so it is not canonical for that value. This transformation is "locally canonical" in that it is canonical for nonzero values of I. We will ignore this essentially topological problem.

Let's use our polar-canonical transformation C_α to help us solve the harmonic oscillator. We substitute expressions (5.28) and (5.29) for x and p_x in the Hamiltonian, getting our new Hamiltonian:

$$H'(t, \theta, I) = \frac{\alpha I}{m}(\cos\theta)^2 + \frac{kI}{\alpha}(\sin\theta)^2. \qquad (5.30)$$

If we choose $\alpha = \sqrt{km}$ then we obtain

$$H'(t, \theta, I) = \sqrt{\frac{k}{m}} I = \omega I, \qquad (5.31)$$

and the new Hamiltonian no longer depends on the coordinate. Hamilton's equation for I is

$$DI(t) = -\partial_1 H'(t, \theta(t), I(t)) = 0, \qquad (5.32)$$

so I is constant. The equation for θ is

$$D\theta(t) = \partial_2 H'(t, \theta(t), I(t)) = \omega, \qquad (5.33)$$

so

$$\theta(t) = \omega t + \varphi. \qquad (5.34)$$

In the original variables,

$$\begin{aligned} x(t) &= \sqrt{2I(t)/\alpha} \sin\theta(t) \\ &= A\sin(\omega t + \varphi), \end{aligned} \qquad (5.35)$$

with the constant $A = \sqrt{2I(t)/\alpha}$. So we have found the solution to the problem by making a canonical transformation to new phase-space variables for which the solution is easy and then transforming the solutions back to the original variables.

Exercise 5.3: Trouble in Lagrangian world

Is there a Lagrangian L' that corresponds to the harmonic oscillator Hamiltonian $H'(t, \theta, I) = \omega I$? What could this possibly mean?

Exercise 5.4: Group properties

If we say that C_H is canonical with respect to Hamiltonians H and H' if and only if $D_s H \circ C_H = DC_H \cdot D_s H'$, then:

a. Show that the composition of canonical transformations is canonical.

b. Show that composition of canonical transformations is associative.

c. Show that the identity transformation is canonical.

d. Show that there is an inverse for a canonical transformation and the inverse is canonical.

5.2.1   Time-Dependent Transformations

We have seen that for time-dependent point transformations the Hamiltonian appropriate for the transformed system is the original Hamiltonian composed with the transformation and augmented with an additive correction. Here we find a similar decomposition for general time-dependent canonical transformations.

The key to this decomposition is to separate the time part and the phase-space part of the Hamiltonian state derivative:[8]

$$D_s H(s) = (1, +\partial_2 H(s), -\partial_1 H(s))$$
$$= T(s) + \mathcal{D}H(s) \tag{5.36}$$

where

$$T(s) = (1, 0, 0), \tag{5.37}$$
$$\mathcal{D}H(s) = (0, +\partial_2 H(s), -\partial_1 H(s)), \tag{5.38}$$

as code:[9]

```
(define (T-func s)
  (up 1
      (zero-like (coordinates s))
      (zero-like (momenta s))))

(define ((D-phase-space H) s)
  (up 0 (((partial 2) H) s) (- (((partial 1) H) s))))
```

If we assume that $H' = H \circ C_{\mathrm{H}} + K$, then the canonical condition (5.20) becomes

$$D_s H \circ C_{\mathrm{H}} = DC_{\mathrm{H}} \cdot D_s(H \circ C_{\mathrm{H}} + K). \tag{5.39}$$

Expanding the state derivative, the canonical condition is

$$(T + \mathcal{D}H) \circ C_{\mathrm{H}} = DC_{\mathrm{H}} \cdot (T + \mathcal{D}(H \circ C_{\mathrm{H}} + K)). \tag{5.40}$$

---

[8] Unlike D_s, $\mathcal{D}$ is linear and can be a derivative operator.

[9] The procedure `zero-like` produces a structure of zeros with the shape of its argument.

Equation (5.40) is satisfied if the following conditions are met:

$$\mathcal{D}H \circ C_{\mathrm{H}} = DC_{\mathrm{H}} \cdot \mathcal{D}(H \circ C_{\mathrm{H}}) \qquad (5.41)$$

$$T \circ C_{\mathrm{H}} = DC_{\mathrm{H}} \cdot (T + \mathcal{D}K). \qquad (5.42)$$

The value of $T \circ C_{\mathrm{H}}$ does not depend on C_{H}, so this term is really very simple. Notice that equation (5.41) does not depend upon K and that equation (5.42) does not depend upon H.

These can be implemented as follows:

```
(define (canonical-H? C H)
  (- (compose (D-phase-space H) C)
     (* (D C)
        (D-phase-space (compose H C)))))

(define (canonical-K? C K)
  (- (compose T-func C)
     (* (D C)
        (+ T-func (D-phase-space K)))))
```

Rotating coordinates

Consider a time-dependent transformation to uniformly rotating coordinates:[10]

$$q = R(\Omega)(t, q'), \qquad (5.43)$$

with components

$$x = x' \cos(\Omega t) - y' \sin(\Omega t)$$
$$y = x' \sin(\Omega t) + y' \cos(\Omega t). \qquad (5.44)$$

As a program this is

```
(define ((rotating Omega) state)
  (let ((t (time state)) (qp (coordinate state)))
    (let ((xp (ref qp 0)) (yp (ref qp 1)) (zp (ref qp 2)))
      (up (- (* (cos (* Omega t)) xp)
             (* (sin (* Omega t)) yp))
          (+ (* (sin (* Omega t)) xp)
             (* (cos (* Omega t)) yp))
          zp))))
```

The extension of this transformation to a phase-space transformation is

---

[10]This is just a rearrangement of the arguments of R_z: $R(\Omega)(t, q') = R_z(\Omega t)(q')$.

```
(define (C-rotating Omega) (F->CH (rotating Omega)))
```

We first verify that this time-dependent transformation satisfies
equation (5.41). We will try it for an arbitrary Hamiltonian with
three degrees of freedom:

```
(define H-arbitrary
  (literal-function 'H
    (-> (UP Real (UP Real Real Real) (DOWN Real Real Real))
        Real)))
```

```
((canonical-H? (C-rotating 'Omega) H-arbitrary)
 (up 't (up 'xp 'yp 'zp) (down 'pp_x 'pp_y 'pp_z)))
(up 0 (up 0 0 0) (down 0 0 0))
```

And it works. Note that this result did not depend on any details
of the Hamiltonian, suggesting that we might be able to make a
test that does not require a Hamiltonian. We will see that shortly.

Since we have a point transformation, we can compute the required adjustment to the Hamiltonian:

```
((F->K (rotating 'Omega))
 (up 't (up 'xp 'yp 'zp) (down 'pp_x 'pp_y 'pp_z)))
(+ (* Omega pp_x yp) (* -1 Omega pp_y xp))
```

So, for this transformation an appropriate correction to the Hamiltonian is

$$K(\Omega)(t; x', y', z'; p'_x, p'_y, p'_z) = -\Omega(x'p'_y - y'p'_x), \qquad (5.45)$$

which is minus the rate of rotation of the coordinate system multiplied by the angular momentum. We implement K as a procedure

```
(define ((K Omega) s)
  (let ((qp (coordinate s)) (pp (momentum s)))
    (let ((xp (ref qp 0)) (yp (ref qp 1))
          (ppx (ref pp 0)) (ppy (ref pp 1)))
      (* -1 Omega (- (* xp ppy) (* yp ppx))))))
```

and apply the test. We find:

```
((canonical-K? (C-rotating 'Omega) (K 'Omega))
 (up 't (up 'xp 'yp 'zp) (down 'pp_x 'pp_y 'pp_z)))
(up 0 (up 0 0 0) (down 0 0 0))
```

The residuals are zero so this K correctly completes the canonical
transformation.

5.2.2   Abstracting the Canonical Condition

We just saw that for the case of rotating coordinates the truth of equation (5.41) did not depend on the details of the Hamiltonian. If C_H satisfies equation (5.41) for any H then we can derive a condition on C_H that is independent of H.

Let's start with an expanded version of equation (5.41):

$$\mathcal{D}H \circ C_H = DC_H \cdot ((\mathcal{D}H \circ C_H) \cdot DC_H), \qquad (5.46)$$

using the chain rule.

We introduce a shuffle function:

$$\tilde{J}([a, b, c]) = (0, c, -b). \qquad (5.47)$$

The argument to $\tilde{J}$ is a down tuple of components of the derivative of a Hamiltonian-like function. The shuffle function is linear. Using $\tilde{J}$ we can write $\mathcal{D}H = \tilde{J} \circ DH$.

Let J be the multiplier corresponding to the constant linear function $\tilde{J}$:

$$J = (D\tilde{J})(s^\star), \qquad (5.48)$$

where $s^\star$ is an arbitrary argument, shaped like $DH(s)$, that is compatible for multiplication with s. The value of $s^\star$ is irrelevant because $D\tilde{J}$ is a constant function. Then we can rewrite equation (5.46) as

$$J \cdot DH(C_H(s')) = DC_H(s') \cdot J \cdot (DH(C_H(s')) \cdot DC_H(s')). \qquad (5.49)$$

We can move the $DC_H(s')$ to the left of $DH(C_H(s'))$ by taking its transpose:[11]

$$
\begin{aligned}
J \cdot DH(C_H(s')) & \\
= DC_H(s') & \cdot J \cdot ((DC_H(s'))^{\mathrm{T}} \cdot DH(C_H(s'))).
\end{aligned} \qquad (5.50)
$$

---

[11] For each linear transformation $T : A \to A$ of incremental phase-space states there is a unique linear transformation $T^{\mathrm{T}} : A^\star \to A^\star$ of the dual space, called the *transpose* of T, such that for every real-valued linear function $g : A \to \mathbf{R}$ of incremental phase-space states, and for every $a \in A$ we have $(T^{\mathrm{T}}(g))(a) = g(T(a))$. As linear multipliers $(DT(a))^{\mathrm{T}} \cdot Dg(a) \cdot a = Dg(a) \cdot DT(a) \cdot a$. But for arbitrary a this is $(DT(a))^{\mathrm{T}} \cdot Dg(a) = Dg(a) \cdot DT(a)$. In our application, $DT(a)$ is $DC_H(s')$, and $Dg(a)$ is $DH(C_H(s'))$.

Since $(DC_H(s'))^T$ is a linear transformation and multiplication is associative for the multipliers of linear transformations, we can write

$$J \cdot DH(C_H(s')) = DC_H(s') \cdot J \cdot (DC_H(s'))^T \cdot DH(C_H(s')). \quad (5.51)$$

This is true for any H if

$$J = DC_H(s') \cdot J \cdot (DC_H(s'))^T. \quad (5.52)$$

As a program, this is[12],[13]

```
(define (J-func DHs)
  (up 0 (ref DHs 2) (- (ref DHs 1))))

(define ((canonical-transform? C) s)
  (let ((J ((D J-func) (compatible-shape s)))
        (DCs ((D C) s)))
    (- J (* DCs J (transpose DCs s)))))
```

This condition, equation (5.52), on C_H, called the *canonical condition*, does not depend on the details of H. This is a remarkable result: we can decide whether a phase-space transformation preserves the dynamics of Hamilton's equations without further reference to the details of the dynamical system. If the transformation is time dependent we can add a correction to the Hamiltonian to make it canonical.

Examples
The polar-canonical transformation satisfies the canonical condition:

```
((canonical-transform? (polar-canonical 'alpha))
 (up 't 'theta 'I))
(up (up 0 0 0) (up 0 0 0) (up 0 0 0))
```

---

[12] The procedure `compatible-shape` takes any structure and produces another structure that is guaranteed to multiply with the given structure to produce a numerical quantity. For example, the shape of $DH(s)$ is a compatible shape to the shape of s: if they are multiplied the result is a numerical quantity. This is the s^* that appears in equation (5.48).

[13] The procedure `transpose` is simply defined for traditional matrices, but because structures that specify linear transformations may have arbitrary substructure, the procedure needs to be supplied with a template that specifies this structure. So the procedure `transpose` takes two arguments: `(transpose ms rs)`, where `ms` is the structure to be transposed and the template `rs` is a structure that is appropriate for multiplication with `ms` on the right.

But not every transformation we might try satisfies the canonical condition. For example, we might try $x = p \sin \theta$ and $p_x = p \cos \theta$. The implementation is

```
(define (a-non-canonical-transform state)
  (let ((t (time state))
        (theta (coordinate state))
        (p (momentum state)))
    (let ((x (* p (sin theta)))
          (p_x (* p (cos theta))))
      (up t x p_x))))
```

```
((canonical-transform? a-non-canonical-transform)
 (up 't 'theta 'p))
(up (up 0 0 0) (up 0 0 (+ -1 p)) (up 0 (+ 1 (* -1 p)) 0))
```

So this transformation does not satisfy the canonical condition.

Canonical condition and Poisson brackets

The canonical condition can be written simply in terms of Poisson brackets.

The Poisson bracket can be written in terms of $\tilde{J}$:

$$\{f, g\} = (Df) \cdot (\tilde{J} \circ (Dg)) = (Df) \cdot J \cdot (Dg), \tag{5.53}$$

as can be seen by writing out the components.

We break the transformation C_H into position and momentum parts:

$$q = A(t, q', p') \tag{5.54}$$
$$p = B(t, q', p'). \tag{5.55}$$

In terms of the individual component functions, the canonical condition (5.52) is

$$\delta_j^i = \{A^i, B_j\}$$
$$0 = \{A^i, A^j\}$$
$$0 = \{B_i, B_j\} \tag{5.56}$$

where δ_j^i is 1 if $i = j$ and 0 otherwise. These equations are called the *fundamental Poisson brackets*. If a transformation satisfies these Poisson bracket relations then it satisfies the canonical condition.

We have found that a transformation is canonical if its position-momentum part satisfies the canonical condition, but for a time-dependent transformation we may have to modify the Hamiltonian by the addition of a suitable K. We can rewrite these conditions in terms of Poisson brackets. If the Hamiltonian is

$$H'(t, q', p') = H(t, A(t, q', p'), B(t, q', p')) + K(t, q', p'), \qquad (5.57)$$

the transformation will be canonical if the coordinate-momentum transformation satisfies the fundamental Poisson brackets, and K satisfies:

$$\{A^i, K\} + \partial_0 A^i = 0$$
$$\{B_j, K\} + \partial_0 B_j = 0. \qquad (5.58)$$

Exercise 5.5: Poisson bracket conditions

Fill in the details to show that the canonical condition (5.52) is equivalent to the fundamental Poisson brackets (5.56) and that the condition on K (5.42) is equivalent to the Poisson bracket condition on K (5.58).

Symplectic matrices

It is convenient to reformulate the canonical condition in terms of matrices. We can obtain a matrix representation of a structure with the utility `s->m` that takes a structure that represents a multiplier of a linear transformation and returns a matrix representation of the multiplier. The procedure `s->m` takes three arguments: `(s->m ls A rs)`. The `ls` and `rs` specify the shapes of objects that multiply `A` on the left and right to give a numerical value. These specify the basis. So, the matrix representation of the multiplier corresponding to $\tilde{J}$ is

```
(let* ((s (up 't (up 'x 'y) (down 'px 'py)))
       (s* (compatible-shape s))
       (J ((D J-func) s*)))
  (s->m s* J s*))
(matrix-by-rows (list 0 0 0 0 0)
                (list 0 0 0 1 0)
                (list 0 0 0 0 1)
                (list 0 -1 0 0 0)
                (list 0 0 -1 0 0))
```

This matrix, **J**, is useful, so we supply a procedure `J-matrix` so that `(J-matrix n)` gives this matrix for an n degree-of-freedom system.

We can now reexpress the canonical condition (5.52) as a matrix equation:

$$\mathbf{J} = \mathbf{DC}_H(s') \cdot \mathbf{J} \cdot (\mathbf{DC}_H(s'))^\mathsf{T}. \tag{5.59}$$

There is a further simplification available. The elements of the first row and the first column of the matrix representation of $\widetilde{J}$ are all zeros. This has simplifying consequences. Consider a general transformation of phase-space states (for two degrees of freedom):

```
(define C-general
  (literal-function 'C
    (-> (UP Real (UP Real Real) (DOWN Real Real))
        (UP Real (UP Real Real) (DOWN Real Real)))))
```

Consider transformations for which the time does not depend on the coordinates or momenta[14]

```
(define (C-simple-time s)
  (let ((cs (C-general s)))
    (up ((literal-function 'tau) (time s))
        (coordinates cs)
        (momenta cs))))
```

For this kind of transformation the first row and the first column of the residuals of the `canonical-transform?` test are identically zero:

```
(let* ((s (up 't (up 'x 'y) (down 'p_x 'p_y)))
       (s* (compatible-shape s)))
  (m:nth-row
   (s->m s* ((canonical-transform? C-simple-time) s) s*)
   0))
(up 0 0 0 0 0)

(let ((s (up 't (up 'x 'y) (down 'p_x 'p_y)))
      (s* (compatible-shape s)))
  (m:nth-col
   (s->m s* ((canonical-transform? C-simple-time) s) s*)
   0))
(up 0 0 0 0 0)
```

---

[14] Actually, this is more interesting: we allow transformations that arbitrarily distort time, as tau is an arbitrary literal function. The canonical condition is concerned only with the possibly time-dependent transformation of coordinates and momenta.

But for C-general these are not zero. Since the transformations we are considering at most shift time, we need to consider only the submatrix associated with the coordinates and the momenta.

The *qp* submatrix[15] of dimension $2n \times 2n$ of the matrix $\mathbf{J}$ is called the *symplectic unit* for n degrees of freedom:

$$\mathbf{J}_n = \begin{pmatrix} \mathbf{0}_{n \times n} & \mathbf{1}_{n \times n} \\ -\mathbf{1}_{n \times n} & \mathbf{0}_{n \times n} \end{pmatrix}. \tag{5.60}$$

The matrix $\mathbf{J}_n$ satisfies the following identities:

$$\mathbf{J}_n^{\mathrm{T}} = \mathbf{J}_n^{-1} = -\mathbf{J}_n. \tag{5.61}$$

A $2n \times 2n$ matrix $\mathbf{A}$ that satisfies the relation

$$\mathbf{J}_n = \mathbf{A}\mathbf{J}_n\mathbf{A}^{\mathrm{T}} \tag{5.62}$$

is called a *symplectic matrix*. We can determine whether a matrix is symplectic:

```
(define (symplectic-matrix? M)
  (let ((2n (m:dimension M)))
    (let ((J (symplectic-unit (quotient 2n 2))))
      (- J (* M J (transpose M))))))
```

An appropriate symplectic unit matrix of a given size is produced by the procedure symplectic-unit.

If the matrix representation of the derivative of a transformation is a symplectic matrix the transformation is a *symplectic transformation*. Here is a test for whether a transformation is symplectic:[16]

```
(define ((symplectic-transform? C) s)
  (symplectic-matrix? (qp-submatrix ((D-as-matrix C) s))))
```

---

[15]The *qp* submatrix of a square matrix of dimension $2n + 1$ is the $2n$-dimensional matrix obtained by deleting the first row and the first column of the given matrix. This can be computed by:

```
(define (qp-submatrix m)
  (m:submatrix m 1 (m:num-rows m) 1 (m:num-cols m)))
```

[16]The procedure D-as-matrix is defined as:

```
(define ((D-as-matrix F) s)
  (s->m (compatible-shape (F s)) ((D F) s) s))
```

The procedure `symplectic-transform?` returns a zero matrix if and only if the transformation being tested passes the symplectic matrix test.

For example, the point transformations are symplectic. We can show this for a general possibly time-dependent two-degree-of-freedom point transformation:

```
(define (F s)
  ((literal-function 'F
      (-> (X Real (UP Real Real)) (UP Real Real)))
   (time s) (coordinates s)))

((symplectic-transform? (F->CH F))
 (up 't (up 'x 'y) (down 'px 'py)))
(matrix-by-rows (list 0 0 0 0)
                (list 0 0 0 0)
                (list 0 0 0 0)
                (list 0 0 0 0))
```

More generally, the phase-space part of the canonical condition is equivalent to the symplectic condition (for two degrees of freedom) even in the case of an unrestricted phase-space transformation.

```
(let* ((s (up 't (up 'x 'y) (down 'p_x 'p_y)))
       (s* (compatible-shape s)))
  (- (qp-submatrix
      (s->m s* ((canonical-transform? C-general) s) s*))
     ((symplectic-transform? C-general) s)))
(matrix-by-rows (list 0 0 0 0)
                (list 0 0 0 0)
                (list 0 0 0 0)
                (list 0 0 0 0))
```

Exercise 5.6: Symplectic matrices

Let $\mathbf{A}$ be a symplectic matrix: $\mathbf{J}_n = \mathbf{A} \mathbf{J}_n \mathbf{A}^\mathrm{T}$. Show that $\mathbf{A}^\mathrm{T}$ and $\mathbf{A}^{-1}$ are symplectic.

Exercise 5.7: Polar-canonical transformations

Let x, p and θ, I be two sets of canonically conjugate variables. Consider transformations of the form $x = \beta I^\alpha \sin \theta$ and $p = \beta I^\alpha \cos \theta$. Determine all α and β for which this transformation is symplectic.

Exercise 5.8: Standard map

Is the standard map a symplectic transformation? Recall that the standard map is: $I' = I + K \sin \theta$, with $\theta' = \theta + I'$, both modulo 2π.

Exercise 5.9: Whittaker transform

Shew that the transformation $q = \log\left((\sin p')/q'\right)$ with $p = q' \cot p'$ is symplectic.

5.3 Invariants of Canonical Transformations

Canonical transformations allow us to change the phase-space coordinate system that we use to express a problem, preserving the form of Hamilton's equations. If we solve Hamilton's equations in one phase-space coordinate system we can use the transformation to carry the solution to the other coordinate system. What other properties are preserved by a canonical transformation?

Noninvariance of pv

We noted in equation (5.10) that point transformations that are canonical extensions of time-independent coordinate transformations preserve the value of pv. This does not hold for more general canonical transformations. We can illustrate this with the polar-canonical transformation. Along corresponding paths x, p_x and θ, I

$$
x(t) = \sqrt{\frac{2I(t)}{\alpha}} \sin \theta(t)
$$
$$
p_x(t) = \sqrt{2I(t)\alpha} \cos \theta(t), \tag{5.63}
$$

and so Dx is

$$
Dx(t) = D\theta(t)\sqrt{\frac{2I(t)}{\alpha}} \cos \theta(t) + DI(t)\frac{1}{\sqrt{2I(t)\alpha}} \sin \theta(t). \tag{5.64}
$$

The difference of pv and the transformed $p'v'$ is

$$
p_x(t)Dx(t) - I(t)D\theta(t)
$$
$$
= I(t)D\theta(t)\left(2\cos^2 \theta(t) - 1\right) + DI(t)\sin \theta(t) \cos \theta(t). \tag{5.65}
$$

In general this is not zero. So the product pv is not necessarily invariant under general canonical transformations.

Invariance of Poisson brackets

Here is a remarkable fact: the composition of the Poisson bracket of two phase-space state functions with a canonical transformation is the same as the Poisson bracket of each of the two functions composed with the transformation separately. Loosely speaking, the Poisson bracket is invariant under canonical phase-space transformations.

Let f and g be two phase-space state functions. Using the $\tilde{J}$ representation of the Poisson bracket (see section 5.2.2), we deduce

$$
\begin{aligned}
\{f \circ C_{\mathrm{H}}, g \circ C_{\mathrm{H}}\} & \\
&= (D(f \circ C_{\mathrm{H}})) \cdot (\tilde{J} \circ D(g \circ C_{\mathrm{H}})) \\
&= (Df \circ C_{\mathrm{H}}) \cdot DC_{\mathrm{H}} \cdot (\tilde{J} \circ ((Dg \circ C_{\mathrm{H}}) \cdot DC_{\mathrm{H}})) \\
&= (Df \circ C_{\mathrm{H}}) \cdot (\tilde{J} \circ Dg \circ C_{\mathrm{H}}) \\
&= (Df \cdot (\tilde{J} \circ Dg)) \circ C_{\mathrm{H}} \\
&= \{f, g\} \circ C_{\mathrm{H}},
\end{aligned}
\tag{5.66}
$$

where the fact that C_{H} satisfies equation (5.41) was used in the middle. This is

$$
\{f \circ C_{\mathrm{H}}, g \circ C_{\mathrm{H}}\} = \{f, g\} \circ C_{\mathrm{H}}.
\tag{5.67}
$$

Volume preservation

Consider a canonical transformation C_{H}. Let $\hat{C}_t$ be a function with parameter t such that $(q, p) = \hat{C}_t(q', p')$ if $(t, q, p) = C_{\mathrm{H}}(t, q', p')$. The function $\hat{C}_t$ maps phase-space coordinates to alternate phase-space coordinates at a given time. Consider regions R in (q, p) and R' in (q', p') such that $R = \hat{C}_t(R')$. The volume of region R' is

$$
V(R) = \int_R \hat{1} = \int_{R'} \det(D\hat{C}_t),
\tag{5.68}
$$

where $\hat{1}$ is the function whose value is one for every input. Now if C_{H} is symplectic then the determinant of $D\hat{C}_t$ is one (see section 4.2.3), so

$$
V(R) = V(R').
\tag{5.69}
$$

Thus, phase-space volume is preserved by symplectic transformations.

Liouville's theorem shows that time evolution preserves phase-space volume. Here we see that canonical transformations also preserve phase volumes. Later, we will find that time evolution actually generates a canonical transformation.

The symplectic 2-form
Define

$$\omega(\zeta_1, \zeta_2) = P(\zeta_2)Q(\zeta_1) - P(\zeta_1)Q(\zeta_2), \tag{5.70}$$

where $Q = I_1$ and $P = I_2$ are the coordinate and momentum selectors, respectively. The arguments ζ_1 and ζ_2 are incremental phase-space states with zero time components.

The ω form can also be written as a sum over degrees of freedom:

$$\omega(\zeta_1, \zeta_2) = \sum_i \left(P_i(\zeta_2)Q^i(\zeta_1) - P_i(\zeta_1)Q^i(\zeta_2) \right). \tag{5.71}$$

Notice that the contributions for each i do not mix components from different degrees of freedom.

This bilinear form is closely related to the symplectic 2-form of differential geometry. It differs in that the symplectic 2-form is formally a function of the phase-space point as well as the incremental vectors.

Under a canonical transformation $s = C_H(s')$, incremental states transform with the derivative

$$\zeta_i = DC_H(s')\zeta_i'. \tag{5.72}$$

We will show that the 2-form is invariant under this transformation

$$\omega(\zeta_1, \zeta_2) = \omega(\zeta_1', \zeta_2'), \tag{5.73}$$

if the time components of the ζ_i' are both zero.

We have shown that condition (5.41) does not depend on the details of the Hamiltonian H. So if a transformation satisfies the canonical condition we can use condition (5.41) with H replaced by an arbitrary function f of phase-space states:

$$\mathcal{D}f(C_H(s')) = (DC_H(s')) \cdot (\mathcal{D}(f \circ C_H)(s')). \tag{5.74}$$

In terms of ω, the Poisson bracket is

$$\{f,g\}(s) = \omega(\mathcal{D}f(s), \mathcal{D}g(s)) \tag{5.75}$$

as can be seen by writing out the components. We use the fact that Poisson brackets are invariant under canonical transformations:

$$(\{f,g\} \circ C_H)(s') = \{f \circ C_H, g \circ C_H\}(s'). \tag{5.76}$$

Using the relation (5.74) to expand the left-hand side of equation (5.76) we obtain:

$$
\begin{aligned}
(\{f,g\} \circ C_H)(s') \\
&= \omega((\mathcal{D}f \circ C_H)(s'), (\mathcal{D}g \circ C_H)(s')) \\
&= \omega((DC_H(s')) \cdot (\mathcal{D}(f \circ C_H)(s')), \\
&\quad\quad (DC_H(s')) \cdot (\mathcal{D}(g \circ C_H)(s'))).
\end{aligned}
\tag{5.77}
$$

The right-hand side of equation (5.76) is

$$\{f \circ C_H, g \circ C_H\}(s') = \omega(\mathcal{D}(f \circ C_H)(s'), \mathcal{D}(g \circ C_H)(s')). \tag{5.78}$$

Now the left-hand side must equal the right-hand side for any f and g, so the equation must also be true for arbitrary ζ_i' of the form

$$
\begin{aligned}
\zeta_1' &= \mathcal{D}(f \circ C_H)(s') \\
\zeta_2' &= \mathcal{D}(g \circ C_H)(s').
\end{aligned}
\tag{5.79}
$$

So the ζ_i' are arbitrary incremental states with zero time components.

We have proven that

$$\omega(\zeta_1', \zeta_2') = \omega(DC_H(s') \cdot \zeta_1', \; DC_H(s') \cdot \zeta_2'). \tag{5.80}$$

for canonical C_H and incremental states ζ_i' with zero time components. Using equation (5.72), we have

$$\omega(\zeta_1', \zeta_2') = \omega(\zeta_1, \zeta_2). \tag{5.81}$$

Thus the bilinear antisymmetric function ω is invariant under even time-varying canonical transformations if the increments are restricted to have zero time component.

As a program, ω is

```
(define (omega zeta1 zeta2)
  (- (* (momentum zeta2) (coordinate zeta1))
     (* (momentum zeta1) (coordinate zeta2))))
```

On page 356 we showed that point transformations are symplectic. Here we can see that the 2-form is preserved under these transformations for two degrees of freedom:

```
(define (F s)
  ((literal-function 'F
    (-> (X Real (UP Real Real)) (UP Real Real)))
   (time s)
   (coordinates s)))
```

```
(let ((s (up 't (up 'x 'y) (down 'p_x 'p_y)))
      (zeta1 (up 0 (up 'dx1 'dy1) (down 'dp1_x 'dp1_y)))
      (zeta2 (up 0 (up 'dx2 'dy2) (down 'dp2_x 'dp2_y))))
  (let ((DCs ((D (F->CH F)) s)))
    (- (omega zeta1 zeta2)
       (omega (* DCs zeta1) (* DCs zeta2)))))
0
```

Alternatively, let $\mathbf{z}_1$ and $\mathbf{z}_2$ be the matrix representations of the qp parts of ζ_1 and ζ_2. The matrix representation of ω is

$$\omega(\zeta_1, \zeta_2) = \mathbf{z}_1^{\mathrm{T}} \cdot \mathbf{J}_n \cdot \mathbf{z}_2. \tag{5.82}$$

Let $\mathbf{A}$ be the matrix representation of the qp part of $DC_H(s')$ Then the invariance of ω is equivalent to

$$\mathbf{z}_1^{\mathrm{T}} \cdot \mathbf{A}^{\mathrm{T}} \cdot \mathbf{J}_n \cdot \mathbf{A} \cdot \mathbf{z}_2 = \mathbf{z}_1^{\mathrm{T}} \cdot \mathbf{J}_n \cdot \mathbf{z}_2. \tag{5.83}$$

But this is true if

$$\mathbf{A}^{\mathrm{T}} \cdot \mathbf{J}_n \cdot \mathbf{A} = \mathbf{J}_n, \tag{5.84}$$

which is equivalent to the condition that $\mathbf{A}$ is symplectic. (If a matrix is symplectic then its transpose is symplectic. See exercise 5.6).

The symplectic condition is symmetrical in that if $\mathbf{A}$ is symplectic then $\mathbf{A}^{\mathrm{T}}$ is symplectic, because the symplectic unit is invertible. The canonical condition

$$\mathbf{J} = DC_H(s') \cdot \mathbf{J} \cdot (DC_H(s'))^{\mathrm{T}}, \tag{5.85}$$

is satisfied by time-varying canonical transformations, and time-varying canonical transformations are symplectic. But if the transformation is time varying then

$$\mathbf{J} = (\mathbf{DC_H}(s'))^{\mathrm{T}} \cdot \mathbf{J} \cdot \mathbf{DC_H}(s'), \tag{5.86}$$

is not satisfied because $\mathbf{J}$ is not invertible. Equation (5.86) is satisfied, however, for time-independent transformations.

Poincaré integral invariant

The invariance of the symplectic 2-form under canonical transformations has a simple interpretation. Consider how the area of an incremental parallelogram in phase space transforms under canonical transformation. Let $(\Delta q, \Delta p)$ and $(\delta q, \delta p)$ be small increments in phase space, originating at (q, p). Consider the incremental parallelogram with vertex at (q, p) with these two phase-space increments as edges. The sum of the areas of the canonical projections of this incremental parallelogram can be written

$$\sum_i \Delta A_i = \sum_i (\Delta q^i \delta p_i - \Delta p_i \delta q^i). \tag{5.87}$$

The right-hand side is the sum of the areas on the canonical planes;[17] for each i the area of a parallelogram is computed from the components of the vectors defining its adjacent sides. Let $\zeta_1 = (0, \Delta q, \Delta p)$ and $\zeta_2 = (0, \delta q, \delta p)$; then the sum of the areas of the incremental parallelograms is just

$$\sum_i \Delta A_i = \omega(\zeta_1, \zeta_2), \tag{5.88}$$

where ω is the bilinear antisymmetric function introduced in equation (5.70). The function ω is invariant under canonical transformations, so the sum of the areas of the incremental parallelograms is invariant under canonical transformations.

There is an integral version of this differential relation. Consider the oriented area of a region R' in phase space (see figure 5.2). Suppose we make a canonical transformation from coordinates

---

[17] The q^i, p_i plane is the *i*th *canonical plane* in these phase-space variables.

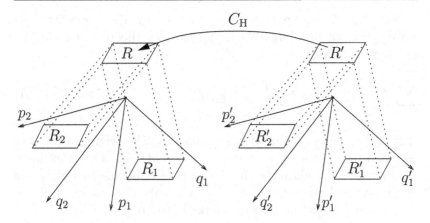

Figure 5.2   A region R' in phase space is mapped by a canonical transformation C_{H} to a region R. The projections of region R onto the planes formed by canonical basis pairs q_j, p_j are R_j. The projections of R' are R'_j. In general, the areas of the regions R and R' are not the same, but the sums of the areas of the canonical plane projections are the same.

(q', p') to (q, p) taking region R' to region R. The boundary of the region in the transformed coordinates is just the image under the canonical transformation of the original boundary. Let R_{q^i, p_i} be the projection of the region R onto the q^i, p_i plane of coordinate q^i and conjugate momentum p_i, and let A_i be its area. Similarly, let $R'_{q'^i, p'_i}$ be the projection of R' onto the q'^i, p'_i plane, and let A'_i be its area.

The area of an arbitrary region is just the limit of the sum of the areas of incremental parallelograms that cover the region, so the sum of oriented areas is preserved by canonical transformations:

$$\sum_i A_i = \sum_i A'_i. \tag{5.89}$$

That is, the sum of the projected areas on the canonical planes is preserved by canonical transformations. Another way to say this is

$$\sum_i \int_{R_{q^i, p_i}} dq^i \, dp_i = \sum_i \int_{R'_{q'^i, p'_i}} dq'^i \, dp'_i. \tag{5.90}$$

The equality-of-areas relation (5.90) can also be written as an equality of line integrals using Stokes's theorem, for simply-connected regions R_{q^i,p_i} and $R'_{q'^i,p'_i}$:

$$\sum_i \oint_{\partial R_{q^i,p_i}} p_i dq^i = \sum_i \oint_{\partial R'_{q'^i,p'_i}} p'_i dq'^i. \qquad (5.91)$$

The canonical planes are disjoint except at the origin, so the projected areas intersect in at most one point. Thus we may independently accumulate the line integrals around the boundaries of the individual projections of the region onto the canonical planes into a line integral around the unprojected region:

$$\oint_{\partial R} \sum_i p_i dq^i = \oint_{\partial R'} \sum_i p'_i dq'^i. \qquad (5.92)$$

Exercise 5.10: Watch out

Consider the canonical transformation C_H:

$$(t, x, p) = C_H(t, \theta, J) = (t, \sqrt{2(J+a)} \sin\theta, \sqrt{2(J+a)} \cos\theta).$$

a. Show that the transformation is symplectic for any a.

b. Show that equation (5.92) is not generally satisfied for the region enclosed by a curve of constant J.

5.4   Generating Functions

We have considered a number of properties of general canonical transformations without having a method for coming up with them. Here we introduce the method of *generating functions*. The generating function is a real-valued function that compactly specifies a canonical transformation through its partial derivatives, as follows.

Consider a real-valued function $F_1(t, q, q')$ mapping configurations expressed in two coordinate systems to the reals. We will use F_1 to construct a canonical transformation from one coordinate system to the other. We will show that the following relations among the coordinates, the momenta, and the Hamiltonians specify a canonical transformation:

$$p = \partial_1 F_1(t, q, q') \tag{5.93}$$
$$p' = -\partial_2 F_1(t, q, q') \tag{5.94}$$
$$H'(t, q', p') - H(t, q, p) = \partial_0 F_1(t, q, q'). \tag{5.95}$$

The transformation will then be explicitly given by solving for one set of variables in terms of the others: To obtain the primed variables in terms of the unprimed ones, let A be the inverse of $\partial_1 F_1$ with respect to the third argument,

$$q' = A(t, q, \partial_1 F_1(t, q, q')); \tag{5.96}$$

then

$$q' = A(t, q, p) \tag{5.97}$$
$$p' = -\partial_2 F_1(t, q, A(t, q, p)). \tag{5.98}$$

Let B be the coordinate part of the phase-space transformation $q = B(t, q', p')$. This B is an inverse function of $\partial_2 F_1$, satisfying

$$q = B(t, q', -\partial_2 F_1(t, q, q')). \tag{5.99}$$

Using B, we have

$$q = B(t, q', p') \tag{5.100}$$
$$p = \partial_1 F_1(t, B(t, q', p'), q'). \tag{5.101}$$

To put the transformation in explicit form requires that the inverse functions A and B exist.

   We can use the above relations to verify that some given transformation from one set of phase-space coordinates (q, p) with Hamiltonian function $H(t, q, p)$ to another set (q', p') with Hamiltonian function $H'(t, q', p')$ is canonical by finding an $F_1(t, q, q')$ such that the above relations are satisfied. We can also use arbitrarily chosen generating functions of type F_1 to generate new canonical transformations.

The polar-canonical transformation
The polar-canonical transformation (5.27) from coordinate and momentum (x, p_x) to new coordinate and new momentum (θ, I),

$$x = \sqrt{\frac{2I}{\alpha}} \sin \theta \tag{5.102}$$

$$p_x = \sqrt{2I\alpha} \cos \theta, \tag{5.103}$$

introduced earlier, is canonical. This can also be demonstrated by finding a suitable F_1 generating function. The generating function satisfies a set of partial differential equations, (5.93) and (5.94):

$$p_x = \partial_1 F_1(t, x, \theta) \tag{5.104}$$

$$I = -\partial_2 F_1(t, x, \theta). \tag{5.105}$$

Using relations (5.102) and (5.103), which specify the transformation, equation (5.104) can be rewritten

$$p_x = x\alpha \cot \theta = \partial_1 F_1(t, x, \theta), \tag{5.106}$$

which is easily integrated to yield

$$F_1(t, x, \theta) = \frac{\alpha}{2} x^2 \cot \theta + \varphi(t, \theta), \tag{5.107}$$

where φ is some integration "constant" with respect to the first integration. Substituting this form for F_1 into the second partial differential equation (5.105), we find

$$I = -\partial_2 F_1(t, x, \theta) = \frac{\alpha}{2} \frac{x^2}{(\sin \theta)^2} - \partial_1 \varphi(t, \theta), \tag{5.108}$$

but if we set $\varphi = 0$ the desired relations are recovered. So the generating function

$$F_1(t, x, \theta) = \frac{\alpha}{2} x^2 \cot \theta \tag{5.109}$$

generates the polar-canonical transformation. This shows that this transformation is canonical.

5.4.1 F_1 Generates Canonical Transformations

We can prove directly that the transformation generated by an F_1 is canonical by showing that if Hamilton's equations are satisfied in one set of coordinates then they will be satisfied in the other set of coordinates. Let F_1 take arguments (t, x, y). The relations among the coordinates are

$$p_x = \partial_1 F_1(t, x, y)$$
$$p_y = -\partial_2 F_1(t, x, y) \tag{5.110}$$

and the Hamiltonians are related by

$$H'(t, y, p_y) = H(t, x, p_x) + \partial_0 F_1(t, x, y). \tag{5.111}$$

Substituting the generating function relations (5.110) into this equation, we have

$$
\begin{aligned}
H'(t, y, &-\partial_2 F_1(t, x, y)) \\
&= H(t, x, \partial_1 F_1(t, x, y)) + \partial_0 F_1(t, x, y).
\end{aligned} \tag{5.112}
$$

Take the partial derivatives of this equality of expressions with respect to the variables x and y:[18]

$$
\begin{aligned}
-(\partial_2 H')^j (\partial_1 (\partial_2 F_1)_j)_i & \\
&= (\partial_1 H)_i + (\partial_2 H)^j (\partial_1 (\partial_1 F_1)_j)_i + (\partial_1 \partial_0 F_1)_i \\
(\partial_1 H')_i - (\partial_2 H')^j (\partial_2 (\partial_2 F_1)_j)_i & \\
&= (\partial_2 H)^j (\partial_2 (\partial_1 F_1)_j)_i + (\partial_2 \partial_0 F_1)_i
\end{aligned} \tag{5.113}
$$

where the arguments are unambiguous and have been suppressed. On solution paths we can use Hamilton's equations for the (x, p_x) system to replace the partial derivatives of H with derivatives of x and p_x, obtaining

$$
\begin{aligned}
-(\partial_2 H')^j (\partial_1 (\partial_2 F_1)_j)_i & \\
&= -(Dp_x)_i + (Dx)^j (\partial_1 (\partial_1 F_1)_j)_i + (\partial_1 \partial_0 F_1)_i \\
(\partial_1 H')_i - (\partial_2 H')^j (\partial_2 (\partial_2 F_1)_j)_i & \\
&= (Dx)^j (\partial_2 (\partial_1 F_1)_j)_i + (\partial_2 \partial_0 F_1)_i.
\end{aligned} \tag{5.114}
$$

Now compute the derivatives of p_x and p_y, from equations (5.110), along consistent paths:

---

[18]The structure $\partial_2 \partial_1 F_1$ is a *down* of *downs*, so it is compatible for contraction with an *up* on either side. But it is not symmetrical, so the associations must be specified. To solve this problem we use index notation (ugh!).

So we use indices to select particular components of structured objects. If an index symbol appears both as a superscript and as a subscript in an expression, the value of the expression is the sum over all possible values of the index symbol of the designated components (Einstein summation convention). Thus, for example, if $\dot{q}$ and p are of dimension n then the indicated product $p_i \dot{q}^i$ is to be interpreted as $\sum_{i=0}^{n-1} p_i \dot{q}^i$.

$$(Dp_x)_i = (\partial_1(\partial_1 F_1)_i)_j(Dx)^j + (\partial_2(\partial_1 F_1)_i)_j(Dy)^j + \partial_0(\partial_1 F_1)_i$$
$$(Dp_y)_i = -(\partial_1(\partial_2 F_1)_i)_j(Dx)^j - (\partial_2(\partial_2 F_1)_i)_j(Dy)^j - \partial_0(\partial_2 F_1)_i.$$
$$(5.115)$$

Using the fact that elementary partials commute, $(\partial_2(\partial_1 F_1)_i)_j = (\partial_1(\partial_2 F_1)_j)_i$, and substituting this expression for $(Dp_x)_i$ into the first of equations (5.114) yields

$$-(\partial_2 H')^j(\partial_1(\partial_2 F_1)_j)_i = -(\partial_1(\partial_2 F_1)_j)_i(Dy)^j. \qquad (5.116)$$

Provided that $\partial_2\partial_1 F_1$ is nonsingular,[19] we have derived one of Hamilton's equations for the (y, p_y) system:

$$Dy(t) = \partial_2 H'(t, y(t), p_y(t)). \qquad (5.117)$$

Hamilton's other equation,

$$Dp_y(t) = -\partial_1 H'(t, y(t), p_y(t)), \qquad (5.118)$$

can be derived in a similar way. So the generating function relations indeed specify a canonical transformation.

5.4.2   Generating Functions and Integral Invariants

Generating functions can be used to specify a canonical transformation by the prescription given above. Here we show how to get a generating function from a canonical transformation, and derive the generating function rules.

The generating function representation of canonical transformations can be derived from the Poincaré integral invariants, as follows. We first show that, given a canonical transformation, the integral invariants imply the existence of a function of phase-space coordinates that can be written as a path-independent line integral. Then we show that partial derivatives of this function, represented in mixed coordinates, give the generating function relations between the old and new coordinates. We need to do this only for time-independent transformations because time-dependent transformations become time independent in the extended phase space (see section 5.5).

---

[19] A structure is nonsingular if the determinant of the matrix representation of the structure is nonzero.

Generating functions of type F_1

Let C be a time-independent canonical transformation, and let C_t be the qp-part of the transformation. The transformation C_t preserves the integral invariant equation (5.90). One way to express the equality of areas is as a line integral (5.92):

$$\oint_{\partial R} \sum_i p_i dq^i = \oint_{\partial R'} \sum_i p_i' dq'^i, \tag{5.119}$$

where R' is a two-dimensional region in (q', p') coordinates at time t, $R = C_t(R')$ is the corresponding region in (q, p) coordinates, and ∂R indicates the boundary of the region R. This holds for any region and its boundary. We will show that this implies there is a function $F(t, q', p')$ that can be defined in terms of line integrals

$$F(t, q', p') - F(t, q_0', p_0')$$
$$= \int_{\gamma = C_t(\gamma')} \sum_i p_i dq^i - \int_{\gamma'} \sum_i p_i' dq'^i, \tag{5.120}$$

where γ' is a curve in phase-space coordinates that begins at $\gamma'(0) = (q_0', p_0')$ and ends at $\gamma'(1) = (q', p')$, and γ is its image under C_t.

Let

$$G_t(\gamma') = \int_{\gamma = C_t(\gamma')} \sum_i p_i dq^i - \int_{\gamma'} \sum_i p_i' dq'^i, \tag{5.121}$$

and let γ_1' and γ_2' be two paths with the same endpoints. Then

$$G_t(\gamma_2') - G_t(\gamma_1') = \oint_{\partial R} \sum_i p_i dq^i - \oint_{\partial R'} \sum_i p_i' dq'^i$$
$$= 0. \tag{5.122}$$

So the value of $G_t(\gamma')$ depends only on the endpoints of γ'.

Let

$$\bar{G}_{t,q_0',p_0'}(q', p') = G_t(\gamma'), \tag{5.123}$$

where γ' is any path from q_0', p_0' to q', p'. Changing the initial point from $q_0'\, p_0'$ to $q_1'\, p_1'$ changes the value of $\bar{G}$ by a constant:

$$\bar{G}_{t,q_1',p_1'}(q', p') - \bar{G}_{t,q_0',p_0'}(q', p') = \bar{G}_{t,q_1',p_1'}(q_0', p_0'). \tag{5.124}$$

If we define F so that

$$F(t, q', p') = \bar{G}_{t, q'_1, p'_1}(q', p'), \tag{5.125}$$

then

$$F(t, q', p') - F(t, q'_0, p'_0) = \bar{G}_{t, q'_0, p'_0}(q', p'), \tag{5.126}$$

demonstrating equation (5.120).

The phase-space point (q, p) in unprimed variables corresponds to (q', p') in primed variables, at an arbitrary time t. Both p and q are determined given q' and p'. In general, given any two of these four quantities, we can solve for the other two. If we can solve for the momenta in terms of the positions we get a particular class of generating functions.[20] We introduce the functions

$$\begin{aligned} p &= f_p(t, q, q') \\ p' &= f_{p'}(t, q, q') \end{aligned} \tag{5.127}$$

that solve the transformation equations $(t, q, p) = C(t, q', p')$ for the momenta in terms of the coordinates at a specified time. With these we introduce a function $F_1(t, q, q')$ such that

$$F_1(t, q, q') = F(t, q, f_p(t, q, q')). \tag{5.128}$$

The function F_1 has the same value as F but has different arguments. We will show that this F_1 is in fact the generating function for canonical transformations introduced in section 5.4. Let's be explicit about the definition of F_1 in terms of a line integral:

$$\begin{aligned} F_1(t, q, q') &- F_1(t, q_0, q'_0) \\ &= \int_{q_0, q'_0}^{q, q'} \left(f_p(t, q, q') dq - f_{p'}(t, q, q') dq' \right). \end{aligned} \tag{5.129}$$

The two line integrals can be combined into this one because they are both expressed as integrals along a curve in (q, q').

---

[20]Point transformations are not in this class: we cannot solve for the momenta in terms of the positions for point transformations, because for a point transformation the primed and unprimed coordinates can be deduced from each other, so there is not enough information in the coordinates to deduce the momenta.

We can use the path independence of F_1 to compute the partial derivatives of F_1 with respect to particular components and consequently derive the generating function relations for the momenta.[21] So we conclude that

$$(\partial_1 F_1(t, q, q'))_i = f_{p_i}(t, q, q') \tag{5.130}$$

$$(\partial_2 F_1(t, q, q'))_i = -f_{p'_i}(t, q, q'). \tag{5.131}$$

These are just the configuration and momentum parts of the generating function relations for canonical transformation. So starting with a canonical transformation, we can find a generating function that gives the coordinate–momentum part of the transformation through its derivatives.

Starting from a general canonical transformation, we have constructed an F_1 generating function from which the canonical transformation may be rederived. So we expect there is a generating function for every canonical transformation.[22]

Generating functions of type F_2

Point transformations were excluded from the previous argument because we could not deduce the momenta from the coordinates. However, a similar derivation allows us to make a generating function for this case. The integral invariants give us an equality of area integrals. There are other ways of writing the equality-of-areas relation (5.90) as a line integral. We can also write

$$\oint_{\partial R} \sum_i p_i dq^i = -\oint_{\partial R'} \sum_i q'_i dp'^i. \tag{5.132}$$

The minus sign arises because by flipping the axes we are traversing the area in the opposite sense. Repeating the argument just

---

[21]Let F be defined as the path-independent line integral

$$F(x) = \int_{x_0}^{x} \sum_i f_i(x) dx^i + F(x_0);$$

then $\partial_i F(x) = f_i(x)$.

[22]There may be some singular cases and topological problems that prevent this from being rigorously true.

given, we can define a function

$$F'(t, q', p') - F'(t, q'_0, p'_0)$$
$$= \int_{\gamma = C(t,\gamma')} \sum_i p_i dq^i + \int_{\gamma'} \sum_i q'_i dp'^i \qquad (5.133)$$

that is independent of the path γ'. If we can solve for q' and p in terms of q and p' we can define the functions

$$q' = f'_{q'}(t, q, p')$$
$$p = f'_p(t, q, p') \qquad (5.134)$$

and define

$$F_2(t, q, p') = F'(t, f'_{q'}(t, q, p'), p'). \qquad (5.135)$$

Then the canonical transformation is given as partial derivatives of F_2:

$$(\partial_1 F_2(t, q, p'))_i = f'_{p_i}(t, q, p') \qquad (5.136)$$

and

$$(\partial_2 F_2(t, q, p'))^i = f'_{q_i'}(t, q, p'). \qquad (5.137)$$

Relationship between F_1 and F_2

For canonical transformations that can be described by both an F_1 and an F_2, there must be a relation between them. The alternative line integral expressions for the area integral are related. Consider the difference

$$(F'(t, q', p') - F'(t, q'_0, p'_0)) - (F(t, q', p') - F(t, q'_0, p'_0))$$
$$= \int_{\gamma'} \sum_i p'_i dq'^i + \int_{\gamma'} \sum_i q'_i dp'^i$$
$$= \int_{\gamma'} \sum_i d(p'_i q'^i)$$
$$= \sum_i (p')_i (q')^i - \sum_i (p'_0)_i (q'_0)^i. \qquad (5.138)$$

The functions F and F' are related by an integrated term

$$F'(t, q', p') - F(t, q', p') = p'q', \qquad (5.139)$$

as are F_1 and F_2:

$$F_2(t, q, p') - F_1(t, q, q') = p'q'. \tag{5.140}$$

The generating functions F_1 and F_2 are related by a Legendre transform:

$$p' = -\partial_2 F_1(t, q, q') \tag{5.141}$$
$$p'q' = -F_1(t, q, q') + F_2(t, q, p') \tag{5.142}$$
$$q' = \partial_2 F_2(t, q, p'). \tag{5.143}$$

We have passive variables q and t:

$$-\partial_1 F_1(t, q, q') + \partial_1 F_2(t, q, p') = 0 \tag{5.144}$$
$$-\partial_0 F_1(t, q, q') + \partial_0 F_2(t, q, p') = 0. \tag{5.145}$$

But $p = \partial_1 F_1(t, q, q')$ from the first transformation, so

$$p = \partial_1 F_2(t, q, p'). \tag{5.146}$$

Furthermore, since $H'(t, q', p') - H(t, q, p) = \partial_0 F_1(t, q, q')$ we can conclude that

$$H'(t, q', p') - H(t, q, p) = \partial_0 F_2(t, q, p'). \tag{5.147}$$

5.4.3 Types of Generating Functions

We have used generating functions of the form $F_1(t, q, q')$ to construct canonical transformations:

$$p = \partial_1 F_1(t, q, q') \tag{5.148}$$
$$p' = -\partial_2 F_1(t, q, q') \tag{5.149}$$
$$H'(t, q', p') - H(t, q, p) = \partial_0 F_1(t, q, q'). \tag{5.150}$$

We can also construct canonical transformations with generating functions of the form $F_2(t, q, p')$, where the third argument of F_2 is the momentum in the primed system.[23]

---

[23]The various generating functions are traditionally known by the names F_1, F_2, F_3, and F_4. Please don't blame us.

$$p = \partial_1 F_2(t, q, p') \tag{5.151}$$
$$q' = \partial_2 F_2(t, q, p') \tag{5.152}$$
$$H'(t, q', p') - H(t, q, p) = \partial_0 F_2(t, q, p') \tag{5.153}$$

As in the F_1 case, to put the transformation in explicit form requires that appropriate inverse functions be constructed to allow the solution of the equations.

Similarly, we can construct two other forms for generating functions, named mnemonically enough F_3 and F_4:

$$q = -\partial_1 F_3(t, p, q') \tag{5.154}$$
$$p' = -\partial_2 F_3(t, p, q') \tag{5.155}$$
$$H'(t, q', p') - H(t, q, p) = \partial_0 F_3(t, p, q') \tag{5.156}$$

and

$$q = -\partial_1 F_4(t, p, p') \tag{5.157}$$
$$q' = \partial_2 F_4(t, p, p') \tag{5.158}$$
$$H'(t, q', p') - H(t, q, p) = \partial_0 F_4(t, p, p') \tag{5.159}$$

These four classes of generating functions are called *mixed-variable generating functions* because the canonical transformations they generate give a mixture of old and new variables in terms of a mixture of old and new variables.

In every case, if the generating function does not depend explicitly on time then the Hamiltonians are obtained from one another purely by composition with the appropriate canonical transformation. If the generating function depends on time, then there are additional terms.

The generating functions presented each treat the coordinates and momenta collectively. One could define more complicated generating functions for which the transformations of different degrees of freedom are specified by generating functions of different types.

5.4.4   Point Transformations

Point transformations can be represented in terms of a generating function of type F_2. Equations (5.6), which define a canonical point transformation derived from a coordinate transformation F, are

$$(t, q, p) = C\left(t, q', p'\right) = \left(t, F(t, q'), p'(\partial_1 F(t, q'))^{-1}\right). \qquad (5.160)$$

Let S be the inverse transformation of F with respect to the second argument

$$q' = S(t, q), \qquad (5.161)$$

so that $q' = S(t, F(t, q'))$. The momentum transformation that accompanies this coordinate transformation is

$$p' = p(\partial_1 S(t, q))^{-1}. \qquad (5.162)$$

We can find the generating function F_2 that gives this transformation by integrating equation (5.152) to get

$$F_2(t, q, p') = p' S(t, q) + \varphi(t, q). \qquad (5.163)$$

Substituting this into equation (5.151), we get

$$p = p' \partial_1 S(t, q) + \partial_1 \varphi(t, q). \qquad (5.164)$$

We do not need the freedom provided by φ, so we can set it equal to zero:

$$F_2(t, q, p') = p' S(t, q), \qquad (5.165)$$

with

$$p = p' \partial_1 S(t, q). \qquad (5.166)$$

So this F_2 gives the canonical transformation of equations (5.161) and (5.162).

The canonical transformation for the coordinate transformation S is the inverse of the canonical transformation for F. By design F and S are inverses on the coordinate arguments. The identity function is $q' = I(q') = S(t, F(t, q'))$. Differentiating yields

$$1 = \partial_1 S(t, F(t, q')) \partial_1 F(t, q'), \tag{5.167}$$

so

$$\partial_1 F(t, q') = (\partial_1 S(t, F(t, q')))^{-1}. \tag{5.168}$$

Using this, the relation between the momenta (5.166) is

$$p = p'(\partial_1 F(t, q'))^{-1}, \tag{5.169}$$

showing that F_2 gives a point transformation equivalent to the point transformation (5.160). So from this other point of view the point transformation is canonical.

The F_1 that corresponds to the F_2 for a point transformation is

$$\begin{aligned}
F_1(t, q, q') &= F_2(t, q, p') - p'q' \\
&= p'S(t, q) - p'q' \\
&= 0. \tag{5.170}
\end{aligned}$$

This is why we could not use generating functions of type F_1 to construct point transformations.

Polar and rectangular coordinates

A commonly required point transformation is the transition between polar coordinates and rectangular coordinates:

$$x = r \cos \theta \tag{5.171}$$
$$y = r \sin \theta.$$

Using the formula for the generating function of a point transformation just derived, we find:

$$F_2(t; r, \theta; p_x, p_y) = \begin{bmatrix} p_x & p_y \end{bmatrix} \begin{pmatrix} r \cos \theta \\ r \sin \theta \end{pmatrix}. \tag{5.172}$$

So the full transformation is derived:

$$\begin{aligned}
(x, y) &= \partial_2 F_2(t; r, \theta; p_x, p_y) \\
&= (r \cos \theta, r \sin \theta) \\
[p_r, p_\theta] &= \partial_1 F_2(t; r, \theta; p_x, p_y) \\
&= [p_x \cos \theta + p_y \sin \theta, -p_x r \sin \theta + p_y r \cos \theta]. \tag{5.173}
\end{aligned}$$

We can isolate the rectangular coordinates to one side of the transformation and the polar coordinates to the other:

$$p_r = \frac{1}{r}(p_x x + p_y y)$$

$$p_\theta = -p_x y + p_y x. \tag{5.174}$$

So, interpreted in terms of Newtonian vectors, $p_r = \hat{r} \cdot \vec{p}$ is the radial component of the linear momentum and $p_\theta = \|\vec{r} \times \vec{p}\|$ is the magnitude of the angular momentum. The point transformation is time independent, so the Hamiltonian transforms by composition.

Rotating coordinates

A useful time-dependent point transformation is the transition to a rotating coordinate system. This is most easily accomplished in polar coordinates. Here we have

$$r' = r$$

$$\theta' = \theta - \Omega t, \tag{5.175}$$

where Ω is the angular velocity of the rotating coordinate system. The generating function is

$$F_2(t; r, \theta; p_r', p_\theta') = [\, p_r' \quad p_\theta' \,] \begin{pmatrix} r \\ \theta - \Omega t \end{pmatrix}. \tag{5.176}$$

This yields the transformation equations

$$r' = r$$

$$\theta' = \theta - \Omega t$$

$$p_r = p_r'$$

$$p_\theta = p_\theta', \tag{5.177}$$

which show that the momenta are the same in both coordinate systems. However, here the Hamiltonian is not a simple composition:

$$H'(t; r', \theta'; p_r', p_\theta') = H(t; r', \theta' + \Omega t; p_r', p_\theta') - p_\theta' \Omega. \tag{5.178}$$

The Hamiltonians differ by the derivative of the generating function with respect to the time argument. In transforming to rotating coordinates, the values of the Hamiltonians differ by the product of the angular momentum and the angular velocity of the coordinate system. Notice that this addition to the Hamiltonian is the same as was found earlier (5.45).

Reducing the two-body problem to the one-body problem

In this example we illustrate how canonical transformations can be used to eliminate some of the degrees of freedom, leaving a problem with fewer degrees of freedom.

Suppose that only certain combinations of the coordinates appear in the Hamiltonian. We make a canonical transformation to a new set of phase-space coordinates such that these combinations of the old phase-space coordinates are some of the new phase-space coordinates. We choose other independent combinations of the coordinates to complete the set. The advantage is that these other independent coordinates do not appear in the new Hamiltonian, so the momenta conjugate to them are conserved quantities.

Let's see how this idea enables us to reduce the problem of two gravitating bodies to the simpler problem of the relative motion of the two bodies. In the process we will discover that the momentum of the center of mass is conserved. This simpler problem is an instance of the Kepler problem. The Kepler problem is also encountered in the formulation of the more general n-body problem.

Consider the motion of two masses m_1 and m_2, subject only to a mutual gravitational attraction described by the potential $V(r)$. This problem has six degrees of freedom. The rectangular coordinates of the particles are x_1 and x_2, with conjugate momenta p_1 and p_2. Each of these is a structure of the three rectangular components. The distance between the particles is $r = \|x_1 - x_2\|$. The Hamiltonian for the two-body problem is

$$H(t; x_1, x_2; p_1, p_2) = \frac{p_1^2}{2m_1} + \frac{p_2^2}{2m_2} + V(r). \tag{5.179}$$

The gravitational potential energy depends only on the relative positions of the two bodies. We do not need to specify V further at this point.

Since the only combination of coordinates that appears in the Hamiltonian is $x_2 - x_1$, we choose new coordinates so that one of the new coordinates is this combination:

$$x = x_2 - x_1. \tag{5.180}$$

To complete the set of new coordinates we choose another to be some independent linear combination

$$X = ax_1 + bx_2, \tag{5.181}$$

where a and b are to be determined. We can use an F_2-type generating function

$$F_2(t; x_1, x_2; p, P) = (x_2 - x_1)p + (ax_1 + bx_2)P, \tag{5.182}$$

where p and P will be the new momenta conjugate to x and X, respectively. We deduce

$$(x, X) = \partial_2 F_2(t; x_1, x_2; p, P) = (x_2 - x_1, ax_1 + bx_2)$$
$$[p_1, p_2] = \partial_1 F_2(t; x_1, x_2; p, P) = [-p + aP, p + bP]. \tag{5.183}$$

We can solve these for the new momenta:

$$P = \frac{p_1 + p_2}{a + b} \tag{5.184}$$

$$p = \frac{ap_2 - bp_1}{a + b}. \tag{5.185}$$

The generating function is not time dependent, so the new Hamiltonian is the old Hamiltonian composed with the transformation:

$$
\begin{aligned}
H'(t; x, X; p, P) &= \frac{(-p + aP)^2}{2m_1} + \frac{(p + bP)^2}{2m_2} + V(\|x\|) \\
&= \frac{p^2}{2m} + \frac{P^2}{2M} + V(\|x\|) \\
&\quad + \left(\frac{b}{m_2} - \frac{a}{m_1} \right) pP,
\end{aligned}
\tag{5.186}
$$

with the definitions

$$\frac{1}{m} = \frac{1}{m_1} + \frac{1}{m_2} \tag{5.187}$$

and

$$\frac{1}{M} = \frac{a^2}{m_1} + \frac{b^2}{m_2}. \tag{5.188}$$

We recognize m as the "reduced mass."

Notice that if the term proportional to pP were not present then the x and X degrees of freedom would not be coupled at all, and furthermore, the X part of the Hamiltonian would be just the Hamiltonian of a free particle, which is easy to solve. The condition that the "cross terms" disappear is

$$\frac{b}{m_2} - \frac{a}{m_1} = 0, \tag{5.189}$$

which is satisfied by

$$a = cm_1$$
$$b = cm_2 \tag{5.190}$$

for any c. For a transformation to be defined, c must be nonzero. So with this choice the Hamiltonian becomes

$$H'(t; x, X; p, P) = H_X(t, X, P) + H_x(t, x, p) \tag{5.191}$$

with

$$H_x(t, x, p) = \frac{p^2}{2m} + V(r) \tag{5.192}$$

and

$$H_X(t, X, P) = \frac{P^2}{2M}. \tag{5.193}$$

The reduced mass is the same as before, and now

$$M = \frac{1}{c^2(m_1 + m_2)}. \tag{5.194}$$

Notice that, without further specifying c, the problem has been separated into the problem of determining the relative motion of the two masses, and the problem of the other degrees of freedom. We did not need a priori knowledge that the center of mass might be important; in fact, only for a particular choice of $c = (m_1 + m_2)^{-1}$ does X become the center of mass.

Epicyclic motion

It is often useful to compose a sequence of canonical transformations to make up the transformation we need for any particular mechanical problem. The transformations we have supplied are especially useful as components in these computations.

We will illustrate the use of canonical transformations to learn about planar motion in a central field. The strategy will be to consider perturbations of circular motion in the central field. The analysis will proceed by transforming to a rotating coordinate system that rides on a circular reference orbit, and then making approximations that restrict the analysis to orbits that differ from the circular orbit only slightly.

In rectangular coordinates we can easily write a Hamiltonian for the motion of a particle of mass m in a field defined by a potential energy that is a function only of the distance from the origin as follows:

$$H(t; x, y; p_x, p_y) = \frac{p_x^2 + p_y^2}{2m} + V(\sqrt{x^2 + y^2}). \tag{5.195}$$

In this coordinate system Hamilton's equations are easy, and they are exactly what is needed to develop trajectories by numerical integration, but the expressions are not very illuminating:

$$Dx = \frac{p_x}{m} \tag{5.196}$$

$$Dy = \frac{p_y}{m} \tag{5.197}$$

$$Dp_x = -DV(\sqrt{x^2 + y^2})\frac{x}{\sqrt{x^2 + y^2}} \tag{5.198}$$

$$Dp_y = -DV(\sqrt{x^2 + y^2})\frac{y}{\sqrt{x^2 + y^2}}. \tag{5.199}$$

We can learn more by converting to polar coordinates centered on the source of our field:

$$x = r \cos \varphi \qquad (5.200)$$

$$y = r \sin \varphi. \qquad (5.201)$$

This coordinate system explicitly incorporates the geometrical symmetry of the potential energy. Extending this coordinate transformation to a point transformation, we can write the new Hamiltonian as:

$$H'(t; r, \varphi; p_r, p_\varphi) = \frac{p_r^2}{2m} + \frac{p_\varphi^2}{2mr^2} + V(r). \qquad (5.202)$$

We can now write Hamilton's equations in these new coordinates, and they are much more illuminating than the equations expressed in rectangular coordinates:

$$Dr = \frac{p_r}{m} \qquad (5.203)$$

$$D\varphi = \frac{p_\varphi}{mr^2} \qquad (5.204)$$

$$Dp_r = \frac{p_\varphi^2}{mr^3} - DV(r) \qquad (5.205)$$

$$Dp_\varphi = 0. \qquad (5.206)$$

The angular momentum p_φ is conserved, and we are free to choose its constant value, so $D\varphi$ depends only on r. We also see that we can establish a circular orbit at any radius R_0: we choose $p_\varphi = p_{\varphi_0}$ so that $p_{\varphi_0}^2/(mR_0^3) - DV(R_0) = 0$. This will ensure that $Dp_r = 0$, and thus $Dr = 0$. The square of the angular velocity of this circular orbit is

$$\Omega^2 = \frac{DV(R_0)}{mR_0}. \qquad (5.207)$$

It is instructive to consider how orbits that are close to the circular orbit differ from the circular orbit. This is best done in rotating coordinates in which a body moving in the circular orbit is a stationary point at the origin. We can do this by converting to coordinates that are rotating with the circular orbit and centered on the orbiting body. We proceed in three stages. First we will transform to a polar coordinate system that is rotating at angular velocity Ω. Then we will return to rectangular coordinates, and finally, we will shift the coordinates so that the origin is on the reference circular orbit.

We start by examining the system in rotating polar coordinates. This is a time-dependent coordinate transformation:

$$r' = r \tag{5.208}$$

$$\varphi' = \varphi - \Omega t \tag{5.209}$$

$$p_r' = p_r \tag{5.210}$$

$$p_\varphi' = p_\varphi. \tag{5.211}$$

Using equation (5.178), we can write the new Hamiltonian directly:

$$H''(t; r', \varphi'; p_r', p_\varphi') = \frac{p_r'^2}{2m} + \frac{p_\varphi'^2}{2mr'^2} + V(r') - p_\varphi'\Omega. \tag{5.212}$$

H'' is not time dependent, and therefore it is conserved. It is not the sum of the potential energy and the kinetic energy. Energy is not conserved in the moving coordinate system, but what is conserved here is a new quantity, the *Jacobi constant*, that combines the energy with the product of the angular momentum of the particle in the new coordinate and the angular velocity of the coordinate system. We will want to keep track of this term.

Next, we return to rectangular coordinates, but they are rotating with the reference circular orbit:

$$x' = r' \cos \varphi' \tag{5.213}$$

$$y' = r' \sin \varphi' \tag{5.214}$$

$$p_x' = p_r' \cos \varphi' - \frac{p_\varphi'}{r'} \sin \varphi' \tag{5.215}$$

$$p_y' = p_r' \sin \varphi' + \frac{p_\varphi'}{r'} \cos \varphi'. \tag{5.216}$$

The Hamiltonian is

$$H'''(t; x', y'; p_x', p_y')$$
$$= \frac{p_x'^2 + p_y'^2}{2m} + \Omega(y'p_x' - x'p_y') + V(\sqrt{x'^2 + y'^2}). \tag{5.217}$$

With one more quick manipulation we shift the coordinate system so that the origin is out on our circular orbit. We define new rectangular coordinates ξ and η with the following simple canonical transformation of coordinates and momenta:

$$\xi = x' - R_0 \tag{5.218}$$
$$\eta = y' \tag{5.219}$$
$$p_\xi = p'_x \tag{5.220}$$
$$p_\eta = p'_y. \tag{5.221}$$

In this final coordinate system the Hamiltonian is

$$H''''(t; \xi, \eta; p_\xi, p_\eta) = \frac{p_\xi^2 + p_\eta^2}{2m} + \Omega(\eta p_\xi - (\xi + R_0)p_\eta)$$
$$+ V(\sqrt{(\xi + R_0)^2 + \eta^2}), \tag{5.222}$$

and Hamilton's equations are uselessly complicated, but the next step is to consider only trajectories for which the coordinates ξ and η are small compared with R_0. Under this assumption we will be able to construct approximate equations of motion for these trajectories that are linear in the coordinates, thus yielding simple analyzable motion. To this point we have made no approximations. The equations above are perfectly accurate for any trajectories in a central field.

The idea is to expand the potential-energy term in the Hamiltonian as a series and to discard any term higher than second-order in the coordinates, thus giving us first-order-accurate Hamilton's equations:

$$U(\xi, \eta) = V(\sqrt{(\xi + R_0)^2 + \eta^2}) \tag{5.223}$$

$$= V(R_0 + \xi + \frac{\eta^2}{2R_0} + \cdots) \tag{5.224}$$

$$= V(R_0) + DV(R_0)(\xi + \frac{\eta^2}{2R_0})$$
$$+ D^2 V(R_0)\frac{\xi^2}{2} + \cdots. \tag{5.225}$$

So the (negated) generalized forces are

$$\partial_0 U(\xi, \eta) = DV(R_0) + D^2 V(R_0)\xi + \cdots \tag{5.226}$$
$$\partial_1 U(\xi, \eta) = DV(R_0)\frac{\eta}{R_0} + \cdots. \tag{5.227}$$

With this expansion we obtain the linearized Hamilton's equations:

$$D\xi = \frac{p_\xi}{m} + \Omega\eta \tag{5.228}$$

$$D\eta = \frac{p_\eta}{m} - \Omega(\xi + R_0) \tag{5.229}$$

$$Dp_\xi = -DV(R_0) - D^2V(R_0)\xi + \cdots + \Omega p_\eta \tag{5.230}$$

$$Dp_\eta = -DV(R_0)\frac{\eta}{R_0} + \cdots - \Omega p_\xi. \tag{5.231}$$

Of course, once we have linear equations we know how to solve them exactly. Because the linearized Hamiltonian is conserved we cannot get exponential expansion or collapse, so the possible solutions are quite limited. It is instructive to convert these equations into a second-order system. We use $\Omega^2 = DV(R_0)/(mR_0)$, equation (5.207), to eliminate the DV terms:

$$D^2\xi - 2\Omega D\eta = (\Omega^2 - \frac{D^2V(R_0)}{m})\xi \tag{5.232}$$

$$D^2\eta + 2\Omega D\xi = 0. \tag{5.233}$$

Combining these, we find

$$D^3\xi + \omega^2 D\xi = 0, \tag{5.234}$$

where

$$\omega^2 = 3\Omega^2 + \frac{D^2V(R_0)}{m}. \tag{5.235}$$

Thus we have a simple harmonic oscillator with frequency ω as one of the components of the solution. The general solution has three parts:

$$\begin{pmatrix} \xi(t) \\ \eta(t) \end{pmatrix} = \eta_0 \begin{pmatrix} 0 \\ 1 \end{pmatrix} \tag{5.236}$$

$$+ \xi_0 \begin{pmatrix} 1 \\ -2At \end{pmatrix} \tag{5.237}$$

$$+ C_0 \begin{pmatrix} \sin(\omega t + \varphi_0) \\ \frac{2\Omega}{\omega}\cos(\omega t + \varphi_0) \end{pmatrix} \tag{5.238}$$

where

$$A = \frac{\Omega^2 m - D^2V(R_0)}{4\Omega m}. \tag{5.239}$$

The constants η_0, ξ_0, C_0, and φ_0 are determined by the initial conditions. If $C_0 = 0$, the particle of interest is on a circular trajectory, but not necessarily the same one as the reference trajectory. If $C_0 = 0$ and $\xi_0 = 0$, we have a "fellow traveler," a particle in the same circular orbit as the reference orbit but with different phase. If $C_0 = 0$ and $\eta_0 = 0$, we have a particle in a circular orbit that is interior or exterior to the reference orbit and shearing away from the reference orbit. The shearing is due to the fact that the angular velocity for a circular orbit varies with the radius. The constant A gives the rate of shearing at each radius. If both $\eta_0 = 0$ and $\xi_0 = 0$ but $C_0 \neq 0$, then we have "epicyclic motion." A particle in a nearly circular orbit may be seen to move in an ellipse around the circular reference orbit. The ellipse will be elongated in the direction of circular motion by the factor $2\Omega/\omega$, and it will rotate in the direction opposite to the direction of the circular motion. The initial phase of the epicycle is φ_0. Of course, any combination of these solutions may exist.

The epicyclic frequency ω and the shearing rate A are determined by the force law (the radial derivative of the potential energy). For a force law proportional to a power of the radius,

$$F \propto r^{1-n}, \tag{5.240}$$

the epicyclic frequency is related to the orbital frequency by

$$\frac{\omega}{\Omega} = 2\sqrt{1 - \frac{n}{4}} \tag{5.241}$$

and the shearing rate is

$$\frac{A}{\Omega} = \frac{n}{4}. \tag{5.242}$$

For a few particular integer force laws we see:

n	0	1	2	3	4	5
$\frac{A}{\Omega}$	0	$\frac{1}{4}$	$\frac{1}{2}$	$\frac{3}{4}$	1	$\frac{5}{4}$
$\frac{\omega}{\Omega}$	2	$\sqrt{3}$	$\sqrt{2}$	1	0	$\pm i$

We can get some insight into the kinds of orbits produced by the epicyclic approximation by looking at a few examples. For

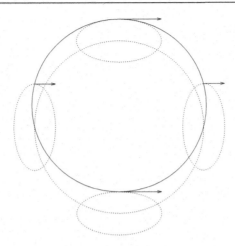

Figure 5.3   Epicyclic construction of an approximate orbit for $F \propto r^{-2}$. The large dotted circle is the reference circular orbit and the dotted ellipses are the epicycles. The epicycles are twice as long as they are wide. The solid ellipse is the approximate trajectory produced by a particle moving on the epicycles. The sense of orbital motion is counterclockwise, and the epicycles are rotating clockwise. The arrows represent the increment of velocity contributed by the epicycle to the circular reference orbit.

some force laws we have integer ratios of epicyclic frequency to orbital frequency. In those cases we have closed orbits. For an inverse-square force law ($n = 3$) we get elliptical orbits with the center of the field at a focus of the ellipse. Figure 5.3 shows how an approximation to such an orbit can be constructed by superposition of the motion on an elliptical epicycle with the motion of the same frequency on a circle. If the force is proportional to the radius ($n = 0$) we get a two-dimensional harmonic oscillator. Here the epicyclic frequency is twice the orbital frequency. Figure 5.4 shows how this yields elliptical orbits that are centered on the source of the central force. An orbit is closed when ω/Ω is a rational fraction. If the force is proportional to the $-3/4$ power of the radius, the epicyclic frequency is $3/2$ the orbital frequency. This yields the three-lobed pattern seen in figure 5.5. For other force laws the orbits predicted by this analysis are multi-lobed patterns produced by precessing approximate ellipses. Most of the cases have incommensurate epicyclic and orbital frequencies, leading to orbits that do not close in finite time.

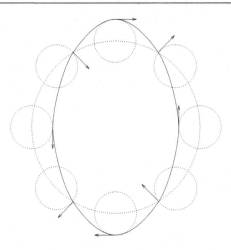

Figure 5.4   Epicyclic construction of an approximate orbit for $F \propto r$. The large dotted circle is the reference circular orbit and the small dotted circles are the epicycles. The solid ellipse is the approximate trajectory produced by a particle moving on the epicycles. The sense of orbital motion is counterclockwise, and the epicycles are rotating clockwise. The arrows represent the increment of velocity contributed by the epicycle to the circular reference orbit.

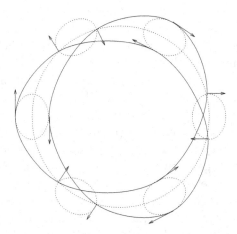

Figure 5.5   Epicyclic construction of an approximate orbit for $F \propto r^{-3/4}$. The large dotted circle is the reference circular orbit and the dotted ellipses are the epicycles. The epicycles have a 4:3 ratio of length to width. The solid trefoil is the approximate trajectory produced by a particle moving on the epicycles. The sense of orbital motion is counterclockwise, and the epicycles are rotating clockwise. The arrows represent the increment of velocity contributed by the epicycle to the circular reference orbit.

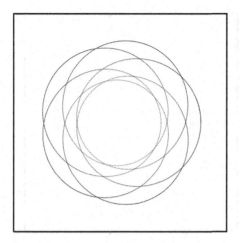

Figure 5.6 The numerically integrated orbit of a particle with a force law $F \propto r^{-2.3}$. For this law the ratio of the epicyclic frequency to the orbital frequency is about .83666—close to 5/6, but not quite. This is manifest in the nearly five-fold symmetry of the rosette-like shape and the fact that one must cross approximately six orbits to get from the inside to the outside of the rosette.

The epicyclic approximation gives a very good idea of what actual orbits look like. Figure 5.6, drawn by numerical integration of the orbit produced by integrating the original rectangular equations of motion for a particle in the field, shows the rosette-type picture characteristic of incommensurate epicyclic and orbital frequencies for an $F = -r^{-2.3}$ force law.

We can directly compare a numerically integrated system with one of our epicyclic approximations. For example, the result of numerically integrating our $F \propto r^{-3/4}$ system is very similar to the picture we obtained by epicycles. (See figure 5.7 and compare it with figure 5.5.)

Exercise 5.11: Collapsing orbits

What exactly happens as the force law becomes steeper? Investigate this by sketching the contours of the Hamiltonian in r, p_r space for various values of the force-law exponent, n. For what values of n are there stable circular orbits? In the case that there are no stable circular orbits, what happens to circular and other noncircular orbits? How are these results consistent with Liouville's theorem and the nonexistence of attractors in Hamiltonian systems?

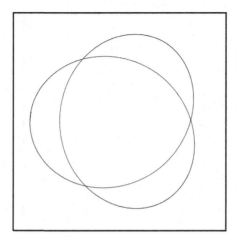

Figure 5.7   The numerically integrated orbit of a particle with a force law $F \propto r^{-3/4}$. For this law the ratio of the epicyclic frequency to the orbital frequency is exactly $3/2$. This is manifest in the three-fold symmetry of the rosette-like shape and the fact that one must cross two orbits to get from the inside to the outside of the rosette.

5.4.5   Total Time Derivatives

The addition of a total time derivative to a Lagrangian leads to the same Lagrange equations. However, the two Lagrangians have different momenta, and they lead to different Hamilton's equations. Here we find out how to represent the corresponding canonical transformation with a generating function.

Let's restate the result about total time derivatives and Lagrangians from the first chapter. Consider some function $G(t, q)$ of time and coordinates. We have shown that if L and L' are related by

$$L'(t, q, \dot{q}) = L(t, q, \dot{q}) + \partial_0 G(t, q) + \partial_1 G(t, q)\dot{q} \qquad (5.243)$$

then the Lagrange equations of motion are the same. The generalized coordinates used in the two Lagrangians are the same, but the momenta conjugate to the coordinates are different. In the usual way, define

$$\mathcal{P}(t, q, \dot{q}) = \partial_2 L(t, q, \dot{q}) \qquad (5.244)$$

and

$$\mathcal{P}'(t,q,\dot{q}) = \partial_2 L'(t,q,\dot{q}). \tag{5.245}$$

So we have

$$\mathcal{P}'(t,q,\dot{q}) = \mathcal{P}(t,q,\dot{q}) + \partial_1 G(t,q). \tag{5.246}$$

Evaluated on a trajectory, we have

$$p'(t) = p(t) + \partial_1 G(t,q(t)). \tag{5.247}$$

This transformation is a special case of an F_2-type transformation. Let

$$F_2(t,q,p') = qp' - G(t,q); \tag{5.248}$$

then the associated transformation is

$$q' = \partial_2 F_2(t,q,p') = q \tag{5.249}$$
$$p = \partial_1 F_2(t,q,p') = p' - \partial_1 G(t,q) \tag{5.250}$$
$$H'(t,q',p') = H(t,q,p) + \partial_0 F_2(t,q,p')$$
$$= H(t,q,p) - \partial_0 G(t,q). \tag{5.251}$$

Explicitly, the new Hamiltonian is

$$H'(t,q',p') = H(t,q',p' - \partial_1 G(t,q')) - \partial_0 G(t,q'), \tag{5.252}$$

where we have used the fact that $q = q'$. The transformation is interesting in that the coordinate transformation is the identity transformation, but the new and old momenta are not the same, even in the case in which G has no explicit time dependence. Suppose we have a Hamiltonian of the form

$$H(t,x,p) = \frac{p^2}{2m} + V(x); \tag{5.253}$$

then the transformed Hamiltonian is

$$H'(t,x',p') = \frac{(p' - \partial_1 G(t,x'))^2}{2m} + V(x') - \partial_0 G(t,x'). \tag{5.254}$$

We see that this transformation may be used to modify terms in the Hamiltonian that are linear in the momenta. Starting from H, the transformation introduces linear momentum terms; starting from H', the transformation eliminates the linear terms.

Driven pendulum

We illustrate the use of this transformation with the driven pendulum. The Hamiltonian for the driven pendulum derived from the $T - V$ Lagrangian (see section 1.6.2) is

$$
H(t, \theta, p_\theta)
$$
$$
= \frac{p_\theta^2}{2ml^2} - glm \cos \theta
$$
$$
+ gm y_s(t) - \frac{p_\theta}{l} \sin \theta D y_s(t) - \frac{m}{2}(\cos \theta)^2 (D y_s(t))^2, \quad (5.255)
$$

where y_s is the drive function. The Hamiltonian is rather messy, and includes a term that is linear in the angular momentum with a coefficient that depends on both the angular coordinate and the time. Let's see what happens if we apply our transformation to the problem to eliminate the linear term. We can identify the transformation function G by requiring that the linear term in momentum be killed:

$$
G(t, \theta) = -ml \cos \theta D y_s(t). \tag{5.256}
$$

The transformed momentum is

$$
p'_\theta = p_\theta + ml \sin \theta D y_s(t), \tag{5.257}
$$

and the transformed Hamiltonian is

$$
H'(t, \theta, p'_\theta) = \frac{(p'_\theta)^2}{2ml^2} - ml(g + D^2 y_s) \cos \theta
$$
$$
+ gm y_s(t) - \frac{m}{2}(y_s(t))^2. \tag{5.258}
$$

Dropping the last two terms, which do not affect the equations of motion, we find

$$
H'(t, \theta, p'_\theta) = \frac{(p'_\theta)^2}{2ml^2} - ml(g + D^2 y_s) \cos \theta. \tag{5.259}
$$

So we have found, by a straightforward canonical transformation, a Hamiltonian for the driven pendulum with the rather simple form of a pendulum with gravitational acceleration that is modified by the acceleration of the pivot. It is, in fact, the Hamiltonian that corresponds to the alternative form of the Lagrangian for the driven pendulum that we found earlier by inspection (see equation 1.120). Here the derivation is by a simple canonical transformation, motivated by a desire to eliminate unwanted terms that are linear in the momentum.

Exercise 5.12: Construction of generating functions

Suppose that canonical transformations

$$(t, q, p) = C_a(t, q', p') \quad \text{and} \quad (t, q', p') = C_b(t, q'', p'')$$

are generated by two F_1-type generating functions, $F_{1a}(t, q, q')$ and $F_{1b}(t, q', q'')$.

a. Show that the generating function for the inverse transformation of C_a is $F_{1c}(t, q', q) = -F_{1a}(t, q, q')$.

b. Define a new kind of generating function,

$$F_x(t, q, q', q'') = F_{1a}(t, q, q') + F_{1b}(t, q', q'').$$

We see that

$$p = \partial_1 F_x(t, q, q', q'') = \partial_1 F_{1a}(t, q, q')$$
$$p'' = -\partial_3 F_x(t, q, q', q'') = -\partial_2 F_{1b}(t, q', q'')$$

Show that $\partial_2 F_x = 0$, allowing a solution to eliminate q'.

c. Using the formulas for p and p'' above, and the result from part **b**, Show that F_x is an appropriate generating function for the composition transformation $C_a \circ C_b$.

Exercise 5.13: Linear canonical transformations

We consider systems with two degrees of freedom and transformations for which the Hamiltonian transforms by composition.

a. Consider the linear canonical transformations that are generated by

$$F_2(t; x_1, x_2; p_1', p_2') = p_1' a x_1 + p_1' b x_2 + p_2' c x_1 + p_2' d x_2.$$

Show that these transformations are just the point transformations, and that the corresponding F_1 is zero.

b. Other linear canonical transformations can be generated by

$$F_1(t; x_1, x_2; x_1', x_2') = x_1' a x_1 + x_1' b x_2 + x_2' c x_1 + x_2' d x_2.$$

Surely we can make even more generators by constructing F_3- and F_4-type transformations analogously. Are all of the linear canonical transformations obtainable in this way? If not, show one that cannot be so generated.

c. Can all linear canonical transformations be generated by compositions of transformations generated by the functions shown in parts a and b above?

d. How many independent parameters are necessary to specify all possible linear canonical transformations for systems with two degrees of freedom?

Exercise 5.14: Integral invariants

Consider the linear canonical transformation for a system with two degrees of freedom generated by the function

$$F_1(t; x_1, x_2; x_1', x_2') = x_1' a x_1 + x_1' b x_2 + x_2' c x_1 + x_2' d x_2,$$

and the general parallelogram with a vertex at the origin and with adjacent sides starting at the origin and extending to the phase-space points $(x_{1a}, x_{2a}, p_{1a}, p_{2a})$ and $(x_{1b}, x_{2b}, p_{1b}, p_{2b})$.

a. Find the area of the given parallelogram and the area of the target parallelogram under the canonical transformation. Notice that the area of the parallelogram is not preserved.

b. Find the areas of the projections of the given parallelogram and the areas of the projections of the target under canonical transformation. Show that the sum of the areas of the projections on the action-like planes is preserved.

Exercise 5.15: Standard-map generating function

Find a generating function for the standard map (see exercise 5.8 on page 357).

5.5   Extended Phase Space

In this section we show that we can treat time as just another coordinate if we wish. Systems described by a time-dependent Hamiltonian may be recast in terms of a time-independent Hamiltonian with an extra degree of freedom. An advantage of this view is that what was a time-dependent canonical transformation can be treated as a time-independent transformation, where there are no additional conditions for adjusting the Hamiltonian.

Suppose that we have some system characterized by a time-dependent Hamiltonian, for example, a periodically driven pendulum. We may imagine that there is some extremely massive oscillator, unperturbed by the motion of the relatively massless pendulum, that produces the drive. Indeed, we may think of time itself as the coordinate of an infinitely massive particle moving uniformly and driving everything else. We often consider the rotation of the Earth as exactly such a stable time reference when performing short-time experiments in the laboratory.

More formally, consider a dynamical system with n degrees of freedom, whose behavior is described by a possibly time-dependent Lagrangian L with corresponding Hamiltonian H. We make a new dynamical system with $n+1$ degrees of freedom by extending the generalized coordinates to include time and introducing a new independent variable. We also extend the generalized velocities to include a velocity for the time coordinate. In this new *extended state space* the coordinates are redundant, so there is a constraint relating the time coordinate to the new independent variable.

We relate the original dynamical system to the extended dynamical system as follows: Let q be a coordinate path. Let $(q_e, t) : \tau \mapsto (q_e(\tau), t(\tau))$ be a coordinate path in the extended system where τ is the new independent variable. Then $q_e = q \circ t$, or $q_e(\tau) = q(t(\tau))$. Consequently, if $v = Dq$ is the velocity along a path then $v_e(\tau) = Dq_e(\tau) = Dq(t(\tau)) \cdot Dt(\tau) = v(t(\tau)) \cdot v_t(\tau)$.

We can find a Lagrangian for the extended system by requiring that the value of the action be unchanged. Introduce the extended Lagrangian action

$$S_e[q_e, t](\tau_1, \tau_2) = \int_{\tau_1}^{\tau_2} (L_e \circ \Gamma[q_e, t]), \qquad (5.260)$$

with

$$L_e(\tau; q_e, t; v_e, v_t) = L(t, q_e, v_e/v_t)v_t. \qquad (5.261)$$

We have

$$S[q](t(\tau_1), t(\tau_2)) = S_e[q \circ t, t](\tau_1, \tau_2). \qquad (5.262)$$

The extended system is subject to a constraint that relates the time to the new independent variable. We assume the constraint is of the form $\varphi(\tau; q_e, t; v_e, v_t) = t - f(\tau) = 0$. The constraint is a holonomic constraint involving the coordinates and time, so we can incorporate this constraint by augmenting the Lagrangian:[24]

$$L'_e(\tau; q_e, t, \lambda; v_e, v_t, v_\lambda)$$
$$= L_e(\tau; q_e, t; v_e, v_t) + v_\lambda(v_t - Df(\tau))$$
$$= L(t, q_e, v_e/v_t)v_t + v_\lambda(v_t - Df(\tau)). \tag{5.263}$$

The Lagrange equations of L'_e for q_e are satisfied for the paths $q \circ t$ where q is any path that satisfies the original Lagrange equations of L.

The momenta conjugate to the coordinates are

$$\mathcal{P}_e(\tau; q_e, t, \lambda; v_e, v_t, v_\lambda)$$
$$= \partial_{2,0}L'_e(\tau; q_e, t, \lambda; v_e, v_t, v_\lambda)$$
$$= \partial_2 L(t, q_e, v_e/v_t)$$
$$= \mathcal{P}(t, q_e, v_e/v_t) \tag{5.264}$$
$$\mathcal{P}_t(\tau; q_e, t, \lambda; v_e, v_t, v_\lambda)$$
$$= \partial_{2,1}L'_e(\tau; q_e, t, \lambda; v_e, v_t, v_\lambda)$$
$$= L(t, q_e, v_e/v_t) - \partial_2 L(t, q_e, v_e/v_t)(v_e/v_t) + v_\lambda$$
$$= -\mathcal{E}(t, q_e, v_e/v_t) + v_\lambda \tag{5.265}$$
$$\mathcal{P}_\lambda(\tau; q_e, t, \lambda; v_e, v_t, v_\lambda)$$
$$= \partial_{2,2}L'_e(\tau; q_e, t, \lambda; v_e, v_t, v_\lambda)$$
$$= v_t - Df(\tau). \tag{5.266}$$

So the extended momenta have the same values as the original momenta at the corresponding states. The momentum conjugate to the time coordinate is the negation of the energy plus v_λ. The momentum conjugate to λ is the constraint, which must be zero.

Next we carry out the transformation to the corresponding Hamiltonian formulation. First, note that the Lagrangian L_e is a homogeneous form of degree one in the velocities. Thus, by Euler's theorem,

$$\partial_2 L_e(\tau; q_e, t; v_e, v_t) \cdot (v_e, v_t) = L_e(\tau; q_e, t; v_e, v_t). \tag{5.267}$$

---

[24]We augment the Lagrangian with the total time derivative of the constraint so that the Legendre transform will be well defined.

The $p\dot{q}$-part of the Legendre transform of L'_e is

$$\partial_2 L'_e(\tau; q_e, t, \lambda; v_e, v_t, v_\lambda) \cdot (v_e, v_t, v_\lambda)$$
$$= \partial_2 L_e(\tau; q_e, t; v_e, v_t) \cdot (v_e, v_t) + v_\lambda v_t + (v_t - Df(\tau))v_\lambda$$
$$= L_e(\tau; q_e, t; v_e, v_t) + v_\lambda v_t + (v_t - Df(\tau))v_\lambda. \qquad (5.268)$$

So the Hamiltonian H'_e corresponding to L'_e is

$$H'_e(\tau; q_e, t, \lambda; p_e, p_t, p_\lambda) = v_\lambda v_t$$
$$= (p_t + H(t, q_e, p_e))(p_\lambda + Df(\tau)). \quad (5.269)$$

We have used the fact that at corresponding states the momenta have the same values, so on paths $p_e = p \circ t$, and

$$\mathcal{E}(t, q_e, v_e/v_t) = H(t, q_e, p_e). \qquad (5.270)$$

The Hamiltonian H'_e does not depend on λ so we deduce that p_λ is constant. In fact, p_λ must be given the value zero, because it is the constraint. When there is a cyclic coordinate we can form a reduced Hamiltonian for the remaining degrees of freedom by substituting the constant value of conserved momentum conjugate to the cyclic coordinate into the Hamiltonian. The resulting Hamiltonian is

$$H_e(\tau; q_e, t; p_e, p_t) = (p_t + H(t, q_e, p_e))Df(\tau). \qquad (5.271)$$

This extended Hamiltonian governs the evolution of the extended system, for arbitrary f.[25]
   Hamilton's equations reduce to

$$Dq_e(\tau) = \partial_2 H(t(\tau), q_e(\tau), p_e(\tau))Df(\tau)$$
$$Dt(\tau) = Df(\tau)$$
$$Dp_e(\tau) = -\partial_1 H(t(\tau), q_e(\tau), p_e(\tau))Df(\tau)$$
$$Dp_t(\tau) = -\partial_0 H(t(\tau), q_e(\tau), p_e(\tau))Df(\tau). \qquad (5.272)$$

---

[25] Once we have made this reduction, taking p_λ to be zero, we can no longer perform a Legendre transform back to the extended Lagrangian system; we cannot solve for p_t in terms of v_t. However, the Legendre transform in the extended system from H'_e to L'_e, with associated state variables, is well defined.

The second equation gives the required relation between t and τ. The first and third equations are equivalent to Hamilton's equations in the original coordinates, as we can see by using $q_e = q \circ t$ to rewrite them:

$$Dq(t(\tau))Dt(\tau) = \partial_2 H(t(\tau), q(t(\tau)), p(t(\tau)))Df(\tau)$$
$$Dp(t(\tau))Dt(\tau) = -\partial_1 H(t(\tau), q(t(\tau)), p(t(\tau)))Df(\tau). \qquad (5.273)$$

Using $Dt(\tau) = Df(\tau)$ and dividing these factors out, we recover Hamilton's equations.[26]

Now consider the special case for which the time is the same as the independent variable: $f(\tau) = \tau$, $Df(\tau) = 1$. In this case $q = q_e$ and $p = p_e$. The extended Hamiltonian becomes

$$H'_e(\tau; q_e, t; p_e, p_t) = p_t + H(t, q_e, p_e). \qquad (5.274)$$

Hamilton's equation for t becomes $Dt(\tau) = 1$, restating the constraint. Hamilton's equations for Dq_e and Dp_e are directly Hamilton's equations:

$$Dq(\tau) = \partial_2 H(\tau, q(\tau), p(\tau))$$
$$Dp(\tau) = -\partial_1 H(\tau, q(\tau), p(\tau)). \qquad (5.275)$$

The extended Hamiltonian (5.274) does not depend on the independent variable, so it is a conserved quantity. Thus, up to an additive constant p_t is equal to minus the energy. The Hamilton's equation for Dp_t relates the change of the energy to $\partial_0 H$. Note that in the more general case, the momentum conjugate to the time is not the negation of the energy. This choice, $t(\tau) = \tau$, is useful for a number of applications.

The extension transformation is canonical in the sense that the two sets of equations of motion describe equivalent dynamics. However, the transformation is not symplectic; in fact, it does not even have the same number of input and output variables.

Exercise 5.16: Homogeneous extended Lagrangian
Verify that L_e is homogeneous of degree one in the velocities.

---

[26] If f is strictly increasing then Df is never zero.

Exercise 5.17: Lagrange equations

a. Verify that the Lagrange equations for q_e are satisfied for exactly the same trajectories that satisfy the original Lagrange equations for q.

b. Verify that the Lagrange equation for t relates the rate of change of energy to $\partial_0 L$.

Exercise 5.18: Lorentz transformations

Investigate Lorentz transformations as point transformations in the extended phase space.

Restricted three-body problem

An example that shows the utility of reformulating a problem in the extended phase space is the restricted three-body problem: the motion of a low-mass particle subject to the gravitational attraction of two other massive bodies that move in some fixed orbit. The problem is an idealization of the situation where a body with very small mass moves in the presence of two bodies with much larger masses. Any effects of the smaller body on the larger bodies are neglected. In the simplest version, the motion of all three bodies is assumed to be in the same plane, and the orbits of the two massive bodies are circular.

The motion of the bodies with larger masses is not influenced by the small mass, so we model this situation as the small body moving in a time-varying field of the larger bodies undergoing a prescribed motion. This situation can be captured as a time-dependent Hamiltonian:

$$H(t; x, y; p_x, p_y) = \frac{p_x^2 + p_y^2}{2m} - \frac{Gmm_1}{r_1(t)} - \frac{Gmm_2}{r_2(t)}, \qquad (5.276)$$

where $r_1(t)$ and $r_2(t)$ are the distances of the small body to the larger bodies, m is the mass of the small body, and m_1 and m_2 are the masses of the larger bodies. Note that $r_1(t)$ and $r_2(t)$ are quantities that depend both on the position of the small particle and the time-varying position of the massive particles.

The massive bodies are in circular orbits and maintain constant distance from the center of mass. Let a_1 and a_2 be the distances to the center of mass; then the distances satisfy $m_1 a_1 = m_2 a_2$. The angular frequency is $\Omega = \sqrt{G(m_1 + m_2)/a^3}$ where a is the distance between the masses.

In polar coordinates, with the center of mass of the subsystem of massive particles at the origin and with r and θ describing the position of the low-mass particle, the positions of the two massive bodies are $a_2 = m_1a/(m_1+m_2)$ with $\theta_2 = \Omega t$, $a_1 = m_2a/(m_1+m_2)$ with $\theta_1 = \Omega t + \pi$. The distances to the point masses are

$$(r_2(t))^2 = r^2 + a_2^2 - 2a_2r\cos(\theta - \Omega t)$$
$$(r_1(t))^2 = r^2 + a_1^2 - 2a_1r\cos(\theta - \Omega t - \pi). \tag{5.277}$$

In polar coordinates, the Hamiltonian is

$$H(t; r, \theta; p_r, p_\theta) = \frac{1}{2m}\left(p_r^2 + \frac{p_\theta^2}{r^2}\right) - \frac{Gmm_1}{r_1(t)} - \frac{Gmm_2}{r_2(t)}. \tag{5.278}$$

The Hamiltonian can be written in terms of some function f such that

$$H(t; r, \theta; p_r, p_\theta) = f(r, \theta - \Omega t, p_r, p_\theta). \tag{5.279}$$

The essential feature is that θ and t appear in the Hamiltonian only in the combination $\theta - \Omega t$.

One way to get rid of the time dependence is to choose a new set of variables with one coordinate equal to this combination $\theta - \Omega t$, by making a point transformation to a rotating coordinate system. We have shown that

$$r' = r \tag{5.280}$$
$$\theta' = \theta - \Omega t \tag{5.281}$$
$$p'_r = p_r \tag{5.282}$$
$$p'_\theta = p_\theta \tag{5.283}$$

with

$$\begin{aligned} H'(t; r', \theta'; p'_r, p'_\theta) &= H(t; r', \theta' + \Omega t; p'_r, p'_\theta) - \Omega p'_\theta \\ &= f(r', \theta', p'_r, p'_\theta) - \Omega p'_\theta \end{aligned} \tag{5.284}$$

is a canonical transformation. The new Hamiltonian, which is not the energy, is conserved because there is no explicit time dependence. It is a useful conserved quantity—the Jacobi constant.[27]

---

[27] Actually, the traditional Jacobi constant is $C = -2H'$.

We can also eliminate the dependence on the independent time-like variable from the Hamiltonian for the restricted problem by going to the extended phase space, choosing $t = \tau$. The Hamiltonian

$$H_e(\tau; r, \theta, t; p_r, p_\theta, p_t) = H(t; r, \theta; p_r, p_\theta) + p_t$$
$$= f(r, \theta - \Omega t, p_r, p_\theta) + p_t \qquad (5.285)$$

is autonomous and is consequently a conserved quantity. Again, we see that θ and t occur only in the combination $\theta - \Omega t$, which suggests a point transformation to a new coordinate $\theta' = \theta - \Omega t$. This point transformation is independent of the new independent variable τ. The transformation is specified in equations (5.280–5.283), augmented by relations specifying how the time coordinate and its conjugate momentum are handled:

$$t = t' \qquad (5.286)$$
$$p_t = -\Omega p'_\theta + p'_t. \qquad (5.287)$$

The new Hamiltonian is obtained by composing the old Hamiltonian with the transformation:

$$H'_e(\tau; r', \theta', t'; p'_r, p'_\theta, p'_t)$$
$$= H_e(\tau; r', \theta' + \Omega t', t'; p'_r, p'_\theta, p'_t - \Omega p'_\theta)$$
$$= f(r', \theta', p'_r, p'_\theta) + p'_t - \Omega p'_\theta. \qquad (5.288)$$

We recognize that the new Hamiltonian in the extended phase space, which has the same value as the original Hamiltonian in the extended phase space, is just the Jacobi constant plus p'_t. The new Hamiltonian does not depend on t', so p'_t is a constant of the motion. In fact, its value is irrelevant to the rest of the dynamical evolution, so we may set the value of p'_t to zero if we like. Thus, we have found that the Hamiltonian in the extended phase space, which is conserved, is just the Jacobi constant plus an additive arbitrary constant. We have two routes to the Jacobi constant: (1) transform the original system to a rotating coordinate system to eliminate the time dependence, but in the process add extra terms to the Hamiltonian, and (2) go to the extended phase space and immediately get a conserved quantity, and by going to rotating coordinates recognize that this Hamiltonian is the same as the

Jacobi constant. So sometimes the Hamiltonian in the extended phase space is a useful conserved quantity.

Exercise 5.19: Transformations in the extended phase space

In section 5.2.1 we found that time-dependent transformations for which the derivative of the coordinate–momentum part is symplectic are canonical only if the Hamiltonian is modified by adding a function K subject to certain constraints (equation 5.42). Show that the constraints on K follow from the symplectic condition in the extended phase space, using the choice $t = \tau$.

5.5.1   Poincaré–Cartan Integral Invariant

The Poincaré invariant (section 5.3) is especially useful in the extended phase space with $t = \tau$. In the extended phase space the extended Hamiltonian does not depend on the independent variable. In the extended phase space canonical transformations are symplectic and the Hamiltonian transforms by composition.

For the special choice of $t = \tau$, equation (5.90) can be rephrased in an interesting way. Let E be the value of the Hamiltonian in the original unextended phase space. Using $q^n = t$ and $p_n = p_t = -E$, we can write

$$\sum_{i=0}^{n-1} \int_{R_i} dq^i dp_i - \int_{R_n} dt dE = \sum_{i=0}^{n-1} \int_{R_i'} dq'^i dp_i' - \int_{R_n'} dt' dE' \quad (5.289)$$

and

$$\oint_{\partial R} \left(\sum_{i=0}^{n-1} p_i dq^i - E dt \right) = \oint_{\partial R'} \left(\sum_{i=0}^{n-1} p_i' dq'^i - E' dt' \right). \quad (5.290)$$

The relations (5.289) and (5.290) are two formulations of the *Poincaré–Cartan integral invariant.*

5.6   Reduced Phase Space

Suppose we have a system with $n+1$ degrees of freedom described by a time-independent Hamiltonian in a $(2n + 2)$-dimensional phase space. Here we can play the converse game: we can choose any generalized coordinate to play the role of "time" and the negation of its conjugate momentum to play the role of a new n-degree-of-freedom time-dependent Hamiltonian in a *reduced phase space* of $2n$ dimensions.

   More precisely, let

$$q = (q^0, ..., q^n)$$
$$p = [p_0, ..., p_n], \tag{5.291}$$

and suppose we have a system described by a time-independent Hamiltonian

$$H(t, q, p) = f(q, p) = E. \tag{5.292}$$

For each solution path there is a conserved quantity E. Let's choose a coordinate q^n to be the time in a reduced phase space. We define the dynamical variables for the n-degree-of-freedom reduced phase space:

$$q_r = (q_r^0, ..., q_r^{n-1})$$
$$p^r = [p_0^r, ..., p_{n-1}^r]. \tag{5.293}$$

In the original phase space a coordinate such as q^n maps time to a coordinate. In the formulation of the reduced phase space we will have to use the inverse function $\tau = (q^n)^{-1}$ to map the coordinate to the time, giving the new coordinates in terms of the new time

$$q_r^i = q^i \circ \tau$$
$$p_i^r = p_i \circ \tau, \tag{5.294}$$

and thus

$$Dq_r^i = D(q^i \circ \tau) = (Dq^i \circ \tau)(D\tau) = (Dq^i \circ \tau)/(Dq^n \circ \tau)$$
$$Dp_i^r = D(p_i \circ \tau) = (Dp_i \circ \tau)(D\tau) = (Dp_i \circ \tau)/(Dq^n \circ \tau). \tag{5.295}$$

   We propose that a Hamiltonian in the reduced phase space is the negative of the inverse of $f(q^0, ..., q^n; p_0, ..., p_n) = E$ with respect to the p_n argument:

$$H_r(x, q_r, p^r) = -(\text{the } p_x \text{ such that } f(q_r, x; p^r, p_x) = E). \tag{5.296}$$

Note that in the reduced phase space we will have indices for the structured variables in the range $0 \dots n-1$, whereas in the original phase space the indices are in the range $0 \dots n$. We will show that H_r is an appropriate Hamiltonian for the given dynamical system in the reduced phase space. To compute Hamilton's equations we must expand the implicit definition of H_r. We define an auxiliary function

$$g(x, q_r, p^r) = f(q_r, x; p^r, -H_r(x, q_r, p^r)). \tag{5.297}$$

Note that *by construction* this function is identically a constant $g = E$. Thus all of its partial derivatives are zero:

$$\partial_0 g = (\partial_0 f)^n - (\partial_1 f)^n \partial_0 H_r = 0$$
$$(\partial_1 g)_i = (\partial_0 f)_i - (\partial_1 f)^n (\partial_1 H_r)_i = 0$$
$$(\partial_2 g)^i = (\partial_1 f)^i - (\partial_1 f)^n (\partial_2 H_r)^i = 0, \tag{5.298}$$

where we have suppressed the arguments. Solving for partials of H_r, we get

$$(\partial_1 H_r)_i = (\partial_0 f)_i / (\partial_1 f)^n = (\partial_1 H)_i / (\partial_2 H)^n$$
$$(\partial_2 H_r)^i = (\partial_1 f)^i / (\partial_1 f)^n = (\partial_2 H)^i / (\partial_2 H)^n . \tag{5.299}$$

Using these relations, we can deduce the Hamilton's equations in the reduced phase space from the Hamilton's equations in the original phase space:

$$Dq_r^i(x) = \frac{Dq^i(\tau(x))}{Dq^n(\tau(x))}$$
$$= \frac{(\partial_2 H(\tau(x), q(\tau(x)), p(\tau(x))))^i}{(\partial_2 H(\tau(x), q(\tau(x)), p(\tau(x))))^n}$$
$$= (\partial_2 H_r(x, q_r(x), p^r(x)))^i \tag{5.300}$$
$$Dp_i^r(x) = \frac{Dp_i(\tau(x))}{Dq^n(\tau(x))}$$
$$= \frac{-(\partial_1 H(\tau(x), q(\tau(x)), p(\tau(x))))_i}{(\partial_2 H(\tau(x), q(\tau(x)), p(\tau(x))))^n}$$
$$= -(\partial_1 H_r(x, q_r(x), p^r(x)))_i. \tag{5.301}$$

Orbits in a central field

Consider planar motion in a central field. We have already seen this expressed in polar coordinates in equation (3.100):

$$H(t; r, \varphi; p_r, p_\varphi) = \frac{p_r^2}{2m} + \frac{p_\varphi^2}{2mr^2} + V(r). \tag{5.302}$$

There are two degrees of freedom and the Hamiltonian is time independent. Thus the energy, the value of the Hamiltonian, is conserved on realizable paths. Let's forget about time and reparameterize this system in terms of the orbital radius r.[28] To do this we solve

$$H(t; r, \varphi; p_r, p_\varphi) = E \tag{5.303}$$

for p_r, obtaining

$$H'(r, \varphi, p_\varphi) = -p_r = -\left(2m(E - V(r)) - \frac{p_\varphi^2}{r^2}\right)^{\frac{1}{2}}, \tag{5.304}$$

which is the Hamiltonian in the reduced phase space.

Hamilton's equations are now quite simple:

$$\frac{d\varphi}{dr} = \frac{\partial H'}{\partial p_\varphi} = \frac{p_\varphi}{r^2}\left(2m(E - V(r)) - \frac{p_\varphi^2}{r^2}\right)^{-\frac{1}{2}} \tag{5.305}$$

$$\frac{dp_\varphi}{dr} = -\frac{\partial H'}{\partial \varphi} = 0. \tag{5.306}$$

The momentum p_φ is independent of r (as it was with t), so for any particular orbit we may define a constant angular momentum L. Thus our problem ends up as a simple quadrature:

$$\varphi(r) = \int^r \frac{L}{r^2}\left(2m(E - V(r)) - \frac{L^2}{r^2}\right)^{-\frac{1}{2}} dr + \varphi_0. \tag{5.307}$$

---

[28]We could have chosen to reparameterize in terms of φ, but then both p_r and r would occur in the resulting time-independent Hamiltonian. The path we have chosen takes advantage of the fact that φ does not appear in our Hamiltonian, so p_φ is a constant of the motion. This structure suggests that to solve this kind of problem we need to look ahead, as in playing chess.

To see the utility of this procedure, we continue our example with a definite potential energy—a gravitating point mass:

$$V(r) = -\frac{\mu}{r}. \tag{5.308}$$

When we substitute this into equation (5.307) we obtain a mess that can be simplified to

$$\varphi(r) = L \int^r \frac{dr}{r\sqrt{2mEr^2 + 2m\mu r - L^2}} + \varphi_0. \tag{5.309}$$

Integrating this, we obtain another mess, which can be simplified and rearranged to obtain the following:

$$\frac{1}{r} = \frac{m\mu}{L^2}\left(1 - \sqrt{1 + \frac{2EL^2}{m\mu^2}}\sin(\varphi(r) - \varphi_0)\right). \tag{5.310}$$

This can be recognized as the polar-coordinate form of the equation of a conic section with eccentricity e and parameter p:

$$\frac{1}{r} = \frac{1 + e\cos\theta}{p} \tag{5.311}$$

where

$$e = \sqrt{1 + \frac{2EL^2}{m\mu^2}}, \quad p = \frac{L^2}{m\mu} \quad \text{and} \quad \theta = \varphi_0 - \varphi(r) - \frac{\pi}{2}. \tag{5.312}$$

In fact, if the orbit is an ellipse with semimajor axis a, we have

$$p = a(1 - e^2) \tag{5.313}$$

and so we can identify the role of energy and angular momentum in shaping the ellipse:

$$E = -\frac{\mu}{2a} \quad \text{and} \quad L = \sqrt{m\mu a(1 - e^2)}. \tag{5.314}$$

What we get from analysis in the reduced phase space is the geometry of the trajectory, but we lose the time-domain behavior. The reduction is often worth the price.

Although we have treated time in a special way so far, we have found that time is not special. It can be included in the coordinates to make a driven system autonomous. And it can be eliminated from any autonomous system in favor of any other coordinate. This leads to numerous strategies for simplifying problems, by removing time variation and then performing canonical transforms on the resulting conservative autonomous system to make a nice coordinate that we can then dump back into the role of time.

Generating functions in extended phase space

We can represent canonical transformations with mixed-variable generating functions. We can extend these to represent transformations in the extended phase space. Let F_2 be a generating function with arguments (t, q, p'). Then, the corresponding F_2^e in the extended phase space can be taken to be

$$F_2^e(\tau; q, t; p', p_t') = tp_t' + F_2(t, q, p). \tag{5.315}$$

The relations between the coordinates and the momenta are the same as before. We also have

$$p_t = (\partial_1 F_2^e)_n(\tau; q, t; p', p_t') = p_t' + \partial_0 F_2(t, q, p)$$
$$t' = (\partial_2 F_2^e)^n(\tau; q, t; p', p_t') = t. \tag{5.316}$$

The first equation gives the relationship between the original Hamiltonians:

$$H'(t, q', p') = H(t, q, p) + \partial_0 F_2(t, q, p), \tag{5.317}$$

as required. Time-independent canonical transformations, where $H' = H \circ C_H$, have symplectic qp part. The generating-function representation of a time-dependent transformation does not depend on the independent variable in the extended phase space. So, in extended phase space the qp part of the transformation, which includes the time and the momentum conjugate to time, is symplectic.

Exercise 5.20: Rotating coordinates in extended phase space

In the extended phase space the time is one of the coordinates. Carry out the transformation to rotating coordinates using an F_2-type generating function in the extended phase space. Compare Hamiltonian (5.178) to the Hamiltonian obtained by composition with the transformation.

5.7 Summary

Canonical transformations can be used to reformulate a problem in coordinates that are easier to understand or that expose some symmetry of a problem.

In this chapter we have investigated different representations of a dynamical system. We have found that different representations will be equivalent if the coordinate–momentum part of the transformation has a symplectic derivative, and if the Hamiltonian transforms in a specified way. If the phase-space transformation is time independent, then the Hamiltonian transforms by composition with the phase-space transformation. The symplectic condition can be equivalently expressed in terms of the fundamental Poisson brackets. The Poisson bracket and the ω function are invariant under canonical transformations. The invariance of ω implies that the sum of the areas of the projections onto fundamental coordinate–momentum planes is preserved (Poincaré integral invariant) by canonical transformations.

A generating function is a real-valued function of the phase-space coordinates and time that represents a canonical transformation through its partial derivatives. We found that every canonical transformation can be represented by a generating function. The proof depends on the Poincaré integral invariant.

We can formulate an extended phase space in which time is treated as another coordinate. Time-dependent transformations are simple in the extended phase space. In the extended phase space the Poincaré integral invariant is the Poincaré–Cartan integral invariant. We can also reformulate a time-independent problem as a time-dependent problem with fewer degrees of freedom, with one of the original coordinates taking on the role of time; this is the reduced phase space.

5.8 Projects

Exercise 5.21: Hierarchical Jacobi coordinates
A Hamiltonian for the n-body problem is

$$H = T + V \tag{5.318}$$

with

$$T(t; x_0, x_1, \ldots, x_{n-1}; p_0, p_1, \ldots, p_{n-1}) = \sum_{i=0}^{n-1} \frac{p_i^2}{2m_i} \qquad (5.319)$$

and

$$V(t; x_0, x_1, \ldots, x_{n-1}; p_0, p_1, \ldots, p_{n-1}) = \sum_{i<j} f_{ij}(\|x_i - x_j\|), \qquad (5.320)$$

where x_i is the tuple of rectangular coordinates for body i and p_i is the tuple of conjugate linear momenta for body i.

The potential energy of the system depends only on the relative positions of the bodies, so the relative motion decouples from the center of mass motion. In this problem we explore canonical transformations that achieve this decoupling.

a. Canonical heliocentric coordinates. The coordinates transform as follows:

$$x_0' = X, \qquad (5.321)$$

where X is the center of mass of the system, and

$$x_i' = x_i - x_0, \qquad (5.322)$$

for $i > 0$, the differences of the position of body i and the body with index 0 (which might be the Sun). Find the associated canonical momenta using an F_2-type generating function. Show that the potential energy can be written solely in terms of the coordinates for $i > 0$. Show that the kinetic energy is not in the form of a sum of squares of momenta divided by mass constants.

b. Jacobi coordinates. The Jacobi coordinates isolate the center of mass motion, without spoiling the usual diagonal quadratic form of the kinetic energy. Define X_i to be the center of mass of the bodies with indices less than or equal to i:

$$X_i = \frac{\sum_{j=0}^{i} m_j x_j}{\sum_{j=0}^{i} m_j}. \qquad (5.323)$$

The Jacobi coordinates are defined by

$$x_{i-1}' = x_i - X_{i-1}, \qquad (5.324)$$

for $0 < i < n$, and

$$x_{n-1}' = X_{n-1}. \qquad (5.325)$$

The coordinates x'_i for $0 < i < n$ are the difference of the position of body $i - 1$ and the center of mass of bodies with lower indices; the coordinate x'_{n-1} is the center of mass of the system. Complete the canonical transformation by finding the conjugate momenta using an F_2-type generating function. Show that the kinetic energy can still be written in the form

$$T(t; x'_0, x'_1, \ldots, x'_{n-1}; p'_0, p'_1, \ldots, p'_{n-1}) = \sum_{i=0}^{n-1} \frac{p'^2_i}{2m'_i}, \qquad (5.326)$$

for some constants m'_i, and that the potential V can be written solely in terms of the Jacobi coordinates x'_i with indices $i > 0$.

c. Hierarchical Jacobi coordinates. Define a "body" as a tuple of a mass and a rectangular position tuple. An n-body "system" is a tuple of n bodies: $(b_0, b_1, \ldots, b_{n-1})$. Define a "linking" transformation $\mathcal{L}_{jk}$ for bodies j and k that takes an n-body system and returns a new linked system:

$$(b'_0, \ldots, b'_{n-1}) = \mathcal{L}_{jk}(b_0, \ldots, b_{n-1}). \qquad (5.327)$$

The bodies in the new system are the same as the bodies in the old system $b_i{}' = b_i$ except for bodies j and k:

$$(m'_j, x'_j) = (m_j m_k/(m_j + m_k), x_k - x_j)$$
$$(m'_k, x'_k) = (m_j + m_k, (m_j x_j + m_k x_k)/(m_j + m_k)). \qquad (5.328)$$

This is a transformation to relative coordinates and center of mass for bodies j and k. Extend this transformation to phase space and show that it preserves the form of the kinetic energy

$$\sum_i \frac{(p_i)^2}{2m_i} = \sum_i \frac{(p'_i)^2}{2m'_i}. \qquad (5.329)$$

Show that the transformation to Jacobi coordinates of part **b** is generated by a composition of linking transformations:

$$\mathcal{L}_{n-2,n-1} \circ \cdots \circ \mathcal{L}_{1,2} \circ \mathcal{L}_{0,1}. \qquad (5.330)$$

Interpret the coordinate transformation produced by such a succession of linking transformations; why do we call this a "linking" transformation? What requirement has to be satisfied for a composition of linking transformations to isolate the center of mass of the system (make it one of the coordinates)? Taking this constraint into account, find hierarchical Jacobi coordinates for a system with six bodies, arranged as two triple systems, each of which is a binary plus a third body. Verify that one of the coordinates is the center of mass of the system, and that the kinetic energy remains a sum of squares of the momenta divided by an appropriate mass constant.

6
Canonical Evolution

> What, then, is time? I know well enough what it
> is, provided that nobody asks me; but if I am
> asked what it is and try to explain, I am baffled.
> All the same I can confidently say that I know that
> if nothing passed, there would be no past time; if
> nothing were going to happen, there would be no
> future time; and if nothing *were* there would be no
> present time.
>
> Augustine of Hippo, from *Confessions*, Book XI,
> Section 14. Translation by R.S. Pine-Coffin, 1961.

Time evolution generates a canonical transformation: if we consider all possible initial states of a Hamiltonian system and follow
all the trajectories for the same time interval, then the map from
the initial state to the final state of each trajectory is a canonical
transformation. Hamilton–Jacobi theory gives a mixed-variable
generating function that generates this time-evolution transformation. For the few integrable systems for which we can solve
the Hamilton–Jacobi equation this transformation gets us action-
angle coordinates, which form a starting point to study perturbations.

6.1   Hamilton–Jacobi Equation

If we could find a canonical transformation so that the transformed
Hamiltonian was identically zero, then by Hamilton's equations
the new coordinates and momenta would be constants. All of the
time variation of the solution would be captured in the canonical
transformation, and there would be nothing more to the solution.
A mixed-variable generating function that does this job satisfies
a partial differential equation called the Hamilton–Jacobi equation. In most cases, a Hamilton–Jacobi equation cannot be solved
explicitly. When it can be solved, however, a Hamilton–Jacobi

equation provides a means of reducing a problem to a useful simple form.

Recall the relations satisfied by an F_2-type generating function:

$$q' = \partial_2 F_2(t, q, p') \tag{6.1}$$

$$p = \partial_1 F_2(t, q, p') \tag{6.2}$$

$$H'(t, q', p') = H(t, q, p) + \partial_0 F_2(t, q, p'). \tag{6.3}$$

If we require the new Hamiltonian to be zero, then F_2 must satisfy the equation

$$0 = H(t, q, \partial_1 F_2(t, q, p')) + \partial_0 F_2(t, q, p'). \tag{6.4}$$

So the solution of the problem is "reduced" to the problem of solving an n-dimensional partial differential equation for F_2 with unspecified new (constant) momenta p'. This is a Hamilton–Jacobi equation, and in some cases we can solve it.

We can also attempt a somewhat less drastic method of solution. Rather than try to find an F_2 that makes the new Hamiltonian identically zero, we can seek an F_2-shaped function W that gives a new Hamiltonian that is solely a function of the new momenta. A system described by this form of Hamiltonian is also easy to solve. So if we set

$$H''(t, q'', p'') = H(t, q, \partial_1 W(t, q, p'')) + \partial_0 W(t, q, p'')$$
$$= E(p'') \tag{6.5}$$

and are able to solve for W, then the problem is essentially solved. In this case, the primed momenta are all constant and the primed positions are linear in time. This is an alternate form of the Hamilton–Jacobi equation.

These forms are related. Suppose that we have a W that satisfies the second form of the Hamilton–Jacobi equation (6.5). Then the F_2 constructed from W

$$F_2(t, q, p') = W(t, q, p') - E(p')t \tag{6.6}$$

satisfies the first form of the Hamilton–Jacobi equation (6.4). Furthermore,

$$p = \partial_1 F_2(t, q, p') = \partial_1 W(t, q, p'), \tag{6.7}$$

so the primed momenta are the same in the two formulations. But

$$q' = \partial_2 F_2(t, q, p')$$
$$= \partial_2 W(t, q, p') - DE(p')t$$
$$= q'' - DE(p')t, \tag{6.8}$$

so we see that the primed coordinates differ by a term that is linear in time—both $p'(t) = p'_0$ and $q'(t) = q'_0$ are constant. Thus we can use either W or F_2 as the generating function, depending on the form of the new Hamiltonian we want.

Note that if H is time independent then we can often find a time-independent W that does the job. For time-independent W the Hamilton–Jacobi equation simplifies to

$$E(p') = H(t, q, \partial_1 W(t, q, p')). \tag{6.9}$$

The corresponding F_2 is then linear in time. Notice that an implicit requirement is that the energy can be written as a function of the new momenta alone. This excludes the possibility that the transformed phase-space coordinates q' and p' are simply initial conditions for q and p.

It turns out that there is flexibility in the choice of the function E. With an appropriate choice the phase-space coordinates obtained through the transformation generated by W are action-angle coordinates.

Exercise 6.1: Hamilton–Jacobi with F_1

We have used an F_2-type generating function to carry out the Hamilton–Jacobi transformations. Carry out the equivalent transformations with an F_1-type generating function. Find the equations corresponding to equations (6.4), (6.5), and (6.9).

6.1.1   Harmonic Oscillator

Consider the familiar time-independent Hamiltonian

$$H(t, x, p) = \frac{p^2}{2m} + \frac{kx^2}{2}. \tag{6.10}$$

We form the Hamilton–Jacobi equation for this problem:

$$0 = H(t, x, \partial_1 F_2(t, x, p')) + \partial_0 F_2(t, x, p'). \tag{6.11}$$

Using $F_2(t, x, p') = W(t, x, p') - E(p')t$, we find

$$E(p') = H(t, x, \partial_1 W(t, x, p')). \tag{6.12}$$

Writing this out explicitly yields

$$E(p') = \frac{(\partial_1 W(t, x, p'))^2}{2m} + \frac{kx^2}{2}, \tag{6.13}$$

and solving for $\partial_1 W$ gives

$$\partial_1 W(t, x, p') = \sqrt{2m\left(E(p') - \frac{kx^2}{2}\right)}. \tag{6.14}$$

Integrating gives the desired W:

$$W(t, x, p') = \int^x \sqrt{2m\left(E(p') - \frac{kz^2}{2}\right)}\, dz. \tag{6.15}$$

We can use either W or the corresponding F_2 as the generating function. First, take W to be the generating function. We obtain the coordinate transformation by differentiating:

$$\begin{aligned} x' &= \partial_2 W(t, x, p') \\ &= \int^x \frac{mDE(p')}{\sqrt{2m\left(E(p') - \frac{kz^2}{2}\right)}}\, dz \end{aligned} \tag{6.16}$$

and then integrating to get

$$x' = \sqrt{\frac{m}{k}}DE(p')\arcsin\left(\sqrt{\frac{k}{2E(p')}}x\right) + C(p'), \tag{6.17}$$

with some integration constant $C(p')$. Inverting this, we get the unprimed coordinate in terms of the primed coordinate and momentum:

$$x = \sqrt{\frac{2E(p')}{k}}\sin\left[\frac{1}{DE(p')}\sqrt{\frac{k}{m}}(x' - C(p'))\right]. \tag{6.18}$$

The new Hamiltonian H' depends only on the momentum:

$$H'(t, x', p') = E(p'). \tag{6.19}$$

The equations of motion are just

$$Dx'(t) = \partial_2 H'(t, x'(t), p'(t)) = DE(p')$$
$$Dp'(t) = -\partial_1 H'(t, x'(t), p'(t)) = 0, \qquad (6.20)$$

with solution

$$x'(t) = DE(p')t + x_0'$$
$$p'(t) = p_0' \qquad (6.21)$$

for initial conditions x_0' and p_0'. If we plug these expressions for $x'(t)$ and $p'(t)$ into equation (6.18) we find

$$
\begin{aligned}
x(t) &= \sqrt{\frac{2E(p')}{k}} \sin\left[\frac{1}{DE(p')}\sqrt{\frac{k}{m}}(DE(p')t + x_0' - C(p'))\right] \\
&= \sqrt{\frac{2E(p')}{k}} \sin\left[\sqrt{\frac{k}{m}}(t - t_0)\right] \\
&= A\sin(\omega t + \varphi), \qquad (6.22)
\end{aligned}
$$

where the angular frequency is $\omega = \sqrt{k/m}$, the amplitude is $A = \sqrt{2E(p')/k}$, and the phase is $\varphi = -\omega t_0 = \omega(x_0' - C(p'))/DE(p')$.

We can also use $F_2 = W - Et$ as the generating function. The new Hamiltonian is zero, so both x' and p' are constant, but the relationship between the old and new variables is

$$
\begin{aligned}
x' &= \partial_2 F_2(t, x, p') \\
&= \partial_2 W(t, x, p') - DE(p')t \\
&= \int^x \frac{mDE(p')}{\sqrt{2m\left(E(p') - \frac{kz^2}{2}\right)}} - DE(p')t \\
&= \sqrt{\frac{m}{k}} DE(p') \arcsin\left(\sqrt{\frac{k}{2E(p')}}x\right) + C(p') - DE(p')t. \quad (6.23)
\end{aligned}
$$

Plugging in the solution $x' = x_0'$ and $p' = p_0'$ and solving for x, we find equation (6.22). So once again we see that the two approaches are equivalent.

It is interesting to note that the solution depends upon the constants $E(p')$ and $DE(p')$, but otherwise the motion is not dependent in any essential way on what the function E actually is. The momentum p' is constant and the values of the constants are set by the initial conditions. Given a particular function E, the initial conditions determine p', but the solution can be obtained without further specifying the E function.

If we choose particular functions E we can get particular canonical transformations. For example, a convenient choice is simply

$$E(p') = \alpha p', \tag{6.24}$$

for some constant α that will be chosen later. We find

$$x = \sqrt{\frac{2\alpha p'}{k}} \sin \frac{\omega}{\alpha} x'. \tag{6.25}$$

So we see that a convenient choice is $\alpha = \omega = \sqrt{k/m}$, so

$$x = \sqrt{\frac{2p'}{\beta}} \sin x', \tag{6.26}$$

with $\beta = \sqrt{km}$. The new Hamiltonian is

$$H'(t, x', p') = E(p') = \omega p'. \tag{6.27}$$

The solution is just $x' = \omega t + x_0'$ and $p' = p_0'$. Substituting the expression for x in terms of x' and p' into $H(t, x, p) = H'(t, x', p')$, we derive

$$p = \left[2m \left(p'\alpha - \frac{k}{2} x^2 \right) \right]^{1/2}$$
$$= \sqrt{2p'\beta} \cos x'. \tag{6.28}$$

The two transformation equations (6.26) and (6.28) are what we have called the polar-canonical transformation (equation 5.29). We have already shown that this transformation is canonical and that it solves the harmonic oscillator, but it was not derived. Here we have derived this transformation as a particular case of the solution of the Hamilton–Jacobi equation.

We can also explore other choices for the E function. For example, we could choose

$$E(p') = \tfrac{1}{2}\alpha p'^2. \tag{6.29}$$

Following the same steps as before, we find

$$x = \sqrt{\frac{\alpha p'^2}{k}} \, \sin \frac{\omega}{\alpha} \frac{x'}{p'}. \tag{6.30}$$

So a convenient choice is again $\alpha = \omega$, leaving

$$x = \frac{p'}{\beta} \sin \frac{x'}{p'}$$

$$p = \beta p' \cos \frac{x'}{p'}, \tag{6.31}$$

with $\beta = (km)^{1/4}$. By construction, this transformation is also canonical and also brings the harmonic oscillator problem into an easily solvable form:

$$H'(t, x', p') = \tfrac{1}{2}\omega p'^2. \tag{6.32}$$

The harmonic oscillator Hamiltonian has been transformed to what looks a lot like the Hamiltonian for a free particle. This is very interesting. Notice that whereas Hamiltonian (6.27) does not have a well defined Legendre transform to an equivalent Lagrangian, the "free particle" harmonic oscillator has a well defined Legendre transform:

$$L'(t, x', \dot{x}') = \frac{\dot{x}'^2}{2\omega}. \tag{6.33}$$

Of course, there may be additional properties that make one choice more useful than others for particular applications.

Exercise 6.2: Pendulum

Formulate and solve a Hamilton–Jacobi equation for the pendulum; investigate both the circulating and oscillating regions of phase space. (Note: This is a long story and requires some knowledge of elliptic functions.)

6.1.2   Hamilton–Jacobi Solution of the Kepler Problem

We can use the Hamilton–Jacobi equation to find canonical coordinates that solve the Kepler problem. This is an essential first step in doing perturbation theory for orbital problems.

In rectangular coordinates (x, y, z), the Kepler Hamiltonian is

$$H_r(t; x, y, z; p_x, p_y, p_z) = \frac{p^2}{2m} - \frac{\mu}{r}, \tag{6.34}$$

where $r^2 = x^2 + y^2 + z^2$ and $p^2 = p_x^2 + p_y^2 + p_z^2$.

We try a generating function of the form $W(t; x, y, z; p'_x, p'_y, p'_z)$. The Hamilton–Jacobi equation is then[1]

$$\begin{aligned}
E(p') = \frac{1}{2m} \Big[&\big(\partial_{1,0} W(t; x, y, z; p'_x, p'_y, p'_z)\big)^2 \\
+ &\big(\partial_{1,1} W(t; x, y, z; p'_x, p'_y, p'_z)\big)^2 \\
+ &\big(\partial_{1,2} W(t; x, y, z; p'_x, p'_y, p'_z)\big)^2 \Big] - \frac{\mu}{r}.
\end{aligned} \tag{6.35}$$

This is a partial differential equation in the three partial derivatives of W. We stare at it a while and give up.

Next we try converting to spherical coordinates. This is motivated by the fact that the potential energy depends only on r. The Hamiltonian in spherical coordinates (r, θ, φ), where θ is the colatitude and φ is the longitude, is

$$H_s(t; r, \theta, \varphi; p_r, p_\theta, p_\varphi) = \frac{1}{2m} \left[p_r^2 + \frac{p_\theta^2}{r^2} + \frac{p_\varphi^2}{r^2(\sin\theta)^2} \right] - \frac{\mu}{r}. \tag{6.36}$$

The Hamilton–Jacobi equation is

$$\begin{aligned}
E(p'_0, &p'_1, p'_2) \\
&= \frac{1}{2m} \Big[\big(\partial_{1,0} W(t; r, \theta, \varphi; p'_0, p'_1, p'_2)\big)^2 \\
&\quad + \frac{1}{r^2} \big(\partial_{1,1} W(t; r, \theta, \varphi; p'_0, p'_1, p'_2)\big)^2 \\
&\quad + \frac{1}{r^2(\sin\theta)^2} \big(\partial_{1,2} W(t; r, \theta, \varphi; p'_0, p'_1, p'_2)\big)^2 \Big] - \frac{\mu}{r}.
\end{aligned} \tag{6.37}$$

---

[1] Remember that $\partial_{1,0}$ means the derivative with respect to the first coordinate position.

We can solve this Hamilton–Jacobi equation by successively isolating the dependence on the various variables. Looking first at the φ dependence, we see that, outside of W, φ appears only in one partial derivative. If we write

$$W(t; r, \theta, \varphi; p'_0, p'_1, p'_2) = f(r, \theta, p'_0, p'_1, p'_2) + p'_2 \, \varphi, \tag{6.38}$$

then $\partial_{1,2} W(t; r, \theta, \varphi; p'_0, p'_1, p'_2) = p'_2$, and then φ does not appear in the remaining equation for f:

$$
\begin{aligned}
E(p'_0, &p'_1, p'_2) \\
&= \frac{1}{2m} \left\{ \left(\partial_0 f(r, \theta, p'_0, p'_1, p'_2) \right)^2 \right. \\
&\quad + \frac{1}{r^2} \left[\left(\partial_1 f(r, \theta, p'_0, p'_1, p'_2) \right)^2 + \frac{(p'_2)^2}{(\sin\theta)^2} \right] \left. \right\} - \frac{\mu}{r}.
\end{aligned} \tag{6.39}
$$

Any function of the p'_i could have been used as the coefficient of φ in the generating function. This particular choice has the nice feature that p'_2 is the z component of the angular momentum.

We can eliminate the θ dependence if we choose

$$f(r, \theta, p'_0, p'_1, p'_2) = R(r, p'_0, p'_1, p'_2) + \Theta(\theta, p'_0, p'_1, p'_2) \tag{6.40}$$

and require that Θ be a solution to

$$\left(\partial_0 \Theta(\theta, p'_0, p'_1, p'_2) \right)^2 + \frac{(p'_2)^2}{(\sin\theta)^2} = (p'_1)^2. \tag{6.41}$$

We are free to choose the right-hand side to be any function of the new momenta. This choice reflects the fact that the left-hand side is non-negative. It turns out that p'_1 is the total angular momentum. This equation for Θ can be solved by quadrature.

The remaining equation that determines R is

$$E(p'_0, p'_1, p'_2) = \frac{1}{2m} \left[\left(\partial_0 R(r, p'_0, p'_1, p'_2) \right)^2 + \frac{1}{r^2} (p'_1)^2 \right] - \frac{\mu}{r}, \tag{6.42}$$

which also can be solved by quadrature.

Altogether the solution of the Hamilton–Jacobi equation reads

$$W(r, \theta, \varphi, p_0', p_1', p_2') = \int^r \left(2mE(p_0', p_1', p_2') + \frac{2m\mu}{r} - \frac{(p_1')^2}{r^2} \right)^{1/2} dr$$
$$+ \int^\theta \left((p_1')^2 - \frac{(p_2')^2}{(\sin\theta)^2} \right)^{1/2} d\theta.$$
$$+ p_2'\varphi. \qquad (6.43)$$

It is interesting that our solution to the Hamilton–Jacobi partial differential equation is of the form

$$W(t; r, \theta, \varphi; p_0', p_1', p_2')$$
$$= R(r, p_0', p_1', p_2') + \Theta(\theta, p_0', p_1', p_2') + \Phi(\varphi, p_0', p_1', p_2'). \qquad (6.44)$$

Thus we have a separation-of-variables technique that involves writing the solution as a sum of functions of the individual variables. This might be contrasted with the separation-of-variables technique encountered in elementary quantum mechanics and classical electrodynamics, which uses products of functions of individual variables.

The coordinates $q' = (q'^0, q'^1, q'^2)$ conjugate to the momenta $p' = [p_0', p_1', p_2']$ are

$$q'^0 = \partial_{2,0} W(t; r, \theta, \varphi; p_0', p_1', p_2')$$
$$= m\partial_0 E(p') \int^r \left(2mE(p') + \frac{2m\mu}{r} - \frac{(p_1')^2}{r^2} \right)^{-1/2} dr$$
$$q'^1 = \partial_{2,1} W(t; r, \theta, \varphi; p_0', p_1', p_2')$$
$$= p_1' \int^\theta \left((p_1')^2 - \frac{(p_2')^2}{(\sin\theta)^2} \right)^{-1/2} d\theta$$
$$+ \int^r \left(m\partial_1 E(p') - \frac{p_1'}{r^2} \right) \left(2mE(p') + \frac{2m\mu}{r} - \frac{(p_1')^2}{r^2} \right)^{-1/2} dr$$
$$q'^2 = \partial_{2,2} W(t; r, \theta, \varphi; p_0', p_1', p_2')$$
$$= \varphi - \frac{p_2'}{(\sin\theta)^2} \int^\theta \left((p_1')^2 - \frac{(p_2')^2}{(\sin\theta)^2} \right)^{-1/2} d\theta$$
$$+ m\partial_2 E(p') \int^r \left(2mE(p') + \frac{2m\mu}{r} - \frac{(p_1')^2}{r^2} \right)^{-1/2} dr.$$

We are still free to choose the functional form of E. A convenient (and conventional) choice is

$$E(p_0', p_1', p_2') = -\frac{m\mu^2}{2(p_0')^2}.$$ (6.45)

With this choice the momentum p_0' has dimensions of angular momentum, and the conjugate coordinate is an angle.

The Hamiltonian for the Kepler problem is reduced to

$$H'(t, q', p') = E(p') = -\frac{m\mu^2}{2(p_0')^2}.$$ (6.46)

Thus

$$q'^0 = nt + \beta^0$$ (6.47)
$$q'^1 = \beta^1$$ (6.48)
$$q'^2 = \beta^2,$$ (6.49)

where $n = m\mu^2/(p_0')^3$ and where β^0, β^1, and β^2 are the initial values of the components of q'. Only one of the new variables changes with time.

The canonical phase-space coordinates can be written in terms of the parameters that specify an orbit. We merely summarize the results; for further explanation see [36] or [38].

Assume we have a bound orbit with semimajor axis a, eccentricity e, inclination i, longitude of ascending node Ω, argument of pericenter ω, and mean anomaly M. The three canonical momenta are $p_0' = \sqrt{m\mu a}$, $p_1' = \sqrt{m\mu a(1 - e^2)}$, and $p_2' = \sqrt{m\mu a(1 - e^2)}\cos i$. The first momentum is related to the energy, the second momentum is the total angular momentum, and the third momentum is the component of the angular momentum in the $\hat{z}$ direction. The conjugate canonical coordinates are $(q')^0 = M$, $(q')^1 = \omega$, and $(q')^2 = \Omega$.

6.1.3   *F₂ and the Lagrangian*

The solution to the Hamilton–Jacobi equation, the mixed-variable generating function that generates time evolution, is related to the action used in the variational principle. In particular, the time derivative of the generating function along realizable paths has the same value as the Lagrangian.

Let $\widetilde{F_2}(t) = F_2(t, q(t), p'(t))$ be the value of F_2 along the paths q and p' at time t. The derivative of $\widetilde{F_2}$ is

$$
\begin{aligned}
D\widetilde{F_2}(t) = {}& \partial_0 F_2(t, q(t), p'(t)) \\
&+ \partial_1 F_2(t, q(t), p'(t)) Dq(t) \\
&+ \partial_2 F_2(t, q(t), p'(t)) Dp'(t).
\end{aligned} \tag{6.50}
$$

Using the Hamilton–Jacobi equation (6.4), this becomes

$$
\begin{aligned}
D\widetilde{F_2}(t) = {}& -H(t, q(t), \partial_1 F_2(t, q(t), p'(t))) \\
&+ \partial_1 F_2(t, q(t), p'(t)) Dq(t) \\
&+ \partial_2 F_2(t, q(t), p'(t)) Dp'(t).
\end{aligned} \tag{6.51}
$$

Now, using equation (6.2), we get

$$
\begin{aligned}
D\widetilde{F_2}(t) = {}& -H(t, q(t), p(t)) \\
&+ p(t) Dq(t) \\
&+ \partial_2 F_2(t, q(t), p'(t)) Dp'(t).
\end{aligned} \tag{6.52}
$$

But $p(t)Dq(t) - H(t, q(t), p(t)) = L(t, q(t), Dq(t))$, so

$$
D\widetilde{F_2}(t) = L(t, q(t), Dq(t)) + \partial_2 F_2(t, q(t), p'(t)) Dp'(t). \tag{6.53}
$$

On realizable paths we have $Dp'(t) = 0$, so along realizable paths the time derivative of F_2 is the same as the Lagrangian along the path. The time integral of the Lagrangian along any path is the action along that path. This means that, up to an additive term that is constant on realizable paths but may be a function of the transformed phase-space coordinates q' and p', the F_2 that solves the Hamilton–Jacobi equation has the same value as the Lagrangian action for realizable paths.

The same conclusion follows for the Hamilton–Jacobi equation formulated in terms of F_1. Up to an additive term that is constant on realizable paths but may be a function of the transformed phase-space coordinates q' and p', the F_1 that solves the corresponding Hamilton–Jacobi equation has the same value as the Lagrangian action for realizable paths.

Recall that a transformation given by an F_2-type generating function is also given by an F_1-type generating function related to it by a Legendre transform (see equation 5.142):

$$F_1(t, q, q') = F_2(t, q, p') - q'p',   \tag{6.54}$$

provided the transformations are nonsingular. In this case, both q' and p' are constant on realizable paths, so the additive constants that make F_1 and F_2 equal to the Lagrangian action differ by $q'p'$.

Exercise 6.3: Harmonic oscillator

Let's check this for the harmonic oscillator (of course).

a. Finish the integral (6.15):

$$W(t, x, p') = \int^x \sqrt{2m\left(E(p') - \frac{kz^2}{2}\right)}\, dz.$$

Write the result in terms of the amplitude $A = \sqrt{2E(p')/k}$.

b. Check that this generating function gives the transformation

$$x' = \partial_2 W(t, x, p') = \sqrt{\frac{m}{k}} DE(p') \arcsin\left(\frac{x}{\sqrt{2E(p')/k}}\right),$$

which is the same as equation (6.17) for a particular choice of the integration constant. The other part of the transformation is

$$p = \partial_1 W(t, x, p') = \sqrt{mk}\sqrt{A^2 - x^2},$$

with the same definition of A as before.

c. Compute the time derivative of the associated F_2 along realizable paths $(Dp'(t) = 0)$, and compare it to the Lagrangian along realizable paths.

6.1.4   The Action Generates Time Evolution

We define the function $\bar{F}(t_1, q_1, t_2, q_2)$ to be the value of the action for a realizable path q such that $q(t_1) = q_1$ and $q(t_2) = q_2$. So $\bar{F}$ satisfies

$$\bar{F}(t_1, q(t_1), t_2, q(t_2)) = S[q](t_1, t_2) = \int_{t_1}^{t_2} L \circ \Gamma[q].   \tag{6.55}$$

For variations η that are not necessarily zero at the end times and for realizable paths q, the variation of the action is

$$\delta_\eta S[q](t_1, t_2) = (\partial_2 L \circ \Gamma[q])\eta|_{t_1}^{t_2}$$
$$= p(t_2)\eta(t_2) - p(t_1)\eta(t_1). \qquad (6.56)$$

Alternatively, the variation of $S[q]$ in equation (6.55) gives

$$\delta_\eta S[q](t_1, t_2) = \partial_1 \bar{F}(t_1, q(t_1), t_2, q(t_2))\eta(t_1)$$
$$+ \partial_3 \bar{F}(t_1, q(t_1), t_2, q(t_2))\eta(t_2). \qquad (6.57)$$

Comparing equations (6.56) and (6.57) and using the fact that the variation η is arbitrary, we find

$$\partial_1 \bar{F}(t_1, q(t_1), t_2, q(t_2)) = -p(t_1)$$
$$\partial_3 \bar{F}(t_1, q(t_1), t_2, q(t_2)) = p(t_2). \qquad (6.58)$$

The partial derivatives of $\bar{F}$ with respect to the coordinate arguments give the momenta. Abstracting off paths, we have

$$\partial_1 \bar{F}(t_1, q_1, t_2, q_2) = -p_1$$
$$\partial_3 \bar{F}(t_1, q_1, t_2, q_2) = p_2. \qquad (6.59)$$

This looks a bit like the F_1-type generating function relations, but here there are two times. Solving equations (6.59) for q_2 and p_2 as functions of t_2 and the initial state t_1, q_1, p_1, we get the time evolution of the system in terms of $\bar{F}$. The function $\bar{F}$ generates time evolution.

If we vary the lower limit of the action integral we get

$$\partial_0(S[q])(t_1, t_2) = -L(t_1, q(t_1), Dq(t_1)). \qquad (6.60)$$

Using equation (6.55), and given a realizable path q such that $q(t_1) = q_1$ and $q(t_2) = q_2$, we get the partial derivatives with respect to the time slots:

$$\partial_0(S[q])(t_1, t_2) = \partial_0 \bar{F}(t_1, q_1, t_2, q_2) + \partial_1 \bar{F}(t_1, q_1, t_2, q_2)Dq(t_1)$$
$$= \partial_0 \bar{F}(t_1, q_1, t_2, q_2) - p(t_1)Dq(t_1). \qquad (6.61)$$

Rearranging the terms of equation (6.61) and using equation (6.60) we get

$$\partial_0 \bar{F}(t_1, q_1, t_2, q_2) = H(t_1, q_1, p_1)$$
$$= H(t_1, q_1, -\partial_1 \bar{F}(t_1, q_1, t_2, q_2)), \tag{6.62}$$

and similarly

$$\partial_2 \bar{F}(t_1, q_1, t_2, q_2) = -H(t_2, q_2, p_2)$$
$$= -H(t_2, q_2, \partial_3 \bar{F}(t_1, q_1, t_2, q_2)). \tag{6.63}$$

These are a pair of Hamilton–Jacobi equations, computed at the endpoints of the path.

The function $\bar{F}$ can be written in terms of an F_1 that satisfies a Hamilton–Jacobi equation for H. We can compute time evolution by using the F_1 solution of the Hamilton–Jacobi equation to express the state (t_1, q_1, p_1) in terms of the constants q' and p'. Using the same solution we can then perform a subsequent transformation back from q' p' to the original state variables at a different time t_2, giving the state (t_2, q_2, p_2). The composition of these two canonical transformations is canonical (see exercise 5.12).

The generating function for the composition is the difference of the generating functions for each step:

$$\bar{F}_x(t_1, q_1, q', t_2, q_2) = F_1(t_2, q_2, q') - F_1(t_1, q_1, q'), \tag{6.64}$$

with the condition

$$\partial_2 F_1(t_2, q_2, q') - \partial_2 F_1(t_1, q_1, q') = 0, \tag{6.65}$$

which allows us to eliminate q' in terms of t_1, q_1, t_2, and q_2. So we can write

$$\bar{F}(t_1, q_1, t_2, q_2) = F_1(t_2, q_2, q') - F_1(t_1, q_1, q'). \tag{6.66}$$

Exercise 6.4: Uniform acceleration

a. Compute the Lagrangian action, as a function of the endpoints and times, for a uniformly accelerated particle. Use this to construct the canonical transformation for time evolution from a given initial state.

b. Solve the Hamilton–Jacobi equation for the uniformly accelerated particle, obtaining the F_1 that makes the transformed Hamiltonian zero. Show that the Lagrangian action can be expressed as a difference of two applications of this F_1.

6.2   Time Evolution is Canonical

We use time evolution to generate a transformation

$$(t, q, p) = \mathcal{C}_\Delta(t', q', p') \tag{6.67}$$

that is obtained in the following way. Let $\sigma(t) = (t, \bar{q}(t), \bar{p}(t))$ be a solution of Hamilton's equations. The transformation $\mathcal{C}_\Delta$ satisfies

$$\mathcal{C}_\Delta(\sigma(t)) = \sigma(t + \Delta), \tag{6.68}$$

or, equivalently,

$$\mathcal{C}_\Delta(t, \bar{q}(t), \bar{p}(t)) = (t + \Delta, \bar{q}(t + \Delta), \bar{p}(t + \Delta)). \tag{6.69}$$

Notice that $\mathcal{C}_\Delta$ changes the time component. This is the first transformation of this kind that we have considered.[2]

Given a state (t', q', p'), we find the phase-space path σ emanating from this state as an initial condition, satisfying

$$q' = \bar{q}(t')$$
$$p' = \bar{p}(t'). \tag{6.70}$$

The value (t, q, p) of $\mathcal{C}_\Delta(t', q', p')$ is then $(t' + \Delta, \bar{q}(t' + \Delta), \bar{p}(t' + \Delta))$.

Time evolution is canonical if the transformation $\mathcal{C}_\Delta$ is symplectic and if the Hamiltonian transforms in an appropriate manner. The transformation $\mathcal{C}_\Delta$ is symplectic if the bilinear antisymmetric form ω is invariant (see equation 5.73) for a general pair of linearized state variations with zero time component.

Let ζ' be an increment with zero time component of the state (t', q', p'). The linearized increment in the value of $\mathcal{C}_\Delta(t', q', p')$ is $\zeta = D\mathcal{C}_\Delta(t', q', p')\zeta'$: The image of the increment is obtained by multiplying the increment by the derivative of the transformation. On the other hand, the transformation is obtained by time evolution, so the image of the increment can also be found by the time evolution of the linearized variational system. Let

$$\bar{\zeta}(t) = (0, \bar{\zeta}_q(t), \bar{\zeta}_p(t))$$
$$\bar{\zeta}'(t) = (0, \bar{\zeta}'_q(t), \bar{\zeta}'_p(t)) \tag{6.71}$$

---

[2]Our theorems about which transformations are canonical are still valid, because a shift of time does not affect the symplectic condition. See footnote 14 in Chapter 5.

be variations of the state path $\sigma(t) = (t, \bar{q}(t), \bar{p}(t))$; then

$$\bar{\zeta}(t + \Delta) = DC_\Delta(t, q(t), p(t))\bar{\zeta}(t)$$
$$\bar{\zeta}'(t + \Delta) = DC_\Delta(t, q(t), p(t))\bar{\zeta}'(t). \tag{6.72}$$

The symplectic requirement is

$$\omega(\bar{\zeta}(t), \bar{\zeta}'(t)) = \omega(\bar{\zeta}(t + \Delta), \bar{\zeta}'(t + \Delta)). \tag{6.73}$$

This must be true for arbitrary Δ, so it is satisfied if the following quantity is constant:

$$A(t) = \omega(\bar{\zeta}(t), \bar{\zeta}'(t))$$
$$= P(\bar{\zeta}'(t))Q(\bar{\zeta}(t)) - P(\bar{\zeta}(t))Q(\bar{\zeta}'(t))$$
$$= \bar{\zeta}'_p(t)\bar{\zeta}_q(t) - \bar{\zeta}_p(t)\bar{\zeta}'_q(t). \tag{6.74}$$

We compute the derivative:

$$DA(t) = D\bar{\zeta}'_p(t)\bar{\zeta}_q(t) + \bar{\zeta}'_p(t)D\bar{\zeta}_q(t)$$
$$\quad - D\bar{\zeta}_p(t)\bar{\zeta}'_q(t) - \bar{\zeta}_p(t)D\bar{\zeta}'_q(t). \tag{6.75}$$

With Hamilton's equations, the variations satisfy

$$D\bar{\zeta}_q(t) = \partial_1\partial_2 H(t, \bar{q}(t), \bar{p}(t))\bar{\zeta}_q(t)$$
$$\quad + \partial_2\partial_2 H(t, \bar{q}(t), \bar{p}(t))\bar{\zeta}_p(t),$$
$$D\bar{\zeta}_p(t) = -\partial_1\partial_1 H(t, \bar{q}(t), \bar{p}(t))\bar{\zeta}_q(t)$$
$$\quad - \partial_2\partial_1 H(t, \bar{q}(t), \bar{p}(t))\bar{\zeta}_p(t). \tag{6.76}$$

Substituting these into DA and collecting terms, we find[3]

$$DA(t) = 0. \tag{6.77}$$

We conclude that time evolution generates a phase-space transformation with symplectic derivative.

To make a canonical transformation we must specify how the Hamiltonian transforms. The same Hamiltonian describes the evolution of a state and a time-advanced state because the latter is just another state. Thus the transformed Hamiltonian is the same as the original Hamiltonian.

---

[3]Partial derivatives of structured arguments do not generally commute, so this deduction is not as simple as it may appear.

Liouville's theorem, again

We deduced that volumes in phase space are preserved by time evolution by showing that the divergence of the phase flow is zero, using the equations of motion (see section 3.8). We can also show that volumes in phase space are preserved by the evolution using the fact that time evolution is a canonical transformation.

We have shown that phase-space volume is preserved for symplectic transformations. Now we have shown that the transformation generated by time evolution is a symplectic transformation. Therefore, the transformation generated by time evolution preserves phase-space volume. This is an alternate proof of Liouville's theorem.

Another time-evolution transformation

There is another canonical transformation that can be constructed from time evolution. We define the transformation $\mathcal{C}'_\Delta$ such that

$$\mathcal{C}'_\Delta = \mathcal{C}_\Delta \circ S_{-\Delta}, \tag{6.78}$$

where $S_\Delta(a, b, c) = (a + \Delta, b, c)$ shifts the time of a phase-space state.[4] More explicitly, given a state (t, q', p'), we evolve the state that is obtained by subtracting Δ from t; that is, we take the state $(t - \Delta, q', p')$ as an initial state for evolution by Hamilton's equations. The state path σ satisfies

$$\sigma(t - \Delta) = (t - \Delta, \bar{q}(t - \Delta), \bar{p}(t - \Delta))$$
$$= (t - \Delta, q', p'). \tag{6.79}$$

The output of the transformation is the state

$$(t, q, p) = \sigma(t) = (t, \bar{q}(t), \bar{p}(t)). \tag{6.80}$$

The transformation satisfies

$$(t, \bar{q}(t), \bar{p}(t)) = \mathcal{C}'_\Delta(t, \bar{q}(t - \Delta), \bar{p}(t - \Delta)). \tag{6.81}$$

The arguments of $\mathcal{C}'_\Delta$ are not a consistent phase-space state; the time argument must be decremented by Δ to obtain a consis-

---

[4]The transformation S_Δ is an identity on the qp components, so it is symplectic. Although it adjusts the time, it is not a time-dependent transformation in that the qp components do not depend upon the time. Thus, if we adjust the Hamiltonian by composition with S_Δ we have a canonical transformation.

tent state. The transformation is completed by evolution of this consistent state.

Why is this a good idea?   Our usual canonical transformations do not change the time component. The $\mathcal{C}_\Delta$ transformation changes the time component, but $\mathcal{C}'_\Delta$ does not. It is canonical and in the usual form:

$$(t, q, p) = \mathcal{C}'_\Delta(t, q', p'). \tag{6.82}$$

The $\mathcal{C}'_\Delta$ transformation requires an adjustment of the Hamiltonian. The Hamiltonian H'_Δ that gives the correct Hamilton's equations at the transformed phase-space point is the original Hamiltonian composed with a function that decrements the independent variable by Δ:

$$H'_\Delta(t, q, p) = H(t - \Delta, q, p) \tag{6.83}$$

or

$$H'_\Delta = H \circ S_{-\Delta}. \tag{6.84}$$

Notice that if H is time independent then $H'_\Delta = H$.

Assume we have a procedure C such that ((C delta-t) state) implements a time-evolution transformation $\mathcal{C}_\Delta$ of the state state with time interval delta-t; then the procedure Cp such that ((Cp delta-t) state) implements $\mathcal{C}'_\Delta$ of the same state and time interval can be derived from the procedure C by using the procedure

```
(define ((C->Cp C) delta-t)
  (compose (C delta-t) (shift-t (- delta-t))))
```

where shift-t implements S_Δ:

```
(define ((shift-t delta-t) state)
  (up
   (+ (time state) delta-t)
   (coordinate state)
   (momentum state)))
```

To complete the canonical transformation we have a procedure that transforms the Hamiltonian:

```
(define ((H->Hp delta-t) H)
  (compose H (shift-t (- delta-t))))
```

So both $\mathcal{C}$ and $\mathcal{C}'$ can be used to make canonical transformations by specifying how the old and new Hamiltonians are related. For $\mathcal{C}_\Delta$ the Hamiltonian is unchanged. For $\mathcal{C}'_\Delta$ the Hamiltonian is time shifted.

Exercise 6.5: Verification

The condition (5.20) that Hamilton's equations are preserved for $\mathcal{C}_\Delta$ is

$$D_s H \circ \mathcal{C}_\Delta = D\mathcal{C}_\Delta \, D_s H'_\Delta,$$

and the condition that Hamilton's equations are preserved for $\mathcal{C}'_\Delta$ is

$$D_s H \circ \mathcal{C}'_\Delta = D\mathcal{C}'_\Delta \, D_s H'_\Delta.$$

Verify that these conditions are satisfied.

Exercise 6.6: Driven harmonic oscillator

We can use the simple driven harmonic oscillator to illustrate that time evolution yields a symplectic transformation that can be extended to be canonical in two ways. We use the driven harmonic oscillator because its solution can be compactly expressed in explicit form.

Suppose that we have a harmonic oscillator with natural frequency ω_0 driven by a periodic sinusoidal drive of frequency ω and amplitude α. The Hamiltonian we will consider is

$$H(t, q, p) = \tfrac{1}{2}p^2 + \tfrac{1}{2}\omega_0^2 q^2 - \alpha q \cos \omega t.$$

The general solution for a given initial state (t_0, q_0, p_0) evolved for a time Δ is

$$\begin{pmatrix} q(t_0 + \Delta) \\ p(t_0 + \Delta)/\omega_0 \end{pmatrix}$$
$$= \begin{pmatrix} \cos \omega_0 \Delta & \sin \omega_0 \Delta \\ -\sin \omega_0 \Delta & \cos \omega_0 \Delta \end{pmatrix} \begin{pmatrix} q_0 - \alpha' \cos \omega t_0 \\ (1/\omega_0)(p_0 + \alpha' \omega \sin \omega t_0) \end{pmatrix}$$
$$+ \begin{pmatrix} \alpha' \cos \omega(t_0 + \Delta) \\ -\alpha'(\omega/\omega_0) \sin \omega(t_0 + \Delta) \end{pmatrix}$$

where $\alpha' = \alpha/(\omega_0^2 - \omega^2)$.

a. Fill in the details of the procedure

```
(define (((C* alpha omega omega0) delta-t) state)
   ... )
```

that implements the time-evolution transformation of the driven harmonic oscillator. Let C be (C* alpha omega omega0).

b. In terms of C*, the general solution emanating from a given state is

```
(define (((solution alpha omega omega0) state0) t)
   (((C* alpha omega omega0) (- t (time state0))) state0))
```

Check that the implementation of C* is correct by using it to construct the solution and verifying that the solution satisfies Hamilton's equations. Further check the solution by comparing to numerical integration.

c. We know that for any phase-space state function F the rate of change of that function along a solution path σ is

$$D(F \circ \sigma) = \partial_0 F \circ \sigma + \{F, H\} \circ \sigma.$$

Show, by writing a short program to test it, that this is true of the function implemented by (C delta) for the driven oscillator. Why is this interesting?

d. Use the procedure symplectic-transform? to show that both C and Cp are symplectic.

e. Use the procedure canonical? to verify that both C and Cp are canonical with the appropriate transformed Hamiltonian.

6.2.1 Another View of Time Evolution

We can also show that time evolution generates canonical transformations using the Poincaré–Cartan integral invariant.

Consider a two-dimensional region of phase-space coordinates, R', at some particular time t' (see figure 6.1). Let R be the image of this region at time t under time evolution for a time interval of Δ. The time evolution is governed by a Hamiltonian H. Let $\sum_i A_i$ be the sum of the oriented areas of the projections of R onto the fundamental canonical planes.[5] Similarly, let $\sum_i A'_i$ be the sum of oriented projected areas for R'. We will show that $\sum_i A_i = \sum_i A'_i$, and thus the Poincaré integral invariant is preserved by time evolution. By showing that the Poincaré integral invariant is preserved, we will have shown that the qp part of the transformation generated by time evolution is symplectic. From this we can construct canonical transformations from time evolution as before.

In the extended phase space we see that the evolution sweeps out a cylindrical volume with endcaps R' and R, each at a fixed time. Let R'' be the two-dimensional region swept out by the trajectories that map the boundary of region R' to the boundary of region R. The regions R, R', and R'' together form the boundary of a volume of phase-state space.

---

[5] By Stokes's theorem we may compute the area of a region by a line integral around the boundary of the region. We define the positive sense of the area to be the area enclosed by a curve that is traversed in a counterclockwise direction, when drawn on a plane with the coordinate on the abscissa and the momentum on the ordinate.

The Poincaré–Cartan integral invariant on the whole boundary is zero.[6] Thus

$$\sum_{i=0}^{n} A_i - \sum_{i=0}^{n} A_i' + \sum_{i=0}^{n} A_i'' = 0, \tag{6.85}$$

where the n index indicates the t, p_t canonical plane. The second term is negative, because in the extended phase space we take the area to be positive if the normal to the surface is outward pointing.

We will show that the Poincaré–Cartan integral invariant for a region of phase space that is generated by time evolution is zero:

$$\sum_{i=0}^{n} A_i'' = 0. \tag{6.86}$$

This will allow us to conclude

$$\sum_{i=0}^{n} A_i - \sum_{i=0}^{n} A_i' = 0. \tag{6.87}$$

The areas of the projection of R and R' on the t, p_t plane are zero because R and R' are at constant times, so for these regions the Poincaré–Cartan integral invariant is the same as the Poincaré integral invariant. Thus

$$\sum_{i=0}^{n-1} A_i = \sum_{i=0}^{n-1} A_i'. \tag{6.88}$$

We are left with showing that the Poincaré–Cartan integral invariant for the region R'' is zero. This will be zero if the contribution from any small piece of R'' is zero. We will show this by showing that the ω form (see equation 5.70) on a small parallelogram in this region is zero. Let $(0; q, t; p, p_t)$ be a vertex of this parallelogram. The parallelogram is specified by two edges ζ_1 and ζ_2 emanating from this vertex. For edge ζ_1 of the parallelogram,

---

[6]We can see this as follows. Let γ be any closed curve in the boundary. This curve divides the boundary into two regions. By Stokes's theorem the integral invariant over both of these pieces can be written as a line integral along this boundary, but they have opposite signs, because γ is traversed in opposite directions to keep the surface on the left. So we conclude that the integral invariant over the entire surface is zero.

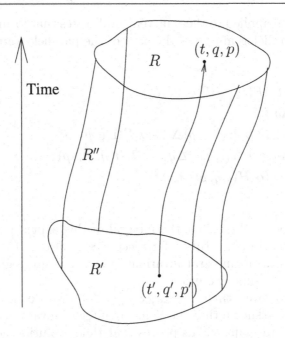

Figure 6.1   All points in some two-dimensional region R' in phase space at time t' are evolved for some time interval Δ. At the time t the set of points define the two-dimensional region R. For example, the state labeled by the phase-space coordinates (t', q', p') evolves to the state labeled by the coordinates (t, q, p).

we take a constant-time phase-space increment with length Δq and Δp in the q and p directions. The first-order change in the Hamiltonian that corresponds to these changes is

$$\Delta H = \partial_1 H(t, q, p)\Delta q + \partial_2 H(t, q, p)\Delta p \tag{6.89}$$

for constant time $\Delta t = 0$. The increment Δp_t is the negative of ΔH. So the extended phase-space increment is

$$\zeta_1 = (0; \Delta q, 0; \Delta p, -\partial_1 H(t, q, p)\Delta q - \partial_2 H(t, q, p)\Delta p). \tag{6.90}$$

The edge ζ_2 is obtained by time evolution of the vertex for a time interval Δt. Using Hamilton's equations, we obtain

$$\zeta_2 = (0; Dq(t)\Delta t, \Delta t; Dp(t)\Delta t, Dp_t(t)\Delta t) \tag{6.91}$$
$$= (0; \partial_2 H(t, q, p)\Delta t, \Delta t; -\partial_1 H(t, q, p)\Delta t, -\partial_0 H(t, q, p)\Delta t).$$

The ω form applied to these incremental states that form the edges of this parallelogram gives the area of the parallelogram:

$$\omega(\zeta_1, \zeta_2)$$
$$= Q(\zeta_1)P(\zeta_2) - P(\zeta_1)Q(\zeta_2)$$
$$= (\Delta q, 0)$$
$$\quad \cdot (-\partial_1 H(t, q, p)\Delta t, -\partial_0 H(t, q, p)\Delta t)$$
$$- (\Delta p, -\partial_1 H(t, q, p)\Delta q - \partial_2 H(t, q, p)\Delta p)$$
$$\quad \cdot (\partial_2 H(t, q, p)\Delta t, \Delta t)$$
$$= 0. \tag{6.92}$$

So we may conclude that the integral of this expression over the entire surface of the tube of trajectories is also zero. Thus the Poincaré–Cartan integral invariant is zero for any region that is generated by time evolution.

Having proven that the trajectory tube provides no contribution, we have shown that the Poincaré integral invariant of the two endcaps is the same. This proves that time evolution generates a symplectic qp transformation.

Area preservation of surfaces of section

We can use the Poincaré–Cartan invariant to prove that for autonomous two-degree-of-freedom systems, surfaces of section (constructed appropriately) preserve area.

To show this we consider a surface of section for one coordinate (say q_2) equal to zero. We construct the section by accumulating the (q_1, p_1) pairs. We assume that all initial conditions have the same energy. We compute the sum of the areas of canonical projections in the extended phase space again. Because all initial conditions have the same $q_2 = 0$ there is no area on the q_2, p_2 plane, and because all the trajectories have the same value of the Hamiltonian the area of the projection on the t, p_t plane is also zero. So the sum of areas of the projections is just the area of the region on the surface of section. Now let each point on the surface of section evolve to the next section crossing. For each point on the section this may take a different amount of time. Compute the sum of the areas again for the mapped region. Again, all points of the mapped region have the same q_2, so the area on the q_2, p_2 plane is zero, and they continue to have the same energy, so the area on the t, p_t plane is zero. So the area of the mapped region is again just the area on the surface of section, the q_1, p_1 plane. Time evolution preserves the sum of areas, so the area on the surface of section is the same as the mapped area.

Thus surfaces of section preserve area provided that the section points are entirely on a canonical plane. For example, to make the Hénon–Heiles surfaces of section (see section 3.6.3) we plotted p_y versus y when $x = 0$ with $p_x \geq 0$. So for all section points the x coordinate has the fixed value 0, the trajectories all have the same energy, and the points accumulated are entirely in the y, p_y canonical plane. So the Hénon–Heiles surfaces of section preserve area.

6.2.2   Yet Another View of Time Evolution

We can show directly from the action principle that time evolution generates a symplectic transformation.

Recall that the Lagrangian action S is

$$S[q](t_1, t_2) = \int_{t_1}^{t_2} L \circ \Gamma[q]. \tag{6.93}$$

We computed the variation of the action in deriving the Lagrange equations. The variation is (see equation 1.33)

$$\delta_\eta S[q](t_1, t_2) = (\partial_2 L \circ \Gamma[q])\eta|_{t_1}^{t_2} - \int_{t_1}^{t_2} (\mathsf{E}\,[L] \circ \Gamma[q])\eta, \tag{6.94}$$

rewritten in terms of the Euler–Lagrange operator E. In the derivation of the Lagrange equations we considered only variations that preserved the endpoints of the path being tested. However, equation (6.94) is true of arbitrary variations. Here we consider variations that are not zero at the endpoints around a realizable path q (one for which $\mathsf{E}\,[L] \circ \Gamma[q] = 0$). For these variations the variation of the action is just the integrated term:

$$\delta_\eta S[q](t_1, t_2) = (\partial_2 L \circ \Gamma[q])\eta|_{t_1}^{t_2} = p(t_2)\eta(t_2) - p(t_1)\eta(t_1). \tag{6.95}$$

Recall that p and η are structures, and the product implies a sum of products of components.

Consider a continuous family of realizable paths; the path for parameter s is $\tilde{q}(s)$ and the coordinates of this path at time t are $\tilde{q}(s)(t)$. We define $\tilde{\eta}(s) = D\tilde{q}(s)$; the variation of the path along the family is the derivative of the parametric path with respect to the parameter. Let

$$\widetilde{S}(s) = S[\tilde{q}(s)](t_1, t_2) \tag{6.96}$$

be the value of the action from t_1 to t_2 for path $\tilde{q}(s)$. The derivative of the action along this parametric family of paths is[7]

$$D\widetilde{S}(s) = \delta_{\tilde{\eta}(s)} S[\tilde{q}(s)]$$
$$= (\partial_2 L \circ \Gamma[\tilde{q}(s)])\tilde{\eta}(s)|_{t_1}^{t_2} - \int_{t_1}^{t_2} (\mathsf{E}[L] \circ \Gamma[\tilde{q}(s)])\tilde{\eta}(s). \tag{6.97}$$

Because $\tilde{q}(s)$ is a realizable path, $\mathsf{E}[L] \circ \Gamma[\tilde{q}(s)] = 0$. So

$$D\widetilde{S}(s) = (\partial_2 L \circ \Gamma[\tilde{q}(s)])\tilde{\eta}(s)|_{t_1}^{t_2}$$
$$= \tilde{p}(s)(t_2)\tilde{\eta}(s)(t_2) - \tilde{p}(s)(t_1)\tilde{\eta}(s)(t_1), \tag{6.98}$$

where $\tilde{p}(s)$ is the momentum conjugate to $\tilde{q}(s)$. The integral of $D\widetilde{S}$ is

$$S[\tilde{q}(s_2)](t_1, t_2) - S[\tilde{q}(s_1)](t_1, t_2) = \int_{s_1}^{s_2} (D\widetilde{S})$$
$$= \int_{s_1}^{s_2} (h(t_2) - h(t_1)), \tag{6.99}$$

where

$$h(t)(s) = \tilde{p}(s)(t)\tilde{\eta}(s)(t) = \tilde{p}(s)(t)D\tilde{q}(s)(t). \tag{6.100}$$

In conventional notation the latter line integral is written

$$\int_{\gamma_2} \sum_i p_i dq^i - \int_{\gamma_1} \sum_i p_i dq^i, \tag{6.101}$$

where $\gamma_1(s) = \tilde{q}(s)(t_1)$ and $\gamma_2(s) = \tilde{q}(s)(t_2)$.

---

[7]Let f be a path-dependent function, $\tilde{\eta}(s) = D\tilde{q}(s)$, and $g(s) = f[\tilde{q}(s)]$. The variation of f at $\tilde{q}(s)$ in the direction $\tilde{\eta}(s)$ is $\delta_{\tilde{\eta}(s)} f[\tilde{q}(s)] = Dg(s)$.

For a loop family of paths (such that $\tilde{q}(s_2) = \tilde{q}(s_1)$), the difference of actions at the endpoints vanishes, so we deduce

$$\oint_{\gamma_2} \sum_i p_i dq^i = \oint_{\gamma_1} \sum_i p_i dq^i, \tag{6.102}$$

which is the line-integral version of the integral invariants.

In terms of area integrals, using Stokes's theorem, this is

$$\sum_i \int_{R_2^i} dp_i dq^i = \sum_i \int_{R_1^i} dp_i dq^i, \tag{6.103}$$

where R_j^i are the regions in the ith canonical plane. We have found that the time evolution preserves the integral invariants, and thus time evolution generates a symplectic transformation.

6.3   Lie Transforms

The evolution of a system under any Hamiltonian generates a continuous family of canonical transformations. To study the behavior of some system governed by a Hamiltonian H, it is sometimes appropriate to use a canonical transformation generated by evolution governed by another Hamiltonian-like function W on the same phase space. Such a canonical transformation is called a *Lie transform*.

The functions H and W are both Hamiltonian-shaped functions defined on the same phase space. Time evolution for an interval Δ governed by H is a canonical transformation $\mathcal{C}_{\Delta,H}$. Evolution by W for an interval ϵ is a canonical transformation $\mathcal{C}'_{\epsilon,W}$:

$$(t, q, p) = \mathcal{C}'_{\epsilon,W}(t, q', p'). \tag{6.104}$$

The independent variable in the H evolution is time, and the independent variable in the W evolution is an arbitrary parameter of the canonical transformation. We chose $\mathcal{C}'$ for the W evolution so that the canonical transformation induced by W does not change the time in the system governed by H.

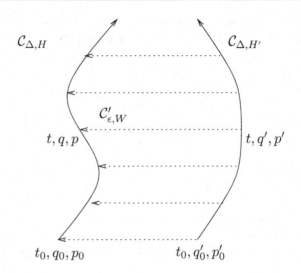

Figure 6.2 Time evolution of a trajectory started at the point (t_0, q_0, p_0), governed by the Hamiltonian H, is transformed by the Lie transform governed by the generator W. The time evolution of the transformed trajectory is governed by the Hamiltonian H'.

Figure 6.2 shows how a Lie transform is used to transform a trajectory. We can see from the diagram that the canonical transformations obey the relation

$$\mathcal{C}'_{\epsilon,W} \circ \mathcal{C}_{\Delta,H'} = \mathcal{C}_{\Delta,H} \circ \mathcal{C}'_{\epsilon,W}. \tag{6.105}$$

For generators W that do not depend on the independent variable, the resulting canonical transformation $\mathcal{C}'_{\epsilon,W}$ is time independent and symplectic. A time-independent symplectic transformation is canonical if the Hamiltonian transforms by composition:[8]

---

[8]In general, the generator W could depend on its independent variable. If so, it would be necessary to specify a rule that gives the initial value of the independent variable for the W evolution. This rule may or may not depend upon the time. If the specification of the independent variable for the W evolution does not depend on time, then the resulting canonical transformation $\mathcal{C}'_{\epsilon,W}$ is time independent and the Hamiltonians transform by composition. If the generator W depends on its independent variable and the rule for specifying its initial value depends on time, then the transformation $\mathcal{C}'_{\epsilon,W}$ is time dependent. In this case there may need to be an adjustment to the relation between the Hamiltonians H and H'. In the extended phase space all these complications disappear: There is only one case. We can assume all generators W are independent of the independent variable.

$$H' = H \circ C'_{\epsilon,W}. \tag{6.106}$$

We will use only Lie transforms that have generators that do not depend on the independent variable.

Lie transforms of functions

The value of a phase-space function F changes if its arguments change. We define the function $E'_{\epsilon,W}$ of a function F of phase-space coordinates (t, q, p) by

$$E'_{\epsilon,W} F = F \circ C'_{\epsilon,W}. \tag{6.107}$$

We say that $E'_{\epsilon,W} F$ is the Lie transform of the function F.

In particular, the Lie transform advances the coordinate and momentum selector functions $Q = I_1$ and $P = I_2$:

$$(E'_{\epsilon,W} Q)(t, q', p') = (Q \circ C'_{\epsilon,W})(t, q', p') = Q(t, q, p) = q$$
$$(E'_{\epsilon,W} P)(t, q', p') = (P \circ C'_{\epsilon,W})(t, q', p') = P(t, q, p) = p. \tag{6.108}$$

So we may restate equation (6.107) as

$$(E'_{\epsilon,W} F)(t, q', p')$$
$$= F(t, (E'_{\epsilon,W} Q)(t, q', p'), (E'_{\epsilon,W} P)(t, q', p')). \tag{6.109}$$

More generally, Lie transforms descend into compositions:

$$(E'_{\epsilon,W}(F \circ G)) = F \circ (E'_{\epsilon,W} G) \tag{6.110}$$

A corollary of the fact that Lie transforms descend into compositions is:

$$E'_{\epsilon_1,W_1} E'_{\epsilon_2,W_2} I = (E'_{\epsilon_1,W_1}(E'_{\epsilon_2,W_2} I)) \circ I$$
$$= (E'_{\epsilon_2,W_2} I) \circ (E'_{\epsilon_1,W_1} I), \tag{6.111}$$

where I is the phase-space identity function: $I(t, q, p) = (t, q, p)$. So the order of application of the operators is reversed from the order of composition of the functions that result from applying the operators.

In terms of $E'_{\epsilon,W}$ we have the canonical transformation

$$q = (E'_{\epsilon,W} Q)(t, q', p')$$
$$p = (E'_{\epsilon,W} P)(t, q', p')$$
$$H' = E'_{\epsilon,W} H. \tag{6.112}$$

We can also say

$$(t, q, p) = (E'_{\epsilon, W} I)(t, q', p'). \tag{6.113}$$

Note that $E'_{\epsilon, W}$ has the property:[9]

$$E'_{\epsilon_1 + \epsilon_2, W} = E'_{\epsilon_1, W} \circ E'_{\epsilon_2, W} = E'_{\epsilon_2, W} \circ E'_{\epsilon_1, W}. \tag{6.114}$$

The identity I is

$$I = E'_{0, W}. \tag{6.115}$$

We can define the inverse function

$$(E'_{\epsilon, W})^{-1} = E'_{-\epsilon, W} \tag{6.116}$$

with the property

$$I = E'_{\epsilon, W} \circ (E'_{\epsilon, W})^{-1} = (E'_{\epsilon, W})^{-1} \circ E'_{\epsilon, W}. \tag{6.117}$$

Simple Lie transforms

For example, suppose we are studying a system for which a rotation would be a helpful transformation. To concoct such a transformation we note that we intend a configuration coordinate to increase uniformly with a given rate. In this case we want an angle to be incremented. The Hamiltonian that consists solely of the momentum conjugate to that configuration coordinate always does the job. So the angular momentum is an appropriate generator for rotations.

The analysis is simple if we use polar coordinates r, θ with conjugate momenta p_r, p_θ. The generator W is just:

$$W(\tau; r, \theta; p_r, p_\theta) = p_\theta \tag{6.118}$$

The family of transformations satisfies Hamilton's equations:

$$\begin{aligned}
Dr &= 0 \\
D\theta &= 1 \\
Dp_r &= 0 \\
Dp_\theta &= 0.
\end{aligned} \tag{6.119}$$

---

[9]The set of transformations $E'_{\epsilon, W}$ with the operation composition and with parameter ϵ is a one-parameter Lie group.

The only variable that appears in W is p_θ, so θ is the only variable that varies as ϵ is varied. In fact, the family of canonical transformations is

$$
\begin{aligned}
r &= r' \\
\theta &= \theta' + \epsilon \\
p_r &= p'_r \\
p_\theta &= p'_\theta.
\end{aligned}
\tag{6.120}
$$

So angular momentum is the generator of a canonical rotation.

The example is simple, but it illustrates one important feature of Lie transformations—they give one set of variables entirely in terms of the other set of variables. This should be contrasted with the mixed-variable generating function transformations, which always give a mixture of old and new variables in terms of a mixture of new and old variables, and thus require an inversion to get one set of variables in terms of the other set of variables. This inverse can be written in closed form only for special cases. In general, there is considerable advantage in using a transformation rule that generates explicit transformations from the start. The Lie transformations are always explicit in the sense that they give one set of variables in terms of the other, but for there to be explicit expressions the evolution governed by the generator must be solvable.

Let's consider another example. This time consider a three-degree-of-freedom problem in rectangular coordinates, and take the generator of the transformation to be the z component of the angular momentum:

$$
W(\tau; x, y, z; p_x, p_y, p_z) = xp_y - yp_x.
\tag{6.121}
$$

The evolution equations are

$$
\begin{aligned}
Dx &= -y \\
Dy &= x \\
Dz &= 0 \\
Dp_x &= -p_y \\
Dp_y &= p_x \\
Dp_z &= 0.
\end{aligned}
\tag{6.122}
$$

We notice that z and p_z are unchanged, and that the equations governing the evolution of x and y decouple from those of p_x and p_y. Each of these pairs of equations represents simple harmonic motion, as can be seen by writing them as second-order systems. The solutions are

$$x = x' \cos \epsilon - y' \sin \epsilon$$
$$y = x' \sin \epsilon + y' \cos \epsilon$$
$$z = z', \tag{6.123}$$
$$p_x = p'_x \cos \epsilon - p'_y \sin \epsilon$$
$$p_y = p'_x \sin \epsilon + p'_y \cos \epsilon$$
$$p_z = p'_z. \tag{6.124}$$

So we see that again a component of the angular momentum generates a canonical rotation. There was nothing special about our choice of axes, so we can deduce that the component of angular momentum about any axis generates rotations about that axis.

Example

Suppose we have a system governed by the Hamiltonian

$$H(t; x, y; p_x, p_y) = \tfrac{1}{2}(p_x^2 + p_y^2) + \tfrac{1}{2}a(x - y)^2 + \tfrac{1}{2}b(x + y)^2. \tag{6.125}$$

Hamilton's equations couple the motion of x and y:

$$Dx = p_x$$
$$Dy = p_y$$
$$Dp_x = -a(x - y) - b(x + y)$$
$$Dp_y = a(x - y) - b(x + y). \tag{6.126}$$

We can decouple the system by performing a coordinate rotation by $\pi/4$. This is generated by

$$W(\tau; x, y; p_x, p_y) = xp_y - yp_x, \tag{6.127}$$

which is similar to the generator for the coordinate rotation above but without the z degree of freedom. Evolving $(\tau; x, y; p_x, p_y)$ by W for an interval of $\pi/4$ gives a canonical rotation:

$$x = x' \cos \pi/4 - y' \sin \pi/4$$
$$y = x' \sin \pi/4 + y' \cos \pi/4$$

$$p_x = p'_x \cos \pi/4 - p'_y \sin \pi/4$$
$$p_y = p'_x \sin \pi/4 + p'_y \cos \pi/4. \tag{6.128}$$

Composing the Hamiltonian H with this time-independent transformation gives the new Hamiltonian

$$H'(t; x', y'; p'_x, p'_y) = (\tfrac{1}{2}(p'_x)^2 + b(x')^2) + (\tfrac{1}{2}(p'_y)^2 + a(y')^2), \tag{6.129}$$

which is a Hamiltonian for two uncoupled harmonic oscillators. ·
So the original coupled problem has been transformed by a Lie transform to a new form for which the solution is easy.

6.4   Lie Series

A convenient way to compute a Lie transform is to approximate it with a series. We develop this technique by extending the idea of a Taylor series.

Taylor's theorem gives us a way of approximating the value of a nice enough function at a point near to a point where the value is known. If we know f and all of its derivatives at t then we can get the value of $f(t + \epsilon)$, for small enough ϵ, as follows:

$$f(t + \epsilon) = f(t) + \epsilon D f(t) + \frac{1}{2}\epsilon^2 D^2 f(t) + \cdots + \frac{1}{n!}\epsilon^n D^n f(t) + \cdots . \tag{6.130}$$

We recall that the power series for the exponential function is

$$e^x = 1 + x + \frac{1}{2}x^2 + \cdots + \frac{1}{n!}x^n + \cdots . \tag{6.131}$$

This suggests that we can formally construct a Taylor-series operator as the exponential of a differential operator[10]

$$e^{\epsilon D} = I + \epsilon D + \frac{1}{2}(\epsilon D)^2 + \cdots + \frac{1}{n!}(\epsilon D)^n + \cdots \tag{6.132}$$

and write

$$f(t + \epsilon) = (e^{\epsilon D} f)(t). \tag{6.133}$$

---

[10]We are playing fast and loose with differential operators here. In a formal treatment it is essential to prove that these games are mathematically well defined and have appropriate convergence properties.

We have to be a bit careful here: $(\epsilon D)^2 = \epsilon D \epsilon D$. We can turn it into $\epsilon^2 D^2$ only because ϵ is a scalar constant, which commutes with every differential operator. But with this caveat in mind we can define the differential operator

$$(e^{\epsilon D} f)(t) = f(t) + \epsilon D f(t) + \frac{1}{2}\epsilon^2 D^2 f(t) + \cdots + \frac{1}{n!}\epsilon^n D^n f(t) + \cdots$$

$$(6.134)$$

Before going on, it is interesting to compute with these a bit. In the code transcripts that follow we develop the series by exponentiation. We can examine the series incrementally by looking at successive elements of the (infinite) sequence of terms of the series. The procedure series:for-each is an incremental traverser that applies its first argument to successive elements of the series given as its second argument. The third argument (when given) specifies the number of terms to be traversed. In each of the following transcripts we print simplified expressions for the successive terms.

The first thing to look at is the general Taylor expansion for an unknown literal function, expanded around t, with increment ϵ. Understanding what we see in this simple problem will help us understand more complex problems later.

```
(series:for-each print-expression
 (((exp (* 'epsilon D))
   (literal-function 'f))
  't)
 6)

(f t)
(* ((D f) t) epsilon)
(* 1/2 (((expt D 2) f) t) (expt epsilon 2))
(* 1/6 (((expt D 3) f) t) (expt epsilon 3))
(* 1/24 (((expt D 4) f) t) (expt epsilon 4))
(* 1/120 (((expt D 5) f) t) (expt epsilon 5))
...
```

We can also look at the expansions of particular functions that we recognize, such as the expansion of sin around 0.

```
(series:for-each print-expression
   (((exp (* 'epsilon D)) sin) 0)
   6)
```

```
0
epsilon
0
(* -1/6 (expt epsilon 3))
0
(* 1/120 (expt epsilon 5))
...
```

It is often instructive to expand functions we usually don't re-member, such as $f(x) = \sqrt{1 + x}$.

```
(series:for-each print-expression
 (((exp (* 'epsilon D))
   (lambda (x) (sqrt (+ x 1))))
  0)
 6)
```

```
1
(* 1/2 epsilon)
(* -1/8 (expt epsilon 2))
(* 1/16 (expt epsilon 3))
(* -5/128 (expt epsilon 4))
(* 7/256 (expt epsilon 5))
...
```

Exercise 6.7: Binomial series

Develop the binomial expansion of $(1 + x)^n$ as a Taylor expansion. Of course, it must be the case that for n a positive integer all of the coeffi-cients except for the first $n + 1$ are zero. However, in the general case, for symbolic n, the coefficients are rather complicated polynomials in n. For example, you will find that the eighth term is

```
(+ (* 1/5040 (expt n 7))
   (* -1/240 (expt n 6))
   (* 5/144 (expt n 5))
   (* -7/48 (expt n 4))
   (* 29/90 (expt n 3))
   (* -7/20 (expt n 2))
   (* 1/7 n))
```

These terms must evaluate to the entries in Pascal's triangle. In partic-ular, this polynomial must be zero for $n < 7$. How is this arranged?

Dynamics

Now, to play this game with dynamical functions we want to pro-vide a derivative-like operator that we can exponentiate, which will give us the time-advance operator. The key idea is to write the derivative of the function in terms of the Poisson bracket. Equation (3.80) shows how to do this in general:

$$D(F \circ \sigma) = (\{F, H\} + \partial_0 F) \circ \sigma. \tag{6.135}$$

We define the operator D_H by

$$D_H F = \partial_0 F + \{F, H\}, \tag{6.136}$$

so

$$D_H F \circ \sigma = D(F \circ \sigma), \tag{6.137}$$

and iterates of this operator can be used to compute higher-order derivatives:

$$D^n(F \circ \sigma) = D_H^n F \circ \sigma. \tag{6.138}$$

We can express the advance of the path function $f = F \circ \sigma$ for an interval ϵ with respect to H as a power series in the derivative operator D_H applied to the phase-space function F and then composed with the path:

$$f(t + \epsilon) = (e^{\epsilon D} f)(t) = (e^{\epsilon D_H} F) \circ \sigma(t). \tag{6.139}$$

Indeed, we can implement the time-advance operator $E_{\epsilon, H}$ with this series, when it converges.

Exercise 6.8: Iterated derivatives

Show that equation (6.138) is correct.

Exercise 6.9: Lagrangian analog

Compare D_H with the total time derivative operator. Recall that

$$D_t F \circ \Gamma[q] = D(F \circ \Gamma[q])$$

abstracts the derivative of a function of a path through state space to a function of the derivatives of the path. Define another derivative operator D_L, analogous to D_H, that would give the time derivative of functions along Lagrangian state paths that are solutions of Lagrange's equations for a given Lagrangian. How might this be useful?

For time-independent Hamiltonian H and time-independent state function F, we can simplify the computation of the advance of F. In this case we define the *Lie derivative operator* L_H such that

$$L_H F = \{F, H\}, \tag{6.140}$$

which reads "the Lie derivative of F with respect to H."[11] So

$$D_H = \partial_0 + L_H \tag{6.141}$$

and for time-independent F

$$D(F \circ \sigma) = L_H F \circ \sigma. \tag{6.142}$$

We can iterate this process to compute higher derivatives. So

$$L_H^2 F = \{\{F, H\}, H\}, \tag{6.143}$$

and successively higher-order Poisson brackets of F with H give successively higher-order derivatives when evaluated on the trajectory.

Let $f = F \circ \sigma$. We have

$$Df = (L_H F) \circ \sigma \tag{6.144}$$
$$D^2 f = (L_H^2 F) \circ \sigma \tag{6.145}$$
$$\cdots .$$

Thus we can rewrite the advance of the path function f for an interval ϵ with respect to H as a power series in the Lie derivative operator applied to the phase-space function F and then composed with the path:

$$f(t + \epsilon) = (e^{\epsilon D} f)(t) = (e^{\epsilon L_H} F) \circ \sigma(t). \tag{6.146}$$

We can implement the time-advance operator $E'_{\epsilon, H}$ with the *Lie series* $e^{\epsilon L_H} F$ when this series converges:

$$E'_{\epsilon, H} F = e^{\epsilon L_H} F. \tag{6.147}$$

---

[11] Our L_H is a special case of what is referred to as a Lie derivative in differential geometry. The more general idea is that a vector field defines a flow. The Lie derivative of an object with respect to a vector field gives the rate of change of the object as it is dragged along with the flow. In our case the flow is the evolution generated by Hamilton's equations, with Hamiltonian H.

We have shown that time evolution is canonical, so the series above are formal representations of canonical transformations as power series in the time. These series may not converge, even if the evolution governed by the Hamiltonian H is well defined.

Computing Lie series

We can use the Lie transform as a computational tool to examine the local evolution of dynamical systems. We define the Lie derivative of F as a derivative-like operator relative to the given Hamiltonian function, H:[12]

```
(define ((Lie-derivative H) F)
  (Poisson-bracket F H))
```

We also define a procedure to implement the Lie transform:[13]

```
(define (Lie-transform H t)
  (exp (* t (Lie-derivative H))))
```

Let's start by examining the beginning of the Lie series for the position of a simple harmonic oscillator of mass m and spring constant k. We can implement the Hamiltonian as

```
(define ((H-harmonic m k) state)
  (+ (/ (square (momentum state)) (* 2 m))
     (* 1/2 k (square (coordinate state)))))
```

We make the Lie transform (series) by passing the `Lie-transform` operator an appropriate Hamiltonian function and an interval to evolve for. The resulting operator is then given the `coordinate` procedure, which selects the position coordinates from the phase-space state. The Lie transform operator returns a procedure that, when given a phase-space state composed of a dummy time, a

---

[12]Actually, we define the Lie derivative slightly differently, as follows:

```
(define ((Lie-derivative-procedure H) F)
  (Poisson-bracket F H))
(define Lie-derivative
  (make-operator Lie-derivative-procedure 'Lie-derivative))
```

The reason is that we want `Lie-derivative` to be an *operator*, which is just like a function except that the product of operators is interpreted as composition, whereas the product of functions is the function computing the product of their values.

[13]The `Lie-transform` procedure here is also defined to be an operator, just like `Lie-derivative`.

position x0, and a momentum p0, returns the position resulting
from advancing that state by the interval dt.

```
(series:for-each print-expression
 (((Lie-transform (H-harmonic 'm 'k) 'dt)
   coordinate)
  (up 0 'x0 'p0))
 6)
```

```
x0
(/ (* dt p0) m)
(/ (* -1/2 (expt dt 2) k x0) m)
(/ (* -1/6 (expt dt 3) k p0) (expt m 2))
(/ (* 1/24 (expt dt 4) (expt k 2) x0) (expt m 2))
(/ (* 1/120 (expt dt 5) (expt k 2) p0) (expt m 3))
...
```

We should recognize the terms of this series. We start with the initial position x_0. The first-order correction $(p_0/m)dt$ is due to the
initial velocity. Next we find an acceleration term $(-kx_0/2m)dt^2$
due to the restoring force of the spring at the initial position.

The Lie transform is just as appropriate for showing us how the
momentum evolves over the interval:

```
(series:for-each print-expression
 (((Lie-transform (H-harmonic 'm 'k) 'dt)
   momentum)
  (up 0 'x0 'p0))
 6)
```

```
p0
(* -1 dt k x0)
(/ (* -1/2 (expt dt 2) k p0) m)
(/ (* 1/6 (expt dt 3) (expt k 2) x0) m)
(/ (* 1/24 (expt dt 4) (expt k 2) p0) (expt m 2))
(/ (* -1/120 (expt dt 5) (expt k 3) x0) (expt m 2))
...
```

In this series we see how the initial momentum p_0 is corrected by
the effect of the restoring force $-kx_0dt$, etc.

What is a bit more fun is to see how a more complex phase-
space function is treated by the Lie series expansion. In the experiment below we examine the Lie series developed by advancing
the harmonic-oscillator Hamiltonian, by means of the transform
generated by the same harmonic-oscillator Hamiltonian:

```
(series:for-each print-expression
 (((Lie-transform (H-harmonic 'm 'k) 'dt)
   (H-harmonic 'm 'k))
  (up 0 'x0 'p0))
 6)

(/ (+ (* 1/2 k m (expt x0 2)) (* 1/2 (expt p0 2))) m)
0
0
0
0
0
...
```

As we would hope, the series shows us the original energy expression $(k/2)x_0^2 + (1/2m)p_0^2$ as the first term. Each subsequent correction term turns out to be zero—because the energy is conserved.

Of course, the Lie series can be used in situations where we want to see the expansion of the motion of a system characterized by a more complex Hamiltonian. The planar motion of a particle in a general central field (see equation 3.100) is a simple problem for which the Lie series is instructive. In the following transcript we can see how rapidly the series becomes complicated. It is worth one's while to try to interpret the additive parts of the third (acceleration) term shown below:

```
(series:for-each print-expression
 (((Lie-transform
    (H-central-polar 'm (literal-function 'U))
    'dt)
   coordinate)
  (up 0
      (up 'r_0 'phi_0)
      (down 'p_r_0 'p_phi_0)))
 4)

(up r_0 phi_0)
(up (/ (* dt p_r_0) m)
    (/ (* dt p_phi_0) (* m (expt r_0 2))))
(up
 (+ (/ (* -1/2 ((D U) r_0) (expt dt 2)) m)
    (/ (* 1/2 (expt dt 2) (expt p_phi_0 2))
       (* (expt m 2) (expt r_0 3))))
 (/ (* -1 (expt dt 2) p_phi_0 p_r_0)
    (* (expt m 2) (expt r_0 3))))
```

```
(up
 (+ (/ (* -1/6 (((expt D 2) U) r_0) (expt dt 3) p_r_0)
       (expt m 2))
    (/ (* -1/2 (expt dt 3) (expt p_phi_0 2) p_r_0)
       (* (expt m 3) (expt r_0 4))))
 (+ (/ (* 1/3 ((D U) r_0) (expt dt 3) p_phi_0)
       (* (expt m 2) (expt r_0 3)))
    (/ (* -1/3 (expt dt 3) (expt p_phi_0 3))
       (* (expt m 3) (expt r_0 6)))
    (/ (* (expt dt 3) p_phi_0 (expt p_r_0 2))
       (* (expt m 3) (expt r_0 4))))))
...
```

Of course, if we know the closed-form Lie transform it is probably a good idea to take advantage of it, but when we do not know the closed form the Lie series representation of it can come in handy.

6.5   Exponential Identities

The composition of Lie transforms can be written as products of exponentials of Lie derivative operators. In general, Lie derivative operators do not commute. If A and B are non-commuting operators, then the exponents do not combine in the usual way:

$$e^A e^B \neq e^{A+B}. \tag{6.148}$$

So it will be helpful to recall some results about exponentials of non-commuting operators.

We introduce the commutator

$$[A, B] = AB - BA. \tag{6.149}$$

The commutator is bilinear and satisfies the Jacobi identity

$$[A, [B, C]] + [B, [C, A]] + [C, [A, B]] = 0, \tag{6.150}$$

which is true for all A, B, and C.

We introduce a notation Δ_A for the commutator with respect to the operator A:

$$\Delta_A B = [A, B]. \tag{6.151}$$

In terms of Δ the Jacobi identity is

$$[\Delta_A, \Delta_B] = \Delta_{[A,B]}. \tag{6.152}$$

An important identity is

$$e^C A e^{-C} = e^{\Delta_C} A$$
$$= A + [C, A] + \frac{1}{2}[C, [C, A]] + \cdots. \tag{6.153}$$

We can check this term by term.
   We see that

$$e^C A^2 e^{-C} = e^C A e^{-C} e^C A e^{-C} = \left(e^C A e^{-C}\right)^2, \tag{6.154}$$

using $e^{-C} e^C = I$, the identity operator. Using the same trick, we find

$$e^C A^n e^{-C} = \left(e^C A e^{-C}\right)^n. \tag{6.155}$$

More generally, if f can be represented as a power series then

$$e^C f(A, B, ...) e^{-C} = f(e^C A e^{-C}, e^C B e^{-C}, ...). \tag{6.156}$$

For instance, applying this to the exponential function yields

$$e^C e^A e^{-C} = e^{e^C A e^{-C}}. \tag{6.157}$$

Using equation (6.153), we can rewrite this as

$$e^{\Delta_C} e^A = e^{e^{\Delta_C} A}. \tag{6.158}$$

Exercise 6.10: Commutators of Lie derivatives

a. Let W and W' be two phase-space state functions. Use the Poisson-bracket Jacobi identity (3.93) to show

$$[L_W, L_{W'}] = -L_{\{W,W'\}}. \tag{6.159}$$

b. Consider the phase-space state functions that give the components of the angular momentum in terms of rectangular canonical coordinates

$$J_x(t; x, y, z; p_x, p_y, p_z) = y p_z - z p_y$$
$$J_y(t; x, y, z; p_x, p_y, p_z) = z p_x - x p_z$$
$$J_z(t; x, y, z; p_x, p_y, p_z) = x p_y - y p_x.$$

Show

$$[L_{J_x}, L_{J_y}] + L_{J_z} = 0. \tag{6.160}$$

c. Relate the Jacobi identity for operators to the Poisson-bracket Jacobi identity.

Exercise 6.11: Baker–Campbell–Hausdorff

Derive the rule for combining exponentials of non-commuting operators:

$$e^A e^B = e^{A+B+\frac{1}{2}[A,B]+\cdots}. \tag{6.161}$$

6.6 Summary

The time evolution of any Hamiltonian system induces a canonical transformation: if we consider all possible initial states of a Hamiltonian system and follow all of the trajectories for the same time interval, then the map from the initial state to the final state of each trajectory is a canonical transformation. This is true for any interval we choose, so time evolution generates a continuous family of canonical transformations.

We generalized this idea to generate continuous canonical transformations other than those generated by time evolution. Such transformations will be especially useful in support of perturbation theory.

In rare cases a canonical transformation can be made to a representation in which the problem is easily solvable: when all coordinates are cyclic and all the momenta are conserved. Here we investigated the Hamilton–Jacobi method for finding such canonical transformations. For problems for which the Hamilton–Jacobi method works, we find that the time evolution of the system is given as a canonical transformation.

6.7 Projects

Exercise 6.12: Symplectic integration

Consider a system for which the Hamiltonian H can be split into two parts, H_0 and H_1, each of which describes a system that can be efficiently evolved:

$$H = H_0 + H_1. \tag{6.162}$$

Symplectic integrators construct approximate solutions for the Hamiltonian H from those of H_0 and H_1.

We construct a map of the phase space onto itself in the following way (see [47, 48, 49]). Define $\delta_{2\pi}(t)$ to be an infinite sum of Dirac delta functions, with interval 2π,

$$\delta_{2\pi}(t) = \sum_{n=-\infty}^{\infty} \delta(t - 2\pi n), \tag{6.163}$$

with representation as a Fourier series

$$2\pi\delta_{2\pi}(t) = \sum_{n=-\infty}^{\infty} \cos(nt). \tag{6.164}$$

Recall that a δ function has the property that $\int_{-a}^{a} f\delta = f(0)$ for any positive a and continuous real-valued function f. It is fruitful to think of the delta function as a limit of a function Δ_h that has the value $\Delta_h(t) = 1/h$ in the interval $-h/2 < t < h/2$ and zero otherwise. Now consider the mapping Hamiltonian

$$H_m(t, q, p) = H_0(t, q, p) + 2\pi\delta_{2\pi}(\Omega t)H_1(t, q, p). \tag{6.165}$$

The evolution of the system between the delta functions is governed solely by H_0. To understand how the system evolves across the delta functions think of the delta functions in terms of Δ_h as h goes to zero. Hamilton's equations contain terms from H_1 with the factor $1/h$, which is large, and terms from H_0 that are independent of h. So as h goes to zero, H_0 makes a negligible contribution to the evolution. The evolution across the delta functions is governed solely by H_1. The evolution of H_m is obtained by alternately evolving the system according to the Hamiltonian H_0 for an interval $\Delta t = 2\pi/\Omega$ and then evolving the system according to the Hamiltonian H_1 for the same time interval. The longer-term evolution of H_m is obtained by iterating this map of the phase space onto itself. Fill in the details to show this.

a. In terms of Lie series, the evolution of H_m for one delta function cycle Δt is generated by

$$e^{\Delta t\, L_{H_m}} I = (e^{\Delta t\, L_{H_1}} I) \circ (e^{\Delta t\, L_{H_0}} I). \tag{6.166}$$

The evolution of H_m approximates the evolution of H. Identify the noncommuting operator A with L_{H_0} and B with L_{H_1}.

Use the Baker–Campbell–Hausdorff identity (equation 6.161) to deduce that the local truncation error (the error in the state after one step Δt) is proportional to $(\Delta t)^2$. This mapping is a first-order integrator.

b. By merely changing the phase of the delta functions, we can reduce the truncation error of the map, and the map becomes a second-order integrator. Instead of making a map by alternating a full step Δt governed by H_0 with a full step Δt governed by H_1, we can make a map by evolving the system for a half step $\Delta t/2$ governed by H_0, then for a full step Δt governed by H_1, and then for another half step $\Delta t/2$ governed by H_0. In terms of Lie series the second-order map is generated by

$$e^{\Delta t \, L_{H_m}} I = (e^{(\Delta t/2) \, L_{H_0}} I) \circ (e^{\Delta t \, L_{H_1}} I) \circ (e^{(\Delta t/2) \, L_{H_0}} I). \qquad (6.167)$$

Confirm that the Hamiltonian governing the evolution of this map is the same as the one above but with the phase of the delta functions shifted. Show that the truncation error of one step of this second-order map is indeed proportional to $(\Delta t)^3$.

c. Consider the Hénon–Heiles system. We can split the Hamiltonian (equation 3.135 on page 252) into two solvable Hamiltonians in the following way:

$$H_0(t; x, y; p_x, p_y) = (p_x^2 + p_y^2)/2 + (x^2 + y^2)/2$$
$$H_1(t; x, y; p_x, p_y) = x^2 y - y^3/3. \qquad (6.168)$$

Hamiltonian H_0 is the Hamiltonian of two uncoupled linear oscillators; Hamiltonian H_1 is a nonlinear coupling. The trajectories of the systems described by each of these Hamiltonians can be expressed in closed form, so we do not need the Lie series for actually integrating each part. The Lie series expansions are used only to determine the order of the integrator.

Write programs that implement first-order and second-order maps for the Hénon–Heiles problem. Note that these maps cannot be of the same form as the Poincaré maps that we used to make surfaces of section, because these maps must take and return the entire state. (Why?) An appropriate template for such a map is (`1st-order-map state dt`). This procedure must return a state.

d. Examine the evolution of the energy for both chaotic and quasiperiodic initial conditions. How does the magnitude of the energy error scale with the step size? Is this consistent with the order of the integrator deduced above? How does the energy error grow with time?

e. The generation of surfaces of section from these maps is complicated by the fact that these maps have to maintain their state even though a plotting point might be required between two samples. The maps you made in part **c** regularly sample the state with the integrator timestep. If we must plot a point between two steps we cannot restart the integrator at the state of the plotted point, because that would lose the phase of the integrator step. To make this work the map must plot points but keep its rhythm, so we have to work around the fact that `explore-map` restarts at each plotted point. Here is some code that can be used to construct a Poincaré-type map that can be used with the explorer:

```
(define ((HH-collector win advance E dt sec-eps n) x y done fail)
  (define (monitor last-crossing-state state)
    (plot-point win
                (ref (coordinate last-crossing-state) 1)
                (ref (momentum last-crossing-state) 1)))
  (define (pmap x y cont fail)
    (find-next-crossing y advance dt sec-eps cont))
  (define collector (default-collector monitor pmap n))
  (cond ((and (up? x) (up? y))              ;passed states
         (collector x y done fail))
        ((and (number? x) (number? y))   ;initialization
         (let ((initial-state (section->state E x y)))
           (if (not initial-state)
               (fail)
               (collector initial-state initial-state done fail))))
        (else (error "bad input to HH-collector" x y))))
```

You will notice that the iteration of the map and the plotting of the points is included in this collector, so the map that this produces must replace these parts of the explorer. The #f argument to explore-map allows us to replace the appropriate parts of the explorer with our combination map iterator and plotter HH-collector.

```
(explore-map win
  (HH-collector win 1st-order-map 0.125 0.1 1.e-10 1000)
  #f)
```

Generate surfaces of section using the second-order map. Does the map preserve the chaotic or quasiperiodic character of trajectories?

7

Canonical Perturbation Theory

> Having treated the motion of the moon about the earth, and having obtained an elliptical orbit, [Newton] considered the effect of the sun on the moon's orbit by taking into account the variations of the latter. However, the calculations caused him great difficulties ... Indeed, the problems he encountered were such that [Newton] was prompted to remark to the astronomer John Machin that "... his head never ached but with his studies on the moon."
>
> June Barrow-Green, *Poincaré and the Three Body Problem* [7], p. 15

Closed-form solutions of dynamical systems can be found only rarely. However, some systems differ from a solvable system by the addition of a small effect. The goal of perturbation theory is to relate aspects of the motion of the given system to those of the nearby solvable system. We can try to find a way to transform the exact solution of this approximate problem into an approximate solution to the original problem. We can also use perturbation theory to try to predict qualitative features of the solutions by describing the characteristic ways in which solutions of the solvable system are distorted by the additional effects. For instance, we might want to predict where the largest resonance regions are located or the locations and sizes of the largest chaotic zones. Being able to predict such features can give insight into the behavior of the particular system of interest.

Suppose, for example, we have a system characterized by a Hamiltonian that breaks up into two parts as follows:

$$H = H_0 + \epsilon H_1, \tag{7.1}$$

where H_0 is solvable and ϵ is a small parameter. The difference between our system and a solvable system is then a small additive complication.

There are a number of strategies for doing this. One strategy is to seek a canonical transformation that eliminates from the Hamiltonian the terms of order ϵ that impede solution—this typically introduces new terms of order ϵ^2. Then one seeks another canonical transformation that eliminates the terms of order ϵ^2 impeding solution, leaving terms of order ϵ^3. We can imagine repeating this process until the part that impedes solution is of such high order in ϵ that it can be neglected. Having reduced the problem to a solvable problem, we can reverse the sequence of transformations to find an approximate solution of the original problem. Does this process converge? How do we know we can ever neglect the remaining terms? Let's follow this path and see where it goes.

7.1   Perturbation Theory with Lie Series

Given a system, we look for a decomposition of the Hamiltonian in the form

$$H(t, q, p) = H_0(t, q, p) + \epsilon H_1(t, q, p), \tag{7.2}$$

where H_0 is solvable. We assume that the Hamiltonian has no explicit time dependence; this can be ensured by going to the extended phase space if necessary. We also assume that a canonical transformation has been made so that H_0 depends solely on the momenta:

$$\partial_1 H_0 = 0. \tag{7.3}$$

We carry out a Lie transformation and find the conditions that the Lie generator W must satisfy to eliminate the order ϵ terms from the Hamiltonian.

The Lie transform and associated Lie series specify a canonical transformation:

$$H' = E'_{\epsilon, W} \cdot H = e^{\epsilon L_W} H$$
$$q = (E'_{\epsilon, W} Q)(t, q', p') = (e^{\epsilon L_W} Q)(t, q', p')$$
$$p = (E'_{\epsilon, W} P)(t, q', p') = (e^{\epsilon L_W} P)(t, q', p')$$
$$(t, q, p) = (E'_{\epsilon, W} I)(t, q', p') = (e^{\epsilon L_W} I)(t, q', p'), \tag{7.4}$$

where $Q = I_1$ and $P = I_2$ are the coordinate and momentum selectors and I is the identity function. Recall the definitions

$$e^{\epsilon L_W} F = F + \epsilon L_W F + \frac{1}{2}\epsilon^2 L_W^2 F + \cdots$$

$$= F + \epsilon\{F, W\} + \frac{1}{2}\epsilon^2\{\{F, W\}, W\} + \cdots, \qquad (7.5)$$

with $L_W F = \{F, W\}$.

Applying the Lie transformation to H gives us

$$H' = e^{\epsilon L_W} H$$

$$= H_0 + \epsilon L_W H_0 + \frac{1}{2}\epsilon^2 L_W^2 H_0 + \cdots$$

$$+ \epsilon H_1 + \epsilon^2 L_W H_1 + \cdots$$

$$= H_0 + \epsilon\left(L_W H_0 + H_1\right) + \epsilon^2\left(\frac{1}{2}L_W^2 H_0 + L_W H_1\right) + \cdots. \qquad (7.6)$$

The first-order term in ϵ is zero if W satisfies the condition

$$L_W H_0 + H_1 = 0, \qquad (7.7)$$

which is a linear partial differential equation for W. The transformed Hamiltonian is

$$H' = H_0 + \epsilon^2\left(\frac{1}{2}L_W^2 H_0 + L_W H_1\right) + \cdots$$

$$= H_0 + \frac{1}{2}\epsilon^2 L_W H_1 + \cdots, \qquad (7.8)$$

where we have used condition (7.7) to simplify the ϵ^2 contribution.

This basic step of perturbation theory has eliminated terms of a certain order (order ϵ) from the Hamiltonian, but in doing so has generated new terms of higher order (here ϵ^2 and higher).

At this point we can find an approximate solution by truncating Hamiltonian (7.8) to H_0, which is solvable. The approximate solution for given initial conditions $s_0 = (t_0, q_0, p_0)$ is obtained by finding the corresponding (t_0, q_0', p_0') using the inverse of transformation (7.4). Then the system is evolved to time t using the solutions of the truncated Hamiltonian H_0, giving the state (t, q', p'). The phase-space coordinates of the evolved point are transformed back to the original variables using the transformation (7.4) to state $s = (t, q, p)$. The approximate solution is

$$s = ((E'_{\epsilon,W} I) \circ (E_{t-t_0, H_0} I) \circ (E'_{-\epsilon, W} I))(s_0)$$
$$= (E'_{-\epsilon, W} E_{t-t_0, H_0} E'_{\epsilon, W} I)(s_0)$$
$$= (e^{-\epsilon L w} e^{(t-t_0) D_{H_0}} e^{\epsilon L w} I)(s_0), \qquad (7.9)$$

using the identity (6.111). Notice that the time evolution of H_0 is expressed in terms of the evolution operator E rather than the Lie-transform operator E', because the time must also be advanced. The power-series expansion for $E_{\Delta t, H_0}$ is expressed in terms of D_{H_0} rather than L_{H_0} (see 6.136). If the Lie transform $E'_{\epsilon,W} = e^{\epsilon L w}$ must be evaluated by summing the series, then we must specify the order to which the sum extends.

Assuming everything goes okay, we can imagine repeating this process to eliminate the order ϵ^2 terms and so on, bringing the transformed Hamiltonian as close as we like to H_0. Unfortunately, there are complications. We can understand some of these complications and how to deal with them by considering some specific applications.

7.2 Pendulum as a Perturbed Rotor

The pendulum is a simple one-degree-of-freedom system, for which the solutions are known. If we consider the pendulum as a free rotor with the added complication of gravity, then we can carry out a perturbation step as just described to see how well it approximates the known motion of the pendulum.

The motion of a pendulum is described by the Hamiltonian

$$H(t, \theta, p) = \frac{p^2}{2\alpha} - \epsilon\beta \cos(\theta), \qquad (7.10)$$

with coordinate θ and conjugate angular momentum p, where $\alpha = ml^2$ and $\beta = mgl$. The parameter ϵ allows us to scale the perturbation; it is 1 for the actual pendulum. We divide the Hamiltonian into the free-rotor Hamiltonian and the perturbation from gravity:

$$H = H_0 + \epsilon H_1, \qquad (7.11)$$

where

$$H_0(t, \theta, p) = \frac{p^2}{2\alpha}$$
$$\epsilon H_1(t, \theta, p) = -\epsilon \beta \cos \theta. \tag{7.12}$$

The Lie generator W satisfies condition (7.7):

$$\{H_0, W\} + H_1 = 0, \tag{7.13}$$

or

$$-\frac{p}{\alpha}\partial_1 W(t, \theta, p) - \beta \cos \theta = 0. \tag{7.14}$$

So

$$W(t, \theta, p) = -\frac{\alpha\beta \sin \theta}{p}, \tag{7.15}$$

where the arbitrary integration constant is ignored.

The transformed Hamiltonian is $H' = H_0 + o(\epsilon^2)$. If we can ignore the ϵ^2 contributions, then the transformed Hamiltonian is simply

$$H'(t, \theta', p') = \frac{(p')^2}{2\alpha}, \tag{7.16}$$

with solutions

$$\theta' = \theta_0' + \frac{p_0'}{\alpha}(t - t_0)$$
$$p' = p_0'. \tag{7.17}$$

To connect these solutions to the solutions of the original problem, we use the Lie series

$$\theta = (e^{\epsilon L_W} Q)(t, \theta', p')$$
$$= \theta' + \epsilon\{Q, W\}(t, \theta', p') + \cdots$$
$$= \theta' + \epsilon \partial_2 W(t, \theta', p') + \cdots$$
$$= \theta' + \epsilon \frac{\alpha\beta \sin \theta'}{(p')^2} + \cdots. \tag{7.18}$$

Similarly,

$$p = p' + \epsilon \frac{\alpha \beta \cos \theta'}{p'} + \cdots. \qquad (7.19)$$

Note that if the Lie series is truncated it is not exactly a canonical transformation; only the infinite series is canonical.

The initial values θ_0' and p_0' are determined from the initial values of θ and p by the inverse Lie transformation:

$$\begin{aligned} \theta' &= (e^{-\epsilon L_W} Q)(t, \theta, p) \\ &= \theta - \epsilon \frac{\alpha \beta \sin \theta}{(p)^2} + \cdots \end{aligned} \qquad (7.20)$$

and

$$p' = p - \epsilon \frac{\alpha \beta \cos \theta}{p} + \cdots. \qquad (7.21)$$

Note that if we truncate the coordinate transformations after the first-order terms in ϵ (or any finite order), then the inverse transformation is not exactly the inverse of the transformation.

The approximate solution for given initial conditions (t_0, θ_0, p_0) is obtained by finding the corresponding (t_0, θ_0', p_0') using the transformation (7.20) and (7.21). Then the system is evolved using the solutions (7.17). The phase-space coordinates of the evolved point are transformed back to the original variables using the transformation (7.18) and (7.19).

We define the two parts of the pendulum Hamiltonian:

```
(define ((H0 alpha) state)
  (let ((p (momentum state)))
    (/ (square p) (* 2 alpha))))

(define ((H1 beta) state)
  (let ((theta (coordinate state)))
    (* -1 beta (cos theta))))
```

The Hamiltonian for the pendulum can be expressed as a series expansion in the parameter ϵ by

```
(define (H-pendulum-series alpha beta epsilon)
  (series (H0 alpha) (* epsilon (H1 beta))))
```

where the `series` procedure is a constructor for a series whose first terms are given and all further terms are zero. The Lie generator that eliminates the order ϵ terms is

```
(define ((W alpha beta) state)
  (let ((theta (coordinate state))
        (p (momentum state)))
    (/ (* -1 alpha beta (sin theta)) p)))
```

We check that W satisfies condition (7.7):

```
((+ ((Lie-derivative (W 'alpha 'beta)) (H0 'alpha))
    (H1 'beta))
 (up 't 'theta 'p))
0
```

and that it has the desired effect on the Hamiltonian:

```
(show-expression
 (series:sum
  (((exp (* 'epsilon (Lie-derivative (W 'alpha 'beta))))
    (H-pendulum-series 'alpha 'beta 'epsilon))
   (up 't 'theta 'p))
  2))
```

$$\frac{\frac{1}{2}p^2}{\alpha} + \frac{\frac{1}{2}\alpha\beta^2\epsilon^2 \left(\sin\left(\theta\right)\right)^2}{p^2}$$

Indeed, the order ϵ term has been removed and an order ϵ^2 term has been introduced.

Ignoring the ϵ^2 terms in the new Hamiltonian, the solution is

```
(define (((solution0 alpha beta) t) state0)
  (let ((t0 (time state0))
        (theta0 (coordinate state0))
        (p0 (momentum state0)))
    (up t
        (+ theta0 (/ (* (- t t0) p0) alpha))
        p0)))
```

The transformation from primed to unprimed phase-space coordinates is, including terms up to order,

```
(define ((C alpha beta epsilon order) state)
  (series:sum
   (((Lie-transform (W alpha beta) epsilon)
     identity)
    state)
   order))
```

To second order in ϵ the transformation generated by W is

```
(show-expression
  ((C 'alpha 'beta 'epsilon 2) (up 't 'theta 'p)))
```

$$
\begin{pmatrix}
t \\[6pt]
-\dfrac{\frac{1}{2}\alpha^2\beta^2\epsilon^2 \cos(\theta)\sin(\theta)}{p^4} + \dfrac{\alpha\beta\epsilon\sin(\theta)}{p^2} + \theta \\[14pt]
-\dfrac{\frac{1}{2}\alpha^2\beta^2\epsilon^2}{p^3} + \dfrac{\alpha\beta\epsilon\cos(\theta)}{p} + p
\end{pmatrix}
$$

The inverse transformation is

```
(define (C-inv alpha beta epsilon order)
  (C alpha beta (- epsilon) order))
```

With these components, the perturbative solution (equation 7.9) is

```
(define (((solution epsilon order) alpha beta) delta-t)
  (compose (C alpha beta epsilon order)
           ((solution0 alpha beta) delta-t)
           (C-inv alpha beta epsilon order)))
```

The resulting procedure maps an initial state to the solution state advanced by `delta-t`.

We can examine the behavior of the perturbative solution and compare it to the true behavior of the pendulum. There are several considerations. We have truncated the Lie series for the phase-space transformation. Does the missing part matter? If the missing part does not matter, how well does this perturbation step work?

Figure 7.1 shows that as we increase the number of terms in the Lie series for the phase-space coordinate transformation the result appears to converge. The lone trajectory includes only terms of first order. The others, including terms of second, third, and fourth order, are closely clustered. On the left edge of the graph (at $\theta = -\pi$), the order of the solution increases from the top to the bottom of the graph. In the middle (at $\theta = 0$), the fourth-order curve is between the second-order curve and the third-order curve. In addition to the error in phase-space path, there is also an error in the period—the higher-order orbits have longer peri-

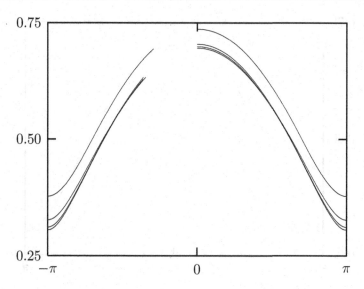

Figure 7.1   The perturbative solution in the phase plane, including terms of first, second, third, and fourth order in the phase-space coordinate transformation. The solutions appear to converge.

ods than the first-order orbit. The parameters are $\alpha = 1.0$ and $\beta = 0.1$. We have set $\epsilon = 1$. Each trajectory was started at $\theta = 0$ with $p = 0.7$. Notice that the initial point on the solution varies between trajectories. This is because the transformation is not perfectly inverted by the truncated Lie series.

Figure 7.2 compares the perturbative solution (with terms up to fourth order) with the actual trajectory of the pendulum. The initial points coincide, to the precision of the graph, because the terms to fourth order are sufficient. The trajectories deviate both in the phase plane and in the period, but they are still quite close.

The trajectories of figures 7.1 and 7.2 are all for the same initial state. As we vary the initial state we find that for trajectories in the circulation region, far from the separatrix, the perturbative solution does quite well. However, if we get close to the separatrix or if we enter the oscillation region, the perturbative solution is nothing like the real solution, and it does not even seem to converge. Figure 7.3 shows what happens when we try to use the perturbative solution inside the oscillation region. Each trajectory was started at $\theta = 0$ with $p = 0.55$. The parameters are $\alpha = 1.0$ and $\beta = 0.1$, as before.

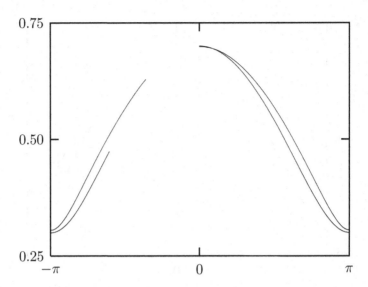

Figure 7.2   The perturbative solution in the phase plane, including terms of fourth order in the phase-space coordinate transformation, is compared with the actual trajectory. The actual trajectory is the lower of the two curves. The parameters are the same as in figure 7.1.

This failure of the perturbation solution should not be surprising. We assumed that the real motion was a distorted version of the motion of the free rotor. But in the oscillation region the assumption is not true—the pendulum is not rotating at all. The perturbative solutions can be valid (if they work at all!) only in a region where the topology of the real orbits is the same as the topology of the perturbative solutions.

We can make a crude estimate of the range of validity of the perturbative solution by looking at the first correction term in the phase-space transformation (7.18). The correction in θ is proportional to $\epsilon\alpha\beta/(p')^2$. This is not a small perturbation if

$$|p'| < \sqrt{\epsilon\alpha\beta}. \tag{7.22}$$

This sets the scale for the validity of the perturbative solution.

We can compare this scale to the size of the oscillation region (see figure 7.4). We can obtain the width of the region of oscillation of the pendulum[1] by considering the separatrix. The value

---

[1]The "width" is measured as the range of momenta.

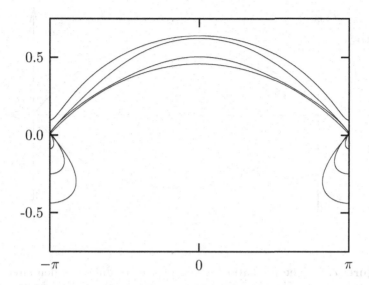

Figure 7.3   The perturbative solution does not converge in the oscillation region. As we include more terms in the Lie series for the phase-space transformation, the resulting trajectory develops loops near the hyperbolic fixed point that increase in size with the order.

of the Hamiltonian on the separatrix is the same as the value at the unstable equilibrium: $H(t, \theta = \pi, p = 0) = \beta\epsilon$. The separatrix has maximum momentum p^{sep} at $\theta = 0$:

$$H(t, 0, p^{\text{sep}}) = H(t, \pi, 0). \tag{7.23}$$

Solving for p^{sep}, the half-width of the region of oscillation, we find

$$p^{\text{sep}} = 2\sqrt{\alpha\beta\epsilon}. \tag{7.24}$$

Comparing equations (7.22) and (7.24), we see that the requirement that the terms in the perturbation solution be small excludes a region of the phase space with the same scale as the region of oscillation of the pendulum.

What the perturbation theory is doing is deforming the phase-space coordinate system so that the problem looks like the free-rotor problem. This deformation is sensible only in the circulating case. So, it is not surprising that the perturbation theory fails in the oscillation region. What may be surprising is how well the perturbation theory works just outside the oscillation region. The

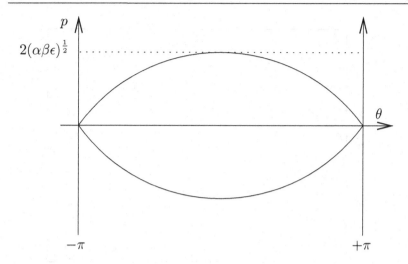

Figure 7.4 The oscillation region of the pendulum is delimited by the separatrix. The maximum momentum occurs at the zero-crossing of the angle. Energy is conserved, so the energy is the same at the point of maximum momentum and at the unstable fixed point. At the unstable fixed point the energy is entirely potential energy, because the momentum is zero. We use this to compute the maximum momentum (where the potential energy is zero and all of the energy is kinetic).

range of p in which the perturbation theory is not valid scales in the same way as the width of the oscillation region. This need not have been the case—the perturbation theory could have failed over a wider range.

Exercise 7.1: Symplectic residual

For the transformation (`C alpha beta epsilon order`), compute the residuals in the symplectic test for various orders of truncation of the Lie series.

7.2.1 Higher Order

We can improve the perturbative solution by carrying out additional perturbation steps. The overall plan is the same as before. We perform a Lie transformation with a new generator that eliminates the desired terms from the Hamiltonian.

After the first step the Hamiltonian is, to second order in ϵ,

$$H'(t, \theta', p') = \frac{(p')^2}{2\alpha} + \epsilon^2 \frac{\alpha \beta^2}{2(p')^2} (\sin \theta')^2 + \cdots$$

$$= \frac{(p')^2}{2\alpha} + \epsilon^2 \frac{\alpha \beta^2}{4(p')^2} (1 - \cos(2\theta')) + \cdots$$

$$= H_0(p') + \epsilon^2 H_2(t, \theta', p') + \cdots. \tag{7.25}$$

Performing a Lie transformation with generator W' yields the Hamiltonian

$$H'' = e^{\epsilon^2 L_{W'}} H'$$

$$= H_0 + \epsilon^2 (L_{W'} H_0 + H_2) + \cdots. \tag{7.26}$$

So the condition on W' that the second-order terms are eliminated is

$$L_{W'} H_0 + H_2 = 0. \tag{7.27}$$

This is

$$-\frac{p'}{\alpha} \partial_1 W'(t, \theta', p') + \frac{\alpha \beta^2}{4(p')^2} (1 - \cos(2\theta')) = 0. \tag{7.28}$$

A generator that satisfies this condition is

$$W'(t, \theta', p') = \frac{\alpha^2 \beta^2}{4(p')^3} \theta' + \frac{\alpha^2 \beta^2}{8(p')^3} \sin(2\theta'). \tag{7.29}$$

There are two contributions to this generator, one proportional to θ' and the other involving a trigonometric function of θ'.

The phase-space coordinate transformation resulting from this Lie transform is found as before. For given initial conditions, we first carry out the inverse transformation corresponding to W, then that for W', solve for the evolution of the system using H_0, then transform back using W' and then W. For initial state $s_0 = (t_0, \theta_0, p_0)$ and advanced state $s = (t, \theta, p)$, the approximate solution is

$$s = (E'_{-\epsilon, W} E'_{-\epsilon^2, W'} E_{(t-t_0), H_0} E'_{\epsilon^2, W'} E'_{\epsilon, W} I)(s_0)$$

$$= (e^{-\epsilon L_W} e^{-\epsilon^2 L_{W'}} e^{(t-t_0) D_{H_0}} e^{\epsilon^2 L_{W'}} e^{\epsilon L_W} I)(s_0). \tag{7.30}$$

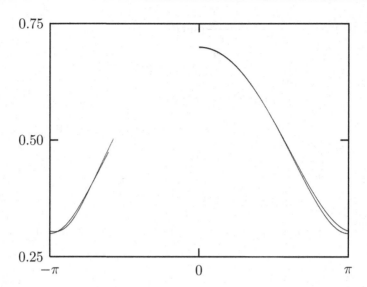

Figure 7.5   The solution using a second perturbation step, eliminating ϵ^2 terms from the Hamiltonian, is compared to the actual solution. The initial agreement is especially good, but the error increases with time.

The solution obtained in this way is compared to the actual evolution of the pendulum in figure 7.5. Terms in all Lie series up to ϵ^4 are included. The perturbative solution, including this second perturbative step, is much closer to the actual solution in the initial segment than the first-order perturbative solution (figure 7.2). The time interval spanned is 10. Over longer times the second-order perturbative solution diverges dramatically from the actual solution, as shown in figure 7.6. These solutions begin at $\theta = 0$ with $p = 0.7$. The parameters are $\alpha = 1.0$ and $\beta = 0.1$. The time interval spanned is 100.

A problem with the perturbative solution is that there are terms in W' and in the corresponding phase-space coordinate transformation that are proportional to θ', and θ' grows linearly with time. So the solution can be valid only for small times; the interval of validity depends on the frequency of the particular trajectory under investigation and the size of the coefficients multiplying the various terms. Such terms in a perturbative representation of the solution that are proportional to time are called *secular terms*. They limit the validity of the perturbation theory to small times.

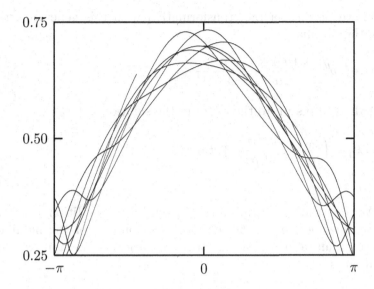

Figure 7.6 The two-step perturbative solution is shown over a longer time. The actual solution is a closed curve in the phase plane; this perturbative solution wanders all over the place and gets worse with time.

7.2.2 Eliminating Secular Terms

A solution to the problem of secular terms was developed by Lindstedt and Poincaré. The goal of each perturbation step is to eliminate terms in the Hamiltonian that prevent solution. However, the term in H' that led to the secular term in the generator W' does not actually impede solution. So a better procedure is to leave that term in the Hamiltonian and find the generator W'' that eliminates only the term that is periodic in θ'. So W'' must satisfy

$$-\frac{p'}{\alpha}\partial_1 W''(t,\theta',p') - \frac{\alpha\beta^2}{4(p')^2}\cos(2\theta') = 0. \tag{7.31}$$

The generator is

$$W''(t,\theta',p') = \frac{\alpha^2\beta^2}{8(p')^3}\sin(2\theta'). \tag{7.32}$$

After we perform a Lie transformation with this generator, the new Hamiltonian is

$$H''(t, \theta'', p'') = \frac{(p'')^2}{2\alpha} + \epsilon^2 \frac{\alpha\beta^2}{4(p'')^2} + \cdots. \tag{7.33}$$

Including terms up to the ϵ^2 term, the solution is

$$\theta'' = \theta_0'' + \left(\frac{p_0''}{\alpha} - \epsilon^2 \frac{\alpha\beta^2}{2(p_0'')^3} \right) (t - t_0)$$

$$p'' = p_0''. \tag{7.34}$$

We construct the solution for a given initial condition as before by composing the transformations, the solution of the modified Hamiltonian, and the inverse transformations. The approximate solution is

$$(t, \theta, p) = (E'_{-\epsilon, W} E'_{-\epsilon^2, W''} E_{(t-t_0), H''} E'_{\epsilon^2, W''} E'_{\epsilon, W} I)(t_0, \theta_0, p_0)$$

$$= (e^{-\epsilon L w} e^{-\epsilon^2 L w''} e^{(t-t_0) D_{H''}} e^{\epsilon^2 L w''} e^{\epsilon L w} I)(t_0, \theta_0, p_0). \tag{7.35}$$

The resulting phase-space evolution is shown in figure 7.7. Now the perturbative solution is a closed curve in the phase plane and is in pretty good agreement with the actual solution.

By modifying the solvable part of the Hamiltonian we are modifying the frequency of the solution. The secular terms appeared because we were trying to approximate a solution with one frequency as a Fourier series with the wrong frequency. As an analogy, consider

$$\sin(\omega + \Delta\omega)t = \sin \omega t \cos \Delta\omega t + \cos \omega t \sin \Delta\omega t$$

$$= \sin \omega t \left(1 - \frac{(\Delta\omega t)^2}{2} + \cdots \right)$$

$$+ \cos \omega t \left(\Delta\omega t + \cdots \right). \tag{7.36}$$

The periodic terms are multiplied by terms that are polynomials in the time. These polynomials are the initial segment of the power series for periodic functions. The infinite series are convergent, but if the series are truncated the error is large at large times.

Continuing the perturbative solution to higher orders is now a straightforward repetition of the steps carried out so far. At each step in the perturbation solution there will be new contributions to

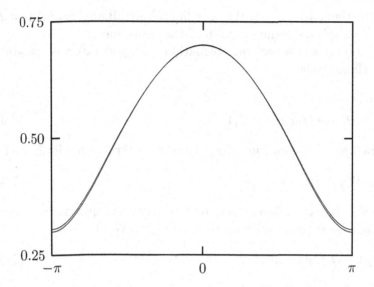

Figure 7.7   The two-step perturbative solution without secular terms is compared to the actual solution. The perturbative solution is now a closed curve and is very close to the actual solution.

the solvable part of the Hamiltonian that absorb potential secular terms. The contribution is just the angle-independent part of the Hamiltonian after the Hamiltonian is written as a Fourier series. The constant part of the Fourier series is the same as the average of the Hamiltonian over the angle. So at each step in the perturbation theory, the average of the perturbation is included with the solvable part of the Hamiltonian and the periodic part is eliminated by a Lie transformation.

7.3   Many Degrees of Freedom

Other problems are encountered in applying perturbation theory to systems with more than a single degree of freedom. Consider an n degrees-of-freedom Hamiltonian of the form

$$H = H_0 + \epsilon H_1, \tag{7.37}$$

where H_0 depends only on the momenta and therefore is solvable. We assume that the Hamiltonian has no explicit time dependence.

We further assume that the coordinates are all angles and that H_1 is a multiply periodic function of the coordinates.

Carrying out a Lie transformation with generator W produces the Hamiltonian

$$H' = e^{\epsilon L_W} H$$
$$= H_0 + \epsilon (L_W H_0 + H_1) + \cdots, \tag{7.38}$$

as before. The condition that the order ϵ terms are eliminated is

$$\{H_0, W\} + H_1 = 0, \tag{7.39}$$

a linear partial differential equation. By assumption, the Hamiltonian H_0 depends only on the momenta. We define

$$\omega_0(p) = \partial_2 H_0(t, \theta, p), \tag{7.40}$$

where $\theta = (\theta^0, \ldots, \theta^{n-1})$, and $p = [p_0, \ldots, p_{n-1}]$. So $\omega_0(p)$ is the up tuple of frequencies of the unperturbed system. The condition on W is

$$\omega_0(p)\partial_1 W(t, \theta, p) = H_1(t, \theta, p). \tag{7.41}$$

As H_1 is a multiply periodic function of the coordinates, we can write it as a Poisson series:[2]

$$H_1(t, \theta, p) = \sum_k A_k(p) \cos(k \cdot \theta), \tag{7.42}$$

where $k = [k_0, \ldots, k_{n-1}]$ ranges over all n-tuples of integers. Similarly, we assume W can be written as a Poisson series:

$$W(t, \theta, p) = \sum_k B_k(p) \sin(k \cdot \theta). \tag{7.43}$$

Substituting these into the condition that order ϵ terms are eliminated, we find

$$\sum_k B_k(p)(k \cdot \omega_0(p)) \cos(k \cdot \theta) = \sum_k A_k(p) \cos(k \cdot \theta). \tag{7.44}$$

---

[2]In general, we need to include sine terms as well, but the cosine expansion is enough for this illustration.

The cosines are orthogonal so the coefficients of corresponding cosine terms must be equal:

$$B_k(p) = \frac{A_k(p)}{k \cdot \omega_0(p)} \tag{7.45}$$

and that the required Lie generator is

$$W(t, \theta, p) = \sum_k \frac{A_k(p)}{k \cdot \omega_0(p)} \sin(k \cdot \theta). \tag{7.46}$$

There are a couple of problems. First, if $A_{0,...,0}$ is nonzero then the expression for $B_{0,...,0}$ involves a division by zero. So the expression for $B_{0,...,0}$ is not correct. The problem is that the corresponding term in H_1 does not involve θ. So the integration for $B_{0,...,0}$ should introduce linear terms in θ. But this is the same situation that led to the secular terms in the perturbation approximation to the pendulum. Having learned our lesson there, we avoid the secular terms by adjoining this term to the solvable Hamiltonian and excluding $k = [0, \ldots, 0]$ from the sum for W. We have

$$H' = H_0 + \epsilon A_{0,...,0} + \cdots, \tag{7.47}$$

and

$$W(t, \theta, p) = \sum_{k \neq [0,...,0]} \frac{A_k(p)}{k \cdot \omega_0(p)} \sin(k \cdot \theta). \tag{7.48}$$

Another problem is that there are many opportunities for small denominators that would make the perturbation large and therefore not a perturbation. As we saw in the perturbation approximation for the pendulum in terms of the rotor, we must exclude certain regions from the domain of applicability of the perturbation approximation. These excluded regions are associated with commensurabilities among the frequencies $\omega_0(p)$. Consider the phase-space transformation of the coordinates

$$\begin{aligned}
\theta &= \left(e^{\epsilon L_W} Q\right)(t, \theta', p') \\
&= \theta' + \epsilon \partial_2 W(t, \theta', p') + \cdots \\
&= \theta' + \epsilon \sum_{k \neq [0,...,0]} \left(\frac{DA_k(p')}{k \cdot \omega_0(p')} - \frac{A_k(p')(k \cdot D\omega(p'))}{(k \cdot \omega_0(p'))^2}\right) \sin(k \cdot \theta).
\end{aligned} \tag{7.49}$$

We must exclude from the domain of applicability all regions for which the coefficients are large. If the second term in the coefficient of sin dominates, the excluded regions satisfy

$$|(k \cdot D\omega(p')) A_k(p)| > (k \cdot \omega_0(p))^2. \tag{7.50}$$

Considering the fact that for any tuple of frequencies $\omega_0(p')$ we can find a tuple of integers k such that $k \cdot \omega(p')$ is arbitrarily small, this problem of small divisors looks very serious.

However, the problem, though serious, is not as bad as it may appear, for a couple of reasons. First, it may be that $A_k \neq 0$ only for certain k. In this case, only the regions for these terms are excluded from the domain of applicability. Second, for analytic functions the magnitude of A_k decreases strongly with the size of k (see [4]):

$$|A_k(p')| \leq Ce^{-\beta|k|_+}, \tag{7.51}$$

for some positive β and C, and where $|k|_+ = |k_0| + |k_1| + \cdots$. At any stage of a perturbation approximation we can limit consideration to just those terms that are larger than a specified magnitude. The size of the excluded region corresponding to a term is of order square root of $|A_k(p')|$ and the inequality (7.51) shows that $|A_k(p')|$ decreases exponentially with the order of the term.

7.3.1 Driven Pendulum as a Perturbed Rotor

More concretely, consider the periodically driven pendulum. We will develop approximate solutions for the driven pendulum as a perturbed rotor.

We use the Hamiltonian

$$H(t, \theta, p) = \frac{p^2}{2ml^2} - \epsilon ml(g - A\omega^2 \cos(\omega t)) \cos \theta. \tag{7.52}$$

For a real driven pendulum $\epsilon = 1$; here it is used to help organize the computation. We will see that it need not be small and can be set to 1 at the end. We can remove the explicit time dependence by going to the extended phase space. The Hamiltonian is

$$H(\tau; \theta, t; p, p_t) \tag{7.53}$$

$$= p_t + \frac{p^2}{2ml^2} - \epsilon ml(g - A\omega^2 \cos(\omega t)) \cos \theta$$

$$= p_t + \frac{p^2}{2\alpha} - \epsilon \beta \cos(\theta) + \epsilon \gamma \cos(\theta + \omega t) + \epsilon \gamma \cos(\theta - \omega t),$$

with the constants $\alpha = ml^2$, $\beta = mlg$, and $\gamma = \frac{1}{2}mlA\omega^2$.

With the intent to approximate the driven pendulum as a perturbed rotor, we choose

$$H_0(\tau; \theta, t; p, p_t) = p_t + \frac{p^2}{2\alpha}$$
$$H_1(\tau; \theta, t; p, p_t) = -\beta \cos \theta + \gamma \cos(\theta + \omega t) + \gamma \cos(\theta - \omega t). \tag{7.54}$$

The perturbation H_1 is particularly simple: it has only three terms, and the coefficients are constants. Because H_1 has only three terms in its Poisson series, only three regions will be excluded from the domain of applicability in the first perturbation step.

The Lie series generator that eliminates the terms in H_1 to first order in ϵ, satisfying

$$\{H_0, W\} + H_1 = 0, \tag{7.55}$$

is

$$W(\tau; \theta, t; p, p_t) = -\frac{\beta}{\omega_r(p)} \sin \theta$$

$$+ \frac{\gamma}{\omega_r(p) + \omega} \sin(\theta + \omega t)$$

$$+ \frac{\gamma}{\omega_r(p) - \omega} \sin(\theta - \omega t), \tag{7.56}$$

where $\omega_r(p) = \partial_{2,0} H_0(\tau; \theta, t; p, p_t) = p/\alpha$ is the unperturbed rotor frequency.

The resulting approximate solution has three regions in which there are small denominators, and so three regions that are excluded from applicability of the perturbative solution. Regions of phase space for which $\omega_r(p)$ is near 0, ω, and $-\omega$ are excluded. Away from these regions the perturbative solution works well, just as in the rotor approximation for the pendulum. Unfortunately, some of the more interesting regions of the phase space of the driven pendulum are excluded: the region in which we find the remnant of the undriven pendulum is excluded, as are the two resonance regions in which the rotation of the pendulum is synchronous with the drive. We need to develop methods for approximating these regions.

7.4   Nonlinear Resonance

We can develop an approximation for an isolated resonance region as follows. We again consider Hamiltonians of the form

$$H = H_0 + \epsilon H_1, \tag{7.57}$$

where $H_0(t, q, p) = \hat{H}_0(p)$ depends only on the momenta and so is solvable. We assume that the Hamiltonian has no explicit time dependence. We further assume that the coordinates are all angles, and that H_1 is a multiply periodic function of the coordinates that can be written

$$H_1(t, \theta, p) = \sum_k A_k(p) \cos(k \cdot \theta). \tag{7.58}$$

Suppose we are interested in a region of phase space for which $n \cdot \omega_0(p)$ is near zero, where n is a tuple of integers, one for each degree of freedom. If we develop the perturbation theory as before with the generator W that eliminates all terms of order ϵ, then the transformed Hamiltonian is H_0, which is analytically solvable, but there would be terms with $n \cdot \omega_0(p)$ in the denominator. The resulting solution is not applicable near this resonance.

Just as the problem of secular terms was solved by grouping more terms with the solvable part of the Hamiltonian, we can develop approximations that are valid in the resonance region by eliminating fewer terms and grouping more terms in the solvable part.

To develop a perturbative approximation in the resonance region for which $n \cdot \omega_0(p)$ is near zero, we take the generator W to be

$$W_n(t, \theta, p) = \sum_{k \neq 0, k \neq n} \frac{A_k(p)}{k \cdot \omega_0(p)} \sin(k \cdot \theta), \tag{7.59}$$

excluding terms in W that lead to small denominators in this region. The transformed Hamiltonian is

$$H'_n(t, \theta, p) = \hat{H}_0(p) + \epsilon A_0(p) + \epsilon A_n(p) \cos(n \cdot \theta) + \cdots, \qquad (7.60)$$

where the additional terms are higher-order in ϵ. Because the term $k = n$ is excluded from the sum in the generating function, that term is left after the transformation.

The transformed Hamiltonian depends only on a single combination of angles, so a change of variables can be made so that the new transformed Hamiltonian is cyclic in all but one coordinate, which is this combination of angles. This transformed Hamiltonian is solvable (reducible to quadratures).

For example, suppose there are two degrees of freedom $\theta = (\theta_1, \theta_2)$ and we are interested in a region of phase space in which $n \cdot \omega_0$ is near zero, with $n = [n_1, n_2]$. The combination of angles $n \cdot \theta$ is slowly varying in the resonance region. The transformed Hamiltonian (7.60) is of the form

$$H'_n(t; \theta_1, \theta_2; p_1, p_2) = \hat{H}_0(p_1, p_2) + \epsilon A_0(p_1, p_2)$$
$$+ \epsilon A_n(p_1, p_2) \cos(n_1 \theta_1 + n_2 \theta_2). \qquad (7.61)$$

We can transform variables to $\sigma = n_1 \theta_1 + n_2 \theta_2$, with second coordinate, say, $\theta' = \theta_2$.[3] Using the F_2-type generating function

$$F_2(t; \theta_1, \theta_2; \Sigma, \Theta') = (n_1 \theta_1 + n_2 \theta_2)\Sigma + \theta_2 \Theta', \qquad (7.62)$$

we find that the transformation is

$$p_1 = n_1 \Sigma$$
$$p_2 = n_2 \Sigma + \Theta'$$
$$\sigma = n_1 \theta_1 + n_2 \theta_2$$
$$\theta' = \theta_2. \qquad (7.63)$$

In these variables the transformed resonance Hamiltonian H'_n becomes

$$H''_n(t; \sigma, \theta'; \Sigma, \Theta') = \hat{H}_0(n_1 \Sigma, n_2 \Sigma + \Theta') + \epsilon A_0(n_1 \Sigma, n_2 \Sigma + \Theta')$$
$$+ \epsilon A_n(n_1 \Sigma, n_2 \Sigma + \Theta') \cos(\sigma). \qquad (7.64)$$

---

[3] Any linearly independent combination will be acceptable here.

This Hamiltonian is cyclic in θ', so Θ' is constant. With this constant momentum, the Hamiltonian for the conjugate pair (σ, Σ) has one degree of freedom. The solutions are level curves of the Hamiltonian. These solutions, reexpressed in terms of the original phase-space coordinates, give the evolution of H'_n. An approximate solution in the resonance region is therefore

$$(t, \theta, p) = (E'_{-\epsilon, W'_n} E_{t-t_0, H'_n} E'_{\epsilon, W'_n} I)(t_0, \theta_0, p_0). \tag{7.65}$$

If the resonance regions are sufficiently separated, then a global solution can be constructed by splicing together such solutions for each resonance region.

7.4.1 Pendulum Approximation

The resonance Hamiltonian (7.64) has a single degree of freedom and is therefore solvable (reducible to quadratures). We can develop an approximate analytic solution in the vicinity of the resonance by making use of the fact that the solution is valid there. The resonance Hamiltonian can be approximated by a generalized pendulum Hamiltonian.

Let

$$\begin{aligned} H''_{n,0}&(t; \sigma, \theta'; \Sigma, \Theta') \\ &= \hat{H}_0(n_1\Sigma, n_2\Sigma + \Theta') + \epsilon A_0(n_1\Sigma, n_2\Sigma + \Theta') \end{aligned} \tag{7.66}$$

and

$$H''_{n,1}(t; \sigma, \theta'; \Sigma, \Theta') = A_n(n_1\Sigma, n_2\Sigma + \Theta')\cos(\sigma); \tag{7.67}$$

then the resonance Hamiltonian is

$$H''_n = H''_{n,0} + \epsilon H''_{n,1}. \tag{7.68}$$

Define the resonance center Σ_n by the requirement that the resonance frequency be zero there:

$$\partial_{2,0} H''_{n,0}(t; \sigma, \theta'; \Sigma_n, \Theta') = 0. \tag{7.69}$$

Now expand both parts of the resonance Hamiltonian about the resonance center:

$$H''_{n,0}(t; \sigma, \theta'; \Sigma, \Theta') = H''_{n,0}(t; \sigma, \theta'; \Sigma_n, \Theta')$$
$$+ \partial_{2,0} H''_{n,0}(t; \sigma, \theta'; \Sigma_n, \Theta') (\Sigma - \Sigma_n)$$
$$+ \frac{1}{2} \partial^2_{2,0} H''_{n,0}(t; \sigma, \theta'; \Sigma_n, \Theta') (\Sigma - \Sigma_n)^2$$
$$+ \cdots, \tag{7.70}$$

and

$$H''_{n,1}(t; \sigma, \theta'; \Sigma, \Theta') = H''_{n,1}(t; \sigma, \theta'; \Sigma_n, \Theta') + \cdots. \tag{7.71}$$

The first term in the expansion of $H''_{n,0}$ is a constant and can be ignored. The coefficient of the second term is zero, from the definition of Σ_n. The third term is the first significant term. We presume here that the first term of $H''_{n,1}$ is a nonzero constant. Now the scale of the separatrix in Σ at resonance is typically proportional to $\sqrt{\epsilon}$. So the third term of $H''_{n,0}$ and the first term of $H''_{n,1}$ are both proportional to ϵ. Subsequent terms are higher-order in ϵ. Keeping only the order ϵ terms, the approximate resonance Hamiltonian is of the form

$$\frac{(\Sigma - \Sigma_n)^2}{2\alpha'} - \epsilon \beta' \cos \sigma, \tag{7.72}$$

which is the Hamiltonian for a pendulum with a shifted center in momentum. This is analytically solvable. The constants are:

$$\alpha' = = 1/(\partial^2_{2,0} H''_{n,0}(t; \sigma, \theta'; \Sigma_n, \Theta'))$$
$$\beta' = = H''_{n,1}(t; \sigma, \theta'; \Sigma_n, \Theta'). \tag{7.73}$$

Driven pendulum resonances

Consider the behavior of the periodically driven pendulum in the vicinity of the resonance $\omega_r(p) = \omega$.

The Hamiltonian (7.54) for the driven pendulum has three resonance terms in H_1. The full generator (7.56) has three terms that are designed to eliminate the corresponding resonance terms in the Hamiltonian. The resulting approximate solution has small denominators close to each of the three resonances, $\omega_r(p) = 0$, $\omega_r(p) = \omega$, and $\omega_r(p) = -\omega$.

To develop a resonance approximation near $\omega_r(p) = \omega$, we do not include the corresponding term in the generator, so that the corresponding term is left in the Hamiltonian. It is helpful to give names to the various terms in the full generator (7.56):

$$W^0(\tau; \theta, t; p, p_t) = -\frac{\beta}{\omega_r(p)} \sin\theta$$

$$W^-(\tau; \theta, t; p, p_t) = \frac{\gamma}{\omega_r(p) + \omega} \sin(\theta + \omega t)$$

$$W^+(\tau; \theta, t; p, p_t) = \frac{\gamma}{\omega_r(p) - \omega} \sin(\theta - \omega t). \tag{7.74}$$

The full generator is $W^0 + W^- + W^+$.

To investigate the motion in the phase space near the resonance $\omega_r(p) = \omega$ (the "+" resonance), we use the generator that excludes the corresponding term

$$W_+ = W^0 + W^-. \tag{7.75}$$

With this generator the transformed Hamiltonian is

$$H_+(\tau; \theta, t; p, p_t) = p_t + \frac{p^2}{2\alpha} + \epsilon\gamma \cos(\theta - \omega t) + \cdots. \tag{7.76}$$

After we exclude the higher-order terms, this Hamiltonian has only a single combination of coordinates, and so can be transformed into a Hamiltonian that is cyclic in all but one degree of freedom. Define the transformation through the mixed-variable generating function

$$F_2(\tau; t, \theta; \Sigma, p_t') = (\theta - \omega t)\Sigma + tp_t', \tag{7.77}$$

giving the transformation

$$\sigma = \theta - \omega t$$
$$t = t'$$
$$p = \Sigma$$
$$p_t = p_t' - \omega\Sigma. \tag{7.78}$$

Expressed in these new coordinates, the resonance Hamiltonian is

$$H_+{}'(\tau; \sigma, t'; \Sigma, p_t') = p_t' - \omega\Sigma + \frac{\Sigma^2}{2\alpha} + \epsilon\gamma \cos\sigma$$

$$= \frac{(\Sigma - \alpha\omega)^2}{2\alpha} + \epsilon\gamma \cos\sigma + p_t' - \frac{1}{2}\alpha\omega^2. \tag{7.79}$$

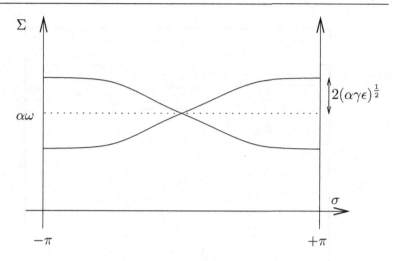

Figure 7.8   Contours of the resonance Hamiltonian H_+' give the motion in the (σ, Σ) plane. In this case the resonance Hamiltonian is a generalized pendulum shifted in momentum and phase. The half-width of the resonance oscillation zone is $2\sqrt{\alpha\gamma\epsilon}$.

This Hamiltonian is cyclic in t', so the solutions are level curves of H_+' in (σ, Σ). Actually, more can be said here because H_+' is already of the form of a pendulum shifted in the Σ direction by $\alpha\omega$ and shifted by π in phase. The shift by π comes about because the sign of the cosine term is positive, rather than negative as in the usual pendulum. A sketch of the level curves is given in figure 7.8.

Exercise 7.2: Resonance width
Verify that the half-width of the resonance region is $2\sqrt{\alpha\gamma\epsilon}$.

Exercise 7.3: With the computer
Verify, with the computer, that with the generator W_+ the transformed Hamiltonian is given by equation (7.76).

An approximate solution of the driven pendulum near the $\omega_r(p) = \omega$ resonance is

$$(\tau; \theta, t; p, p_t) = (E'_{-\epsilon, W_+} E_{\tau - \tau_0, H_+'} E'_{\epsilon, W_+} I)(\tau_0; \theta_0, t_0; p_0, (p_t)_0). \quad (7.80)$$

To find out to what extent the approximate solution models the actual driven pendulum, we make a surface of section using this approximate solution and compare it to a surface of section for the

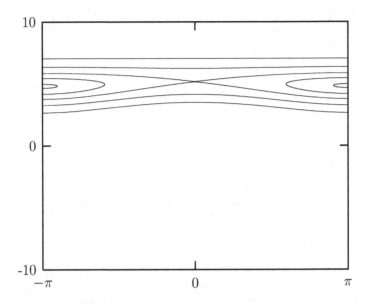

Figure 7.9   Surface of section of the first-order perturbative solution
for the driven pendulum constructed for the region near the resonance
$\omega_r(p) = \omega$. The parameters of the system are $\alpha = 1$, $\beta = 1$, $\gamma = 1/4$,
and $\omega = 5$. Only order ϵ terms were kept in the Lie series for the W
transformation. The perturbative solution captures the essential shape
and position of the resonant island it is designed to approximate.

actual driven pendulum. The surface of section for the approximate solution in the resonance region is shown in figure 7.9. A
surface of section for the actual driven pendulum is shown in the
lower part of figure 7.10. The correspondence is surprisingly good.
Note how the resonance island is not symmetrical about a line of
constant momentum. The resonance Hamiltonian is symmetrical
about $\Sigma = \alpha\omega$, and by itself would give a symmetric resonance
island (see figure 7.8). The necessary distortion is introduced by
the W_+ transformation that eliminates the other resonances. Indeed, in the full section the distortion appears to be generated by
the nearby $\omega_r(p) = 0$ resonance "pushing away" nearby features
so that it has room to fit. However, some features of the actual
section are not represented in figure 7.9: for instance, the small
chaotic zone near the actual separatrix.

The distortion introduced by the transformation generated by
W_+ is small because the terms that it introduces are proportional

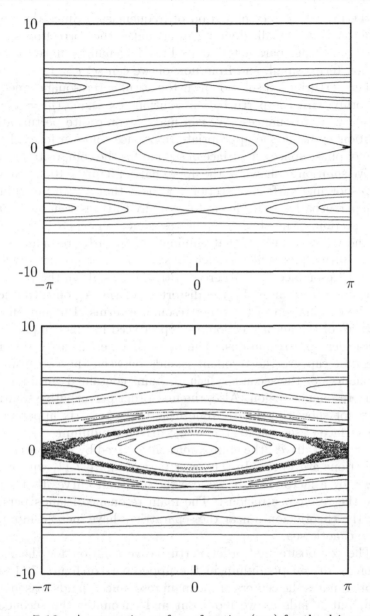

Figure 7.10   A composite surface of section (top) for the driven pendulum is constructed by combining the first-order perturbative solution for the region near the resonance $\omega_r(p) = 0$ and the solutions for the regions near the resonances $\omega_r(p) = \pm\omega$. A corresponding surface of section for the actual driven pendulum is shown below. The parameters of the system are: $\alpha = 1$, $\beta = 1$, $\gamma = 1/4$, and $\omega = 5$.

to the inverse of a combination of frequencies.[4] Since this combination is not small, dividing by it makes the correction small. Thus the "order parameter" ϵ need not be small to make the correction terms small, and from now on we can set $\epsilon = 1$.

The perturbation solution near the $\omega_r(p) = 0$ resonance merges smoothly with the perturbation solutions for the $\omega_r(p) = \omega$ and $\omega_r(p) = -\omega$ resonances. We can make a composite perturbative solution by using the appropriate resonance solution for each region of phase space. A surface of section for the composite perturbative solution is shown in the upper part of figure 7.10, above the corresponding surface of section for the actual driven pendulum. The perturbative solution captures many features seen on the actual section. The shapes of the resonance regions are distorted by the transformations that eliminate the nearby resonances, so the resulting pieces fit together consistently. The predicted width of each resonance region agrees with the actual width: it is not substantially changed by the distortion of the region introduced by the elimination of the other resonance terms. But not all the features of the actual section are reproduced in this composite of first-order approximations. The first-order perturbative solution does not capture the resonant islands between the two primary resonances or the secondary island chains contained within a primary resonance region. Also, the first-order perturbative solution does not show the chaotic zone near the separatrix apparent in the surface of section for the actual driven pendulum.

For larger drives, the approximations derived by first-order perturbations are worse. In the lower part of figure 7.11, with drive larger by a factor of five, we lose the invariant curves that separate the resonance regions. The main resonance islands persist, but the chaotic zones near the separatrices have merged into one large chaotic sea.

The composite first-order perturbative solution for the more strongly driven pendulum in the upper part of figure 7.11 still approximates the centers of the main resonance islands reasonably well, but it fails as we move out and encounter the secondary islands that are visible in the resonance region for $\omega_r(p) = \omega$. Here the approximations for the two regions do not fit together so well. The chaotic sea is found in the region where the perturbative solutions do not match.

---

[4]For W_+ see equations 7.74 and 7.75; for the general relationship between a term in the generator and the coordinate transformation generated see equations 7.48 and 7.49.

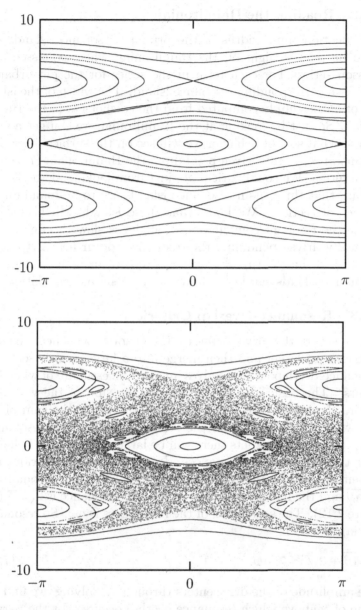

Figure 7.11 Composite surface of section (top) for the driven pendulum constructed by combining the first-order perturbative solution for the region near the resonance $\omega_r(p) = 0$ and the regions near the resonances $\omega_r(p) = \pm\omega$. A corresponding surface of section for the actual driven pendulum is shown below. The parameters of the system are the same as in figure 7.10 except that $\gamma = 5/4$.

7.4.2   Reading the Hamiltonian

The locations and widths of the primary resonance islands can
often be read straight off the Hamiltonian when expressed as a
Poisson series. For each term in the series for the perturbation
there is a corresponding resonance island. The width of the island
can often be simply computed from the coefficients in the Hamiltonian. So just by looking at the Hamiltonian we can get a good
idea of what sort of behavior we will see on the surface of section.
For instance, in the driven pendulum, the Hamiltonian (7.54) has
three terms. We could anticipate, just from looking at the Hamiltonian, that three main resonance islands are to be found on the
surface of section. We know that these islands will be located
where the resonant combination of angles is slow. So for the periodically driven pendulum the resonances occur near $\omega_r(p) = \omega$,
$\omega_r(p) = 0$, and $\omega_r(p) = -\omega$. The approximate widths of the
resonance islands can be computed with a simple calculation.

7.4.3   Resonance-Overlap Criterion

As the size of the drive increases, the chaotic zones near the separatrices get larger and then merge into a large chaotic sea. The
resonance-overlap criterion gives an analytic estimate of when this
occurs. The basic idea is to compare the sum of the widths of
neighboring resonances with their separation. If the sum of the
half-widths is greater than the separation, then the resonance-
overlap criterion predicts there will be large-scale chaotic behavior
near the overlapping resonances. In the case of the periodically
driven pendulum, the half-width of the $\omega_r(p) = 0$ resonance is
$2\sqrt{\alpha\beta}$ and the half-width of the $\omega_r(p) = \omega$ resonance is $2\sqrt{\alpha\gamma}$ (see
figure 7.12). The separation of the resonances is $\alpha\omega$. So resonance
overlap occurs if

$$2\sqrt{\alpha\beta} + 2\sqrt{\alpha\gamma} \geq \alpha\omega. \tag{7.81}$$

The amplitude of the drive enters through γ. Solving, we find the
value of γ above which resonance overlap occurs. For the parameters $\alpha = \beta = 1$, $\omega = 5$ used in figures 7.9–7.11, the resonance
overlap value of γ is 9/4. We see that, in fact, the chaotic zones
have already merged for $\gamma = 5/4$. So in this case the resonance-
overlap criterion overestimates the strength of the resonances required to get large-scale chaotic behavior. This is typical of the
resonance-overlap criterion.

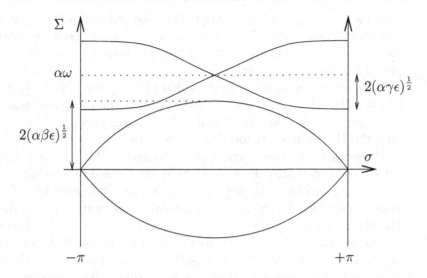

Figure 7.12   Resonance overlap occurs when the sum of the half-widths of adjacent resonances is larger than the spacing between them.

A way of thinking about why the resonance-overlap criterion usually overestimates the strength required to get large-scale chaos is that other effects must be taken into account. For instance, as the drive is increased second-order resonances appear between the primary resonances; these resonances take up space and so resonance overlap occurs for smaller drive than would be expected by considering the primary resonances alone. Also, the chaotic zones at each separatrix have area that must be accounted for.

7.4.4   Higher-Order Perturbation Theory

As the drive is increased, a variety of new islands emerge, which are not evident in the original Hamiltonian. To find approximations for motion in these regions we can use higher-order perturbation theory. The basic plan is the same as before. At any stage the Hamiltonian (which is perhaps a result of earlier stages of perturbation theory) is expressed as a Poisson series (a multiple-angle Fourier series). The terms that are not resonant in a region of interest are eliminated by a Lie transformation. The remaining resonance terms involve only a single combination of angle and are thus solvable by making a canonical transformation to resonance coordinates. We complete the solution and transform back to the original coordinates.

Let's find a perturbative approximation for the second-order islands visible in figure 7.10 between the $\omega_r(p) = 0$ resonance and the $\omega_r(p) = -\omega$ resonance. The details are messy, so we will just give a few intermediate results.

This resonance is not near the three primary resonances, so we can use the full generator (7.56) to eliminate those three primary resonance terms from the Hamiltonian. After this perturbation step the Hamiltonian is too hairy to look at.

We expand the transformed Hamiltonian in Poisson form and divide the terms into those that are resonant and those that are not. The terms that are not resonant can be eliminated by a Lie transform. This Lie transform leaves the resonant terms in the Hamiltonian and introduces an additional distortion to the curves on the surface of section. In this case this additional distortion is small, but very messy to compute, so we will just not include this effect. The resonance Hamiltonian is then (after considerable algebra)

$$
\begin{aligned}
& H_{2:1}'(\tau; \theta, t; p, p_t) \\
& \quad = \frac{p^2}{2\alpha} + p_t + \frac{\alpha\beta\gamma}{4p^2} \frac{\alpha^2\omega^2 + 2\alpha\omega p + 2p^2}{(\alpha\omega + p)^2} \cos\left(2\theta + \omega t\right).
\end{aligned} \tag{7.82}
$$

This is solvable because there is only a single combination of coordinates.

We can get an analytic solution by making the pendulum approximation. The Hamiltonian is already quadratic in the momentum p, so all we need to do is evaluate the coefficient of the potential terms at the resonance center $p_{2:1} = \alpha\omega/2$. The resonance Hamiltonian, in the pendulum approximation, is

$$
H_{2:1}''(\tau; \theta, t; p, p_t) = \frac{p^2}{2\alpha} + \frac{2\beta\gamma}{\alpha\omega^2} \cos\left(2\theta + \omega t\right). \tag{7.83}
$$

Carrying out the transformation to the resonance variable $\sigma = 2\theta - \omega t$ reduces this to a pendulum Hamiltonian with a single degree of freedom. Combining the analytic solution of this pendulum Hamiltonian with the transformations generated by the full W, we get an approximate perturbative solution

$$
(\tau; \theta, t; p, p_t) = (E_{-\epsilon, W}' E_{\tau - \tau_0, H_{2:1}''} E_{\epsilon, W}' I)(\tau_0; \theta_0, t_0; p_0, (p_t)_0). \tag{7.84}
$$

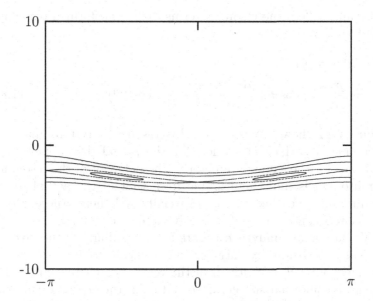

Figure 7.13   Second-order perturbation theory gives an approximation to the second-order islands near the resonance $2\omega_r(p) + \omega = 0$.

A surface of section in the appropriate resonance region using this
solution is shown in figure 7.13. Comparing this to the actual surface of section (figure 7.10), we see that the approximate solution
provides a good representation of this resonance motion.

7.4.5   Stability of the Inverted Vertical Equilibrium

As a second application, we use second-order perturbation theory
to investigate the inverted vertical equilibrium of the periodically
driven pendulum.

Here, the procedure parallels the one just followed, but we focus on a different set of resonance terms. The terms that are
slowly varying for the vertical equilibrium are those that involve
θ but do not involve t, such as $\cos(\theta)$ and $\cos(2\theta)$. So we want
to use the generator $W^+ + W^-$ that eliminates the nonresonant
terms involving combinations of θ and ωt, while leaving the central resonance. After the Lie transform of the Hamiltonian with
this generator, we write the transformed Hamiltonian as a Poisson

series and collect the resonant terms. The transformed resonance Hamiltonian is

$$H'_V(\tau; \theta, t; p, p_t)$$
$$= \frac{p^2}{2\alpha} - \beta \cos\theta + \frac{\alpha\gamma^2(\alpha^2\omega^2 + p^2)}{2(\alpha^2\omega^2 - p^2)^2} \cos(2\theta) + \cdots. \qquad (7.85)$$

Figure 7.14 shows contours of this resonance Hamiltonian H'_V (top) and a surface of section for the actual driven pendulum (bottom) for the same parameters. The behavior of the resonance Hamiltonian is indistinguishable from that of the actual driven pendulum. The theory does especially well here; there are no nearby resonances because the drive frequency is high.

We can get an analytic estimate for the stability of the inverted vertical equilibrium by carrying out a linear stability analysis of the fixed point $\theta = \pi$, $p = 0$ of the resonance Hamiltonian. The algebra is somewhat simpler if we first make the pendulum approximation about the resonance center. The resonance Hamiltonian is then approximately

$$H''_V(\tau; \theta, t; p, p_t) = \frac{p^2}{2\alpha} - \beta \cos\theta + \frac{\gamma^2}{2\alpha\omega^2} \cos(2\theta) + \cdots. \qquad (7.86)$$

Linear stability analysis of the inverted vertical equilibrium indicates stability for

$$\gamma^2 > \alpha\beta\omega^2. \qquad (7.87)$$

In terms of the original physical parameters, the vertical equilibrium is linearly stable if

$$\frac{\omega}{\omega_s} \frac{A}{l} > \sqrt{2}, \qquad (7.88)$$

where $\omega_s = \sqrt{g/l}$, the small-amplitude oscillation frequency. For the vertical equilibrium to be stable, the scaled product of the amplitude of the drive and the drive frequency must be sufficiently large.

This analytic estimate is compared with the behavior of the driven pendulum in figure 7.15. For any given assignment of the parameters, the driven pendulum can be tested for the linear stability of the inverted vertical equilibrium by the methods of chapter 4; this involves determining the roots of the charac-

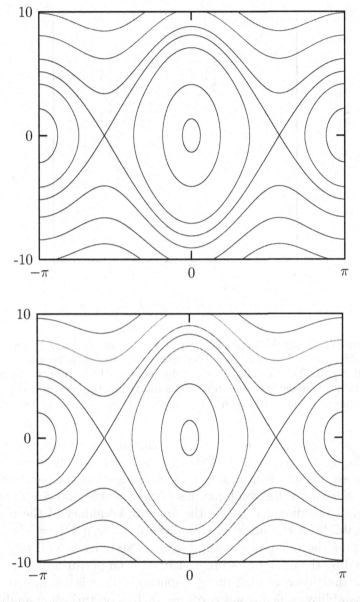

Figure 7.14   Contours of the resonance Hamiltonian H'_V, which has been developed to study the stability of the vertical equilibrium, are shown in the upper plot. A corresponding surface of section for the actual driven pendulum is shown in the lower plot. The parameters are $m = 1\,\text{kg}$, $l = 1\,\text{m}$, $g = 9.8\,\text{m}\,\text{s}^{-2}$, $A = 0.03\,\text{m}$, and $\omega = 100\omega_s$, where $\omega_s = \sqrt{g/l}$.

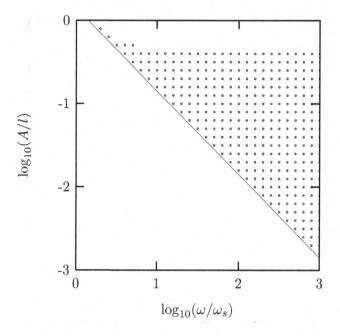

Figure 7.15 Stability of the inverted vertical equilibrium over a range of parameters. The full parameter space displayed was sampled over a regular grid. The dots indicate parameters for which the actual driven pendulum is linearly stable; nothing is plotted in the case of instability. The diagonal line is the locus of points satisfying $(\omega/\omega_s)(A/l) = \sqrt{2}$.

teristic polynomial for a reference orbit at the resonance center. In the figure the stability of the inverted vertical equilibrium was assessed at each point of a grid of assignments of the parameters. A dot is shown for combinations of parameters that are linearly stable. The diagonal line is the analytic boundary of the region of stability of the inverted equilibrium: $(\omega/\omega_s)(A/l) = \sqrt{2}$. We see that the boundary of the region of stability is well approximated by the analytic estimate derived from perturbation theory. Note that for very high drive amplitudes there is another region of instability, which is not captured by this perturbation analysis.

7.5   Summary

The goal of perturbation theory is to relate aspects of the motions of a given system to those of a nearby solvable system. Perturbation theory can be used to predict features such as the size and location of the resonance islands and chaotic zones.

With perturbation analysis we obtain an approximation to the evolution of a system by relating the evolution of the system to that of a different system that, when approximated, can be exactly solved. We can carry this exact solution of the approximate problem back to the original system to obtain an approximate solution of our original problem. The strategy of canonical perturbation theory is to make canonical transformations that eliminate terms in the Hamiltonian that impede solution. Formulation of perturbation theory in terms of Lie series is especially convenient.

We can use first-order perturbation theory to analyze the motion of the undriven pendulum as a free rotor to which gravity is added. In this analysis we find that a small denominator in the series limits the range of applicability of the perturbative solution to regions that are away from the resonant oscillation region.

In higher-order perturbation theory for the pendulum we discover the problem of secular terms, terms that produce error that grow with time. The appearance of secular terms can be avoided by keeping track of how the frequencies change as perturbations are included. In canonical perturbation theory secular terms can be avoided by associating the average part of the perturbation with the solvable part of the Hamiltonian.

In carrying out canonical perturbation theory in higher dimensions we find that the problem of small denominators is more serious. Small denominators arise near every commensurability, and commensurabilities are common. Small denominators can be locally avoided near particular commensurabilities by incorporating the offending terms into the solvable part of the Hamiltonian. If the resonances are isolated, the resulting resonance Hamiltonian is still solvable. In many cases the resonance Hamiltonian is well approximated by a pendulum-like Hamiltonian. A global picture can be constructed by stitching together the solutions for each resonance region constructed separately.

If two resonance regions overlap—that is, if the sum of the half-widths of the resonance regions exceeds their separation— then large-scale chaos ensues. The chaotic regions associated with the separatrices of the overlapping resonances become connected. When the resonances are well approximated by pendulum-like resonances a simple analytic criterion for the appearance of large-scale chaos can be developed.

Higher-order perturbative descriptions can be developed to describe islands that do not correspond to particular terms in the Hamiltonian, secondary resonances, bifurcations, and so on. The theory can be extended to describe as much detail as one wishes.

7.6   Projects

Exercise 7.4: Periodically driven pendulum

a. Work out the details of the perturbation theory for the primary driven pendulum resonances, as displayed in figure 7.10.

b. Work out the details of the perturbation theory for the stability of the inverted vertical equilibrium. Derive the resonance Hamiltonian and plot its contours. Compare these contours to surfaces of section for a variety of parameters.

c. Carry out the linear stability analysis leading to equation (7.88). What is happening in the upper part of figure 7.15? Why is the system unstable when criterion (7.88) predicts stability? Use surfaces of section to investigate this parameter regime.

Exercise 7.5: Spin-orbit coupling

A Hamiltonian for the spin-orbit problem described in section 2.11.2 is

$$
\begin{aligned}
H(t,\theta,p_\theta) &= \frac{p_\theta^2}{2C} - \frac{n^2\epsilon^2 C}{4}\frac{a^3}{R^3(t)}\cos 2(\theta - f(t)) \\
&= \frac{p_\theta^2}{2C} - \frac{n^2\epsilon^2 C}{4}\Big(\cos(2\theta - 2nt) + \frac{7e}{2}\cos(2\theta - 3nt) \\
&\qquad\qquad - \frac{e}{2}\cos(2\theta - nt) + \cdots\Big)
\end{aligned}
\tag{7.89}
$$

where the ignored terms are higher order in eccentricity e. Note that here ϵ is the out-of-roundness parameter.

a. Find the widths and centers of the three primary resonances. Compare the predictions for the widths to the island widths seen on surfaces of section. Write the criterion for resonance overlap and compare to numerical experiments for the transition to large-scale chaos.

b. The fixed point of the synchronous island is offset from the average rate of rotation. This is indicative of a "forced" oscillation of the rotation of the Moon. Develop a perturbative theory for motion in the synchronous island by using a Lie transform to eliminate the two non-synchronous resonances. Predict the location of the fixed point at the center of the synchronous resonance on the surface of section, and thus predict the amplitude of the forced oscillation of the Moon.

8

Appendix: Scheme

> Programming languages should be designed not by piling feature on top of feature, but by removing the weaknesses and restrictions that make additional features appear necessary. Scheme demonstrates that a very small number of rules for forming expressions, with no restrictions on how they are composed, suffice to form a practical and efficient programming language that is flexible enough to support most of the major programming paradigms in use today.
>
> *IEEE Standard for the Scheme Programming Language* [24], p. 3

Here we give an elementary introduction to Scheme.[1] For a more precise explanation of the language see the IEEE standard [24]; for a longer introduction see the textbook [1].

Scheme is a simple programming language based on expressions. An expression names a value. For example, the numeral 3.14 names an approximation to a familiar number. There are primitive expressions, such as numerals, that we directly recognize, and there are compound expressions of several kinds.

Procedure calls

A *procedure call* is a kind of compound expression. A procedure call is a sequence of expressions delimited by parentheses. The first subexpression in a procedure call is taken to name a procedure, and the rest of the subexpressions are taken to name the arguments to that procedure. The value produced by the procedure when applied to the given arguments is the value named by the procedure call. For example,

---

[1]Many of the statements here are valid only assuming that no assignments are used.

```
(+ 1 2.14)
3.14
```

```
(+ 1 (* 2 1.07))
3.14
```

are both compound expressions that name the same number as the numeral 3.14.[2] In these cases the symbols + and * name procedures that add and multiply, respectively. If we replace any subexpression of any expression with an expression that names the same thing as the original subexpression, the thing named by the overall expression remains unchanged. In general, a procedure call is written

(*operator   operand-1* ... *operand-n*)

where *operator* names a procedure and *operand-i* names the *i*th argument.[3]

Lambda expressions

Just as we use numerals to name numbers, we use λ-expressions to name procedures.[4] For example, the procedure that squares its input can be written:

```
(lambda (x) (* x x))
```

This expression can be read: "The procedure of one argument, x, that multiplies x by x." Of course, we can use this expression in any context where a procedure is needed. For example,

```
((lambda (x) (* x x)) 4)
16
```

---

[2]In examples we show the value that would be printed by the Scheme system using slanted characters following the input expression.

[3]In Scheme every parenthesis is essential: you cannot add extra parentheses or remove any.

[4]The logician Alonzo Church [13] invented λ-notation to allow the specification of an anonymous function of a named parameter: λx[expression in x]. This is read, "That function of one argument that is obtained by substituting the argument for x in the indicated expression."

The general form of a λ-expression is

```
(lambda  formal-parameters  body)
```

where *formal-parameters* is a list of symbols that will be the names of the arguments to the procedure and *body* is an expression that may refer to the formal parameters. The value of a procedure call is the value of the body of the procedure with the arguments substituted for the formal parameters.

Definitions

We can use the `define` construct to give a name to any object. For example, if we make the definitions[5]

```
(define pi 3.141592653589793)
```

```
(define square (lambda (x) (* x x)))
```

we can then use the symbols `pi` and `square` wherever the numeral or the λ-expression could appear. For example, the area of the surface of a sphere of radius 5 is

```
(* 4 pi (square 5))
314.1592653589793
```

Procedure definitions may be expressed more conveniently using "syntactic sugar." The squaring procedure may be defined

```
(define (square x) (* x x))
```

which we may read: "To square *x* multiply *x* by *x*."

In Scheme, procedures may be passed as arguments and returned as values. For example, it is possible to make a procedure that implements the mathematical notion of the composition of two functions:[6]

---

[5]The definition of `square` given here is not the definition of `square` in the Scmutils system. In Scmutils, `square` is extended for tuples to mean the sum of the squares of the components of the tuple. However, for arguments that are not tuples the Scmutils `square` does multiply the argument by itself.

[6]The examples are indented to help with readability. Scheme does not care about extra white space, so we may add as much as we please to make things easier to read.

```
(define compose
  (lambda (f g)
    (lambda (x)
      (f (g x)))))

((compose square sin) 2)
.826821810431806

(square (sin 2))
.826821810431806
```

Using the syntactic sugar shown above, we can write the definition more conveniently. The following are both equivalent to the definition above:

```
(define (compose f g)
  (lambda (x)
    (f (g x))))

(define ((compose f g) x)
  (f (g x)))
```

Conditionals

Conditional expressions may be used to choose among several expressions to produce a value. For example, a procedure that implements the absolute value function may be written:

```
(define (abs x)
  (cond ((< x 0) (- x))
        ((= x 0) x)
        ((> x 0) x)))
```

The conditional cond takes a number of clauses. Each clause has a predicate expression, which may be either true or false, and a consequent expression. The value of the cond expression is the value of the consequent expression of the first clause for which the corresponding predicate expression is true. The general form of a conditional expression is

```
(cond ( predicate-1   consequent-1)
      ...
      ( predicate-n   consequent-n))
```

For convenience there is a special predicate expression else that can be used as the predicate in the last clause of a cond.

The if construct provides another way to make a conditional when there is only a binary choice to be made. For example, because we have to do something special only when the argument is negative, we could have defined abs as:

```
(define (abs x)
  (if (< x 0)
      (- x)
      x))
```

The general form of an if expression is

```
(if  predicate  consequent  alternative)
```

If the *predicate* is true the value of the if expression is the value of the *consequent*, otherwise it is the value of the *alternative*.

Recursive procedures

Given conditionals and definitions, we can write recursive procedures. For example, to compute the nth factorial number we may write:

```
(define (factorial n)
  (if (= n 0)
      1
      (* n (factorial (- n 1)))))
```

```
(factorial 6)
720
```

```
(factorial 40)
815915283247897734345611269596115894272000000000
```

Local names

The let expression is used to give names to objects in a local context. For example,

```
(define (f radius)
  (let ((area (* 4 pi (square radius)))
        (volume (* 4/3 pi (cube radius))))
    (/ volume area)))
```

```
(f 3)
1
```

The general form of a `let` expression is

```
(let (( variable-1   expression-1)
      ...
      ( variable-n   expression-n))
   body)
```

The value of the `let` expression is the value of the *body* expression in the context where the variables *variable-i* have the values of the expressions *expression-i*. The expressions *expression-i* may not refer to any of the variables *variable-j* given values in the `let` expression.

A `let*` expression is the same as a `let` expression except that an expression *expression-i* may refer to variables *variable-j* given values earlier in the `let*` expression.

A slight variant of the `let` expression provides a convenient way to express looping constructs. We can write a procedure that implements an alternative algorithm for computing factorials as follows:

```
(define (factorial n)
  (let factlp ((count 1) (answer 1))
    (if (> count n)
        answer
        (factlp (+ count 1) (* count answer)))))

(factorial 6)
720
```

Here, the symbol `factlp` following the `let` is locally defined to be a procedure that has the variables `count` and `answer` as its formal parameters. It is called the first time with the expressions 1 and 1, initializing the loop. Whenever the procedure named `factlp` is called later, these variables get new values that are the values of the operand expressions (+ count 1) and (* count answer).

Compound data—lists and vectors

Data can be glued together to form compound data structures. A list is a data structure in which the elements are linked sequentially. A Scheme vector is a data structure in which the elements are packed in a linear array. New elements can be added to lists, but to access the nth element of a list takes computing time proportional to n. By contrast a Scheme vector is of fixed length, and its elements can be accessed in constant time. All data structures in this book are implemented as combinations of lists and Scheme vectors. Compound data objects are constructed from components by procedures called constructors and the components are accessed by selectors.

The procedure `list` is the constructor for lists. The selector `list-ref` gets an element of the list. All selectors in Scheme are zero-based. For example,

```
(define a-list (list 6 946 8 356 12 620))
```

```
a-list
(6 946 8 356 12 620)
```

```
(list-ref a-list 3)
356
```

```
(list-ref a-list 0)
6
```

Lists are built from pairs. A pair is made using the constructor `cons`. The selectors for the two components of the pair are `car` and `cdr` (pronounced "could-er").[7] A list is a chain of pairs, such that the `car` of each pair is the list element and the `cdr` of each pair is the next pair, except for the last `cdr`, which is a distinguishable value called the empty list and is written (). Thus,

```
(car a-list)
6
```

```
(cdr a-list)
(946 8 356 12 620)
```

```
(car (cdr a-list))
946
```

```
(define another-list
  (cons 32 (cdr a-list)))
```

```
another-list
(32 946 8 356 12 620)
```

---

[7]These names are accidents of history. They stand for "Contents of the Address part of Register" and "Contents of the Decrement part of Register" of the IBM 704 computer, which was used for the first implementation of Lisp in the late 1950s. Scheme is a dialect of Lisp.

```
(car (cdr another-list))
946
```

Both `a-list` and `another-list` share the same tail (their `cdr`).

There is a predicate `pair?` that is true of pairs and false on all other types of data.

Vectors are simpler than lists. There is a constructor `vector` that can be used to make vectors and a selector `vector-ref` for accessing the elements of a vector:

```
(define a-vector
  (vector 37 63 49 21 88 56))

a-vector
#(37 63 49 21 88 56)

(vector-ref a-vector 3)
21

(vector-ref a-vector 0)
37
```

Notice that a vector is distinguished from a list on printout by the character # appearing before the initial parenthesis.

There is a predicate `vector?` that is true of vectors and false for all other types of data.

The elements of lists and vectors may be any kind of data, including numbers, procedures, lists, and vectors. Numerous other procedures for manipulating list-structured data and vector-structured data can be found in the Scheme online documentation.

Symbols

Symbols are a very important kind of primitive data type that we use to make programs and algebraic expressions. You probably have noticed that Scheme programs look just like lists. In fact, they are lists. Some of the elements of the lists that make up programs are symbols, such as + and `vector`.[8] If we are to make programs that can manipulate programs, we need to be able to write an expression that names such a symbol. This is accomplished by the mechanism of *quotation*. The name of the symbol

---

[8]Symbols may have any number of characters. A symbol may not contain whitespace or a delimiter character, such as parentheses, brackets, quotation marks, comma, or #.

+ is the expression '+, and in general the name of an expression
is the expression preceded by a single quote character. Thus the
name of the expression (+ 3 a) is '(+ 3 a).

We can test if two symbols are identical by using the predicate
eq?. For example, we can write a program to determine if an
expression is a sum:

```
(define (sum? expression)
  (and (pair? expression)
       (eq? (car expression) '+)))

(sum? '(+ 3 a))
#t

(sum? '(* 3 a))
#f
```

Here #t and #f are the printed representations of the boolean
values true and false.

Consider what would happen if we were to leave out the quote in
the expression (sum? '(+ 3 a)). If the variable a had the value 4
we would be asking if 7 is a sum. But what we wanted to know
was whether the expression (+ 3 a) is a sum. That is why we
need the quote.

Effects

Sometimes it is necessary to perform some action, such as plot a
point or print a value, in the process of a computation. Such an
action is called an *effect*.[9] For example, to see in more detail how
the factorial program computes its answer we can interpolate a
write-line statement in the body of the factlp internal procedure. This will print out a list of the count and the answer for
each iteration:

```
(define (factorial n)
  (let factlp ((count 1) (answer 1))
    (write-line (list count answer))
    (if (> count n)
        answer
        (factlp (+ count 1) (* count answer)))))
```

---

[9]This is computer-science jargon: An effect is a change to something. For
example, write-line changes the display by printing something to the display.

When we execute the modified `factorial` procedure we can watch the counter incrementing and the answer being built:

```
(factorial 6)
(1 1)
(2 1)
(3 2)
(4 6)
(5 24)
(6 120)
(7 720)
 720
```

The body of every procedure or `let`, as well as the consequent of every `cond` clause, allows statements that have effects to be used. The effect statement generally has no useful value. The final expression in the body or clause produces the value that is returned. In this example the `if` expression produces the value of the `factorial`.

Assignments

Effects like printing a value or plotting a point are pretty benign, but there are more powerful (and thus dangerous) effects, called *assignments*. An assignment *changes* the value of a variable or an entry in a data structure. Almost everything we are computing are mathematical functions: for a particular input they always produce the same result. However, with assignment we can make objects that change their behavior as they are used. For example, we can make a device that counts every time we call it:

```
(define (make-counter)
  (let ((count 0))
    (lambda ()
      (set! count (+ count 1))
      count)))
```

Let's make two counters:

```
(define c1 (make-counter))
(define c2 (make-counter))
```

These two counters have independent local state. Calling a counter causes it to increment its local state variable, `count`, and return its value.

```
(c1)
 1

(c1)
 2

(c2)
 1

(c1)
 3

(c2)
 2
```

Assignment to variables is sometimes useful. For example, it may be useful to accumulate some objects into a list for further analysis. Here is an elegant way to do this:

```
(define (make-collector)
  (let ((lst '()))
    (cons (lambda (new)
            (set! lst (cons new lst))
            new)
          (lambda () lst))))
```

This procedure makes a pair of two procedures. The `car` of the pair is a procedure that adds to a list and the `cdr` of the pair is a procedure that reports the list that has been collected.

Let's make two collectors and play with them:

```
(define c3 (make-collector))
(define c4 (make-collector))

((car c3) 42)
 42

((car c4) 'jerry)
 jerry

((car c3) 28)
 28

((car c3) 14)
 14
```

```
((car c4) 'jack)
 jack

((cdr c3))
 (14 28 42)

((cdr c4))
 (jack jerry)
```

It is also possible to assign to the elements of a data structure, such as a list or vector. This is unnecessary in our work so we won't tell you about how to do it! In general, it is good practice to avoid assignments whenever possible, but if you need them they are available.[10]

---

[10]The discipline of programming without assignments is called *functional programming*. Functional programs are generally easier to understand, and have fewer bugs than *imperative programs*.

9
Appendix: Our Notation

An adequate notation should be understood by at least two people, one of whom may be the author.

Abdus Salam (1950).

We adopt a *functional mathematical notation* that is close to that used by Spivak in his *Calculus on Manifolds* [40]. The use of functional notation avoids many of the ambiguities of traditional mathematical notation; the ambiguities of traditional notation that can impede clear reasoning in classical mechanics. Functional notation carefully distinguishes the function from the value of the function when applied to particular arguments. In functional notation mathematical expressions are unambiguous and self-contained.

We adopt a *generic arithmetic* in which the basic arithmetic operations, such as addition and multiplication, are extended to a wide variety of mathematical types. Thus, for example, the addition operator + can be applied to numbers, tuples of numbers, matrices, functions, etc. Generic arithmetic formalizes the common informal practice used to manipulate mathematical objects.

We often want to manipulate aggregate quantities, such as the collection of all of the rectangular coordinates of a collection of particles, without explicitly manipulating the component parts. Tensor arithmetic provides a traditional way of manipulating aggregate objects: Indices label the parts; conventions, such as the summation convention, are introduced to manipulate the indices. We introduce a *tuple arithmetic* as an alternative way of manipulating aggregate quantities that usually lets us avoid labeling the parts with indices. Tuple arithmetic is inspired by tensor arithmetic but it is more general: not all of the components of a tuple need to be of the same size or type.

The mathematical notation is in one-to-one correspondence with expressions of the computer language *Scheme* [24]. Scheme is based on the λ-calculus [13] and directly supports the manipulation of functions. We augment Scheme with symbolic, nu-

merical, and generic features to support our applications. For a
simple introduction to Scheme, see the Scheme appendix. The
correspondence between the mathematical notation and Scheme
requires that mathematical expressions be unambiguous and self-
contained. Scheme provides immediate feedback in verification
of mathematical deductions and facilitates the exploration of the
behavior of systems.

Functions

The value of the function f, given the argument x, is written $f(x)$.
The expression $f(x)$ denotes the value of the function at the given
argument; when we wish to denote the function we write just f.
Functions may take several arguments. For example, we may have
the function that gives the Euclidean distance between two points
in the plane given by their rectangular coordinates:

$$d(x_1, y_1, x_2, y_2) = \sqrt{(x_2 - x_1)^2 + (y_2 - y_1)^2}. \tag{9.1}$$

In Scheme we can write this as:

```
(define (d x1 y1 x2 y2)
  (sqrt (+ (square (- x2 x1)) (square (- y2 y1)))))
```

Functions may be composed if the range of one overlaps the
domain of the other. The composition of functions is constructed
by passing the output of one to the input of the other. We write
the composition of two functions using the ∘ operation:

$$(f \circ g) : x \mapsto (f \circ g)(x) = f(g(x)). \tag{9.2}$$

A procedure h that computes the cube of the sine of its argument
may be defined by composing the procedures cube and sin:

```
(define h (compose cube sin))
```

```
(h 2)
```
.7518269446689928

which is the same as

```
(cube (sin 2))
```
.7518269446689928

Arithmetic is extended to the manipulation of functions: the usual mathematical operations may be applied to functions. Examples are addition and multiplication; we may add or multiply two functions if they take the same kinds of arguments and if their values can be added or multiplied:

$$(f + g)(x) = f(x) + g(x),$$
$$(fg)(x) = f(x)g(x). \tag{9.3}$$

A procedure g that multiplies the cube of its argument by the sine of its argument is

```
(define g (* cube sin))

(g 2)
7.274379414605454

(* (cube 2) (sin 2))
7.274379414605454
```

Symbolic values

As in usual mathematical notation, arithmetic is extended to allow the use of symbols that represent unknown or incompletely specified mathematical objects. These symbols are manipulated as if they had values of a known type. By default, a Scheme symbol is assumed to represent a real number. So the expression 'a is a literal Scheme symbol that represents an unspecified real number:

```
((compose cube sin) 'a)
(expt (sin a) 3)
```

The default printer simplifies the expression and displays it in a readable form.[1] We can use the simplifier to verify a trigonometric identity:

```
((- (+ (square sin) (square cos)) 1) 'a)
0
```

---

[1] The procedure `print-expression` can be used in a program to print a simplified version of an expression. The default printer in the user interface incorporates the simplifier.

Just as it is useful to be able to manipulate symbolic numbers, it is useful to be able to manipulate symbolic functions. The procedure `literal-function` makes a procedure that acts as a function having no properties other than its name. By default, a literal function is defined to take one real argument and produce one real value. For example, we may want to work with a function $f : \mathbf{R} \to \mathbf{R}$:

```
((literal-function 'f) 'x)
(f x)
```

```
((compose (literal-function 'f) (literal-function 'g)) 'x)
(f (g x))
```

We can also make literal functions of multiple, possibly structured arguments that return structured values. For example, to denote a literal function named g that takes two real arguments and returns a real value ($g : \mathbf{R} \times \mathbf{R} \to \mathbf{R}$) we may write:

```
(define g (literal-function 'g (-> (X Real Real) Real)))
```

```
(g 'x 'y)
(g x y)
```

We may use such a literal function anywhere that an explicit function of the same type may be used.

There is a whole language for describing the type of a literal function in terms of the number of arguments, the types of the arguments, and the types of the values. Here we describe a function that maps pairs of real numbers to real numbers with the expression `(-> (X Real Real) Real)`. Later we will introduce structured arguments and values and show extensions of literal functions to handle these.

Tuples

There are two kinds of tuples: *up* tuples and *down* tuples. We write tuples as ordered lists of their components; a tuple is delimited by parentheses if it is an up tuple and by square brackets if it is a down tuple. For example, the up tuple v of velocity components v^0, v^1, and v^2 is

$$v = \left(v^0, v^1, v^2 \right) . \tag{9.4}$$

The down tuple p of momentum components p_0, p_1, and p_2 is

$$p = [p_0, p_1, p_2] . \tag{9.5}$$

A component of an up tuple is usually identified by a superscript. A component of a down tuple is usually identified by a subscript. We use zero-based indexing when referring to tuple elements. This notation follows the usual convention in tensor arithmetic.

We make tuples with the constructors up and down:

```
(define v (up 'v^0 'v^1 'v^2))

v
(up v^0 v^1 v^2)

(define p (down 'p_0 'p_1 'p_2))

p
(down p_0 p_1 p_2)
```

Tuple arithmetic is different from the usual tensor arithmetic in that the components of a tuple may also be tuples and different components need not have the same structure. For example, a tuple structure s of phase-space states is

$$s = (t, (x, y), [p_x, p_y]). \tag{9.6}$$

It is an up tuple of the time, the coordinates, and the momenta. The time t has no substructure. The coordinates are an up tuple of the coordinate components x and y. The momentum is a down tuple of the momentum components p_x and p_y. This is written:

```
(define s (up 't (up 'x 'y) (down 'p_x 'p_y)))
```

In order to reference components of tuple structures there are selector functions, for example:

$$I(s) = s$$
$$I_0(s) = t$$
$$I_1(s) = (x, y)$$
$$I_2(s) = [p_x, p_y]$$
$$I_{1,0}(s) = x$$
$$\cdots$$
$$I_{2,1}(s) = p_y. \tag{9.7}$$

The sequence of integer subscripts on the selector describes the access chain to the desired component.

The procedure `component` is the general selector procedure that implements the selector functions. For example, $I_{0,1}$ is implemented by (component 0 1):

```
((component 0 1) (up (up 'a 'b) (up 'c 'd)))
b
```

To access a component of a tuple we may also use the selector procedure `ref`, which takes a tuple and an index and returns the indicated element of the tuple:

```
(ref (up 'a 'b 'c) 1)
b
```

We use zero-based indexing everywhere. The procedure `ref` can be used to access any substructure of a tree of tuples:

```
(ref (up (up 'a 'b) (up 'c 'd)) 0 1)
b
```

Two up tuples of the same length may be added or subtracted, elementwise, to produce an up tuple, if the components are compatible for addition. Similarly, two down tuples of the same length may be added or subtracted, elementwise, to produce a down tuple, if the components are compatible for addition.

Any tuple may be multiplied by a number by multiplying each component by the number. Numbers may, of course, be multiplied. Tuples that are compatible for addition form a vector space.

For convenience we define the square of a tuple to be the sum of the squares of the components of the tuple. Tuples can be multiplied, as described below, but the square of a tuple is not the product of the tuple with itself.

The meaning of multiplication of tuples depends on the structure of the tuples. Two tuples are compatible for contraction if they are of opposite types, they are of the same length, and corresponding elements have the following property: either they are both tuples and are compatible for contraction, or one of them is not a tuple. If two tuples are compatible for contraction then generic multiplication is interpreted as contraction: the result is the sum of the products of corresponding components of the tuples. For example, p and v introduced in equations (9.4) and (9.5) above are compatible for contraction; the product is

$$pv = p_0 v^0 + p_1 v^1 + p_2 v^2. \tag{9.8}$$

So the product of tuples that are compatible for contraction is an inner product. Using the tuples p and v defined above gives us

```
(* p v)
(+ (* p_0 v^0) (* p_1 v^1) (* p_2 v^2))
```

Contraction of tuples is commutative: $pv = vp$. Caution: Multiplication of tuples that are compatible for contraction is, in general, not associative. For example, let $u = (5, 2)$, $v = (11, 13)$, and $g = [[3, 5], [7, 9]]$. Then $u(gv) = 964$, but $(ug)v = 878$. The expression ugv is ambiguous. An expression that has this ambiguity does not occur in this book.

The rule for multiplying two structures that are not compatible for contraction is simple. If A and B are not compatible for contraction, the product AB is a tuple of type B whose components are the products of A and the components of B. The same rule is applied recursively in multiplying the components. So if $B = (B^0, B^1, B^2)$, the product of A and B is

$$AB = \left(AB^0, AB^1, AB^2\right). \tag{9.9}$$

If A and C are not compatible for contraction and $C = [C_0, C_1, C_2]$, the product is

$$AC = [AC_0, AC_1, AC_2]. \tag{9.10}$$

Tuple structures can be made to represent linear transformations. For example, the rotation commonly represented by the matrix

$$\begin{bmatrix} \cos\theta & -\sin\theta \\ \sin\theta & \cos\theta \end{bmatrix} \tag{9.11}$$

can be represented as a tuple structure:[2]

$$\left[\begin{pmatrix} \cos\theta \\ \sin\theta \end{pmatrix} \begin{pmatrix} -\sin\theta \\ \cos\theta \end{pmatrix} \right]. \tag{9.12}$$

---

[2]To emphasize the relationship of simple tuple structures to matrix notation we often format up tuples as vertical arrangements of components and down tuples as horizontal arrangements of components. However, we could just as well have written this tuple as $[(\cos\theta, \sin\theta), (-\sin\theta, \cos\theta)]$.

Such a tuple is compatible for contraction with an up tuple that represents a vector. So, for example:

$$\left[\begin{pmatrix} \cos\theta \\ \sin\theta \end{pmatrix} \begin{pmatrix} -\sin\theta \\ \cos\theta \end{pmatrix}\right] \begin{pmatrix} x \\ y \end{pmatrix} = \begin{pmatrix} x\cos\theta - y\sin\theta \\ x\sin\theta + y\cos\theta \end{pmatrix}. \tag{9.13}$$

Two tuples that represent linear transformations, though not compatible for contraction, may also be combined by multiplication. In this case the product represents the composition of the linear transformations. For example, the product of the tuples representing two rotations is

$$\left[\begin{pmatrix} \cos\theta \\ \sin\theta \end{pmatrix} \begin{pmatrix} -\sin\theta \\ \cos\theta \end{pmatrix}\right]\left[\begin{pmatrix} \cos\varphi \\ \sin\varphi \end{pmatrix} \begin{pmatrix} -\sin\varphi \\ \cos\varphi \end{pmatrix}\right]$$
$$= \left[\begin{pmatrix} \cos(\theta+\varphi) \\ \sin(\theta+\varphi) \end{pmatrix} \begin{pmatrix} -\sin(\theta+\varphi) \\ \cos(\theta+\varphi) \end{pmatrix}\right]. \tag{9.14}$$

Multiplication of tuples that represent linear transformations is associative but generally not commutative, just as the composition of the transformations is associative but not generally commutative.

Derivatives

The derivative of a function f is a function, denoted by Df. Our notational convention is that D is a high-precedence operator. Thus D operates on the adjacent function before any other application occurs: $Df(x)$ is the same as $(Df)(x)$. Higher-order derivatives are described by exponentiating the derivative operator. Thus the nth derivative of a function f is notated as $D^n f$.

The procedure for producing the derivative of a function is named D. The derivative of the sin procedure is a procedure that computes cos:

```
(define derivative-of-sine (D sin))

(derivative-of-sine 'x)
(cos x)
```

The derivative of a function f is the function Df whose value for a particular argument is something that can be multiplied by an increment Δx in the argument to get a linear approximation to the increment in the value of f:

$$f(x + \Delta x) \approx f(x) + Df(x)\Delta x. \tag{9.15}$$

For example, let f be the function that cubes its argument ($f(x) = x^3$); then Df is the function that yields three times the square of its argument ($Df(y) = 3y^2$). So $f(5) = 125$ and $Df(5) = 75$. The value of f with argument $x + \Delta x$ is

$$f(x + \Delta x) = (x + \Delta x)^3 = x^3 + 3x^2\Delta x + 3x\Delta x^2 + \Delta x^3 \qquad (9.16)$$

and

$$Df(x)\Delta x = 3x^2\Delta x. \qquad (9.17)$$

So $Df(x)$ multiplied by Δx gives us the term in $f(x + \Delta x)$ that is linear in Δx, providing a good approximation to $f(x + \Delta x) - f(x)$ when Δx is small.

Derivatives of compositions obey the chain rule:

$$D(f \circ g) = ((Df) \circ g) \cdot Dg. \qquad (9.18)$$

So at x,

$$(D(f \circ g))(x) = Df(g(x)) \cdot Dg(x). \qquad (9.19)$$

D is an example of an *operator*. An operator is like a function except that multiplication of operators is interpreted as composition, whereas multiplication of functions is multiplication of the values (see equation 9.3). If D were an ordinary function, then the rule for multiplication would imply that D^2f would just be the product of Df with itself, which is not what is intended. A product of a number and an operator scales the operator. So, for example

```
(((* 5 D) cos) 'x)
(* -5 (sin x))
```

Arithmetic is extended to allow manipulation of operators. A typical operator is $(D+I)(D-I) = D^2-I$, where I is the identity operator, which subtracts a function from its second derivative. Such an operator can be constructed and used as follows:

```
((((* (+ D I) (- D I)) (literal-function 'f)) 'x)
(+ (((expt D 2) f) x) (* -1 (f x)))
```

Derivatives of functions of multiple arguments

The derivative generalizes to functions that take multiple arguments. The derivative of a real-valued function of multiple arguments is an object whose contraction with the tuple of increments in the arguments gives a linear approximation to the increment in the function's value.

A function of multiple arguments can be thought of as a function of an up tuple of those arguments. Thus an incremental argument tuple is an up tuple of components, one for each argument position. The derivative of such a function is a down tuple of the partial derivatives of the function with respect to each argument position.

Suppose we have a real-valued function g of two real-valued arguments, and we want to approximate the increment in the value of g from its value at x, y. If the arguments are incremented by the tuple $(\Delta x, \Delta y)$ we compute:

$$
\begin{aligned}
Dg(x,y) \cdot (\Delta x, \Delta y) &= [\partial_0 g(x,y), \partial_1 g(x,y)] \cdot (\Delta x, \Delta y) \\
&= \partial_0 g(x,y)\Delta x + \partial_1 g(x,y)\Delta y. \quad (9.20)
\end{aligned}
$$

Using the two-argument literal function g defined on page 512, we have:

```
((D g) 'x 'y)
(down (((partial 0) g) x y) (((partial 1) g) x y))
```

In general, partial derivatives are just the components of the derivative of a function that takes multiple arguments (or structured arguments or both; see below). So a partial derivative of a function is a composition of a component selector and the derivative of that function.[3] Indeed:

$$
\partial_0 g = I_0 \circ Dg \qquad\qquad\qquad\qquad (9.21)
$$

$$
\partial_1 g = I_1 \circ Dg. \qquad\qquad\qquad\qquad (9.22)
$$

Concretely, if

$$
g(x,y) = x^3 y^5 \qquad\qquad\qquad\qquad\qquad (9.23)
$$

---

[3]Partial derivative operators such as (partial 2) are operators, so (expt (partial 1) 2) is a second partial derivative.

then

$$Dg(x, y) = [3x^2y^5, 5x^3y^4] \tag{9.24}$$

and the first-order approximation of the increment for changing the arguments by Δx and Δy is

$$g(x + \Delta x, y + \Delta y) - g(x, y) \approx [3x^2y^5, 5x^3y^4] \cdot (\Delta x, \Delta y)$$
$$= 3x^2y^5\Delta x + 5x^3y^4\Delta y. \tag{9.25}$$

Partial derivatives of compositions also obey a chain rule:

$$\partial_i(f \circ g) = ((Df) \circ g) \cdot \partial_i g. \tag{9.26}$$

So if x is a tuple of arguments, then

$$(\partial_i(f \circ g))(x) = Df(g(x)) \cdot \partial_i g(x). \tag{9.27}$$

Mathematical notation usually does not distinguish functions of multiple arguments and functions of the tuple of arguments. Let $h((x, y)) = g(x, y)$. The function h, which takes a tuple of arguments x and y, is not distinguished from the function g that takes arguments x and y. We use both ways of defining functions of multiple arguments. The derivatives of both kinds of functions are compatible for contraction with a tuple of increments to the arguments. Scheme comes in handy here:

```
(define (h s)
  (g (ref s 0) (ref s 1)))

(h (up 'x 'y))
(g x y)

((D g) 'x 'y)
(down (((partial 0) g) x y) (((partial 1) g) x y))

((D h) (up 'x 'y))
(down (((partial 0) g) x y) (((partial 1) g) x y))
```

A phase-space state function is a function of time, coordinates, and momenta. Let H be such a function. The value of H is $H(t, (x, y), [p_x, p_y])$ for time t, coordinates (x, y), and momenta $[p_x, p_y]$. Let s be the phase-space state tuple as in (9.6):

$$s = (t, (x, y), [p_x, p_y]). \tag{9.28}$$

The value of H for argument tuple s is $H(s)$. We use both ways of writing the value of H.

We often show a function of multiple arguments that include tuples by indicating the boundaries of the argument tuples with semicolons and separating their components with commas. If H is a function of phase-space states with arguments t, (x, y), and $[p_x, p_y]$, we may write $H(t; x, y; p_x, p_y)$. This notation loses the up/down distinction, but our semicolon-and-comma notation is convenient and reasonably unambiguous.

The derivative of H is a function that produces an object that can be contracted with an increment in the argument structure to produce an increment in the function's value. The derivative is a down tuple of three partial derivatives. The first partial derivative is the partial derivative with respect to the numerical argument. The second partial derivative is a down tuple of partial derivatives with respect to each component of the up-tuple argument. The third partial derivative is an up tuple of partial derivatives with respect to each component of the down-tuple argument:

$$DH(s) = [\partial_0 H(s), \partial_1 H(s), \partial_2 H(s)] \qquad (9.29)$$
$$= [\partial_0 H(s), [\partial_{1,0} H(s), \partial_{1,1} H(s)], (\partial_{2,0} H(s), \partial_{2,1} H(s))],$$

where $\partial_{1,0}$ indicates the partial derivative with respect to the first component (index 0) of the second argument (index 1) of the function, and so on. Indeed, $\partial_z F = I_z \circ DF$ for any function F and access chain z. So, if we let Δs be an incremental phase-space state tuple,

$$\Delta s = (\Delta t, (\Delta x, \Delta y), [\Delta p_x, \Delta p_y]), \qquad (9.30)$$

then

$$DH(s)\Delta s = \partial_0 H(s)\Delta t$$
$$+ \partial_{1,0} H(s)\Delta x + \partial_{1,1} H(s)\Delta y$$
$$+ \partial_{2,0} H(s)\Delta p_x + \partial_{2,1} H(s)\Delta p_y. \qquad (9.31)$$

Caution: Partial derivative operators with respect to different structured arguments generally do not commute.

In Scheme we must make explicit choices. We usually assume that phase-space state functions are functions of the tuple. For example,

```
(define H
  (literal-function 'H
    (-> (UP Real (UP Real Real) (DOWN Real Real)) Real)))
```

```
(H s)
(H (up t (up x y) (down p_x p_y)))
```

```
((D H) s)
(down
 (((partial 0) H) (up t (up x y) (down p_x p_y)))
 (down (((partial 1 0) H) (up t (up x y) (down p_x p_y)))
       (((partial 1 1) H) (up t (up x y) (down p_x p_y))))
 (up (((partial 2 0) H) (up t (up x y) (down p_x p_y)))
     (((partial 2 1) H) (up t (up x y) (down p_x p_y))))))
```

Structured results

Some functions produce structured outputs. A function whose output is a tuple is equivalent to a tuple of component functions each of which produces one component of the output tuple.

For example, a function that takes one numerical argument and produces a structure of outputs may be used to describe a curve through space. The following function describes a helical path around the z-axis in three-dimensional space:

$$h(t) = (\cos t, \sin t, t) = (\cos, \sin, I)(t). \tag{9.32}$$

The derivative is just the up tuple of the derivatives of each component of the function:

$$Dh(t) = (-\sin t, \cos t, 1). \tag{9.33}$$

We can write

```
(define (helix t)
  (up (cos t) (sin t) t))
```

or just

```
(define helix (up cos sin identity))
```

Its derivative is just the up tuple of the derivatives of each component of the function:

```
((D helix) 't)
(up (* -1 (sin t)) (cos t) 1)
```

In general, a function that produces structured outputs is just treated as a structure of functions, one for each of the components. The derivative of a function of structured inputs that produces structured outputs is an object that when contracted with an incremental input structure produces a linear approximation to the incremental output. Thus, if we define function g by

$$g(x, y) = ((x + y)^2, (y - x)^3, e^{x+y}),\qquad(9.34)$$

then the derivative of g is

$$Dg(x,y) = \left[\begin{pmatrix} 2(x+y) \\ -3(y-x)^2 \\ e^{x+y} \end{pmatrix}, \begin{pmatrix} 2(x+y) \\ 3(y-x)^2 \\ e^{x+y} \end{pmatrix} \right].\qquad(9.35)$$

In Scheme:

```
(define (g x y)
  (up (square (+ x y)) (cube (- y x)) (exp (+ x y))))

((D g) 'x 'y)
(down (up (+ (* 2 x) (* 2 y))
          (+ (* -3 (expt x 2)) (* 6 x y) (* -3 (expt y 2)))
          (* (exp y) (exp x)))
      (up (+ (* 2 x) (* 2 y))
          (+ (* 3 (expt x 2)) (* -6 x y) (* 3 (expt y 2)))
          (* (exp y) (exp x))))
```

Caution must be exercised when taking the derivative of the product of functions that each produce structured results. The problem is that the usual product rule does not hold. Let f and g be functions of x whose results are compatible for contraction to a number. The increment of f for an increment Δx of x is $Df(x)\Delta x$, and similarly for g. The increment of the product fg is $D(fg)(x)\Delta x$, but expanded in terms of the derivative of f and g the increment is $(Df(x)\Delta x)g(x) + f(x)(Dg(x)\Delta x)$. It is not $((Df)(x)g(x) + f(x)(Dg(x)))\Delta x$. The reason is that the shape of the derivative of f is such that $Df(x)$ should be multiplied by Δx rather than $g(x)$.

Exercise 9.1: Chain rule

Let $F(x,y) = x^2y^3$, $G(x,y) = (F(x,y),y)$, and $H(x,y) = F(F(x,y),y)$, so that $H = F \circ G$.

a. Compute $\partial_0 F(x,y)$ and $\partial_1 F(x,y)$.

b. Compute $\partial_0 F(F(x,y),y)$ and $\partial_1 F(F(x,y),y)$.

c. Compute $\partial_0 G(x,y)$ and $\partial_1 G(x,y)$.

d. Compute $DF(a,b)$, $DG(3,5)$ and $DH(3a^2,5b^3)$.

Exercise 9.2: Computing derivatives

We can represent functions of multiple arguments as procedures in several ways, depending upon how we wish to use them. The simplest idea is to identify the procedure arguments with the function's arguments.

For example, we could write implementations of the functions that occur in exercise 9.1 as follows:

```
(define (f x y)
  (* (square x) (cube y)))

(define (g x y)
  (up (f x y) y))

(define (h x y)
  (f (f x y) y))
```

With this choice it is awkward to compose a function that takes multiple arguments, such as f, with a function that produces a tuple of those arguments, such as g. Alternatively, we can represent the function arguments as slots of a tuple data structure, and then composition with a function that produces such a data structure is easy. However, this choice requires the procedures to build and take apart structures.

For example, we may define procedures that implement the functions above as follows:

```
(define (f v)
  (let ((x (ref v 0))
        (y (ref v 1)))
    (* (square x) (cube y))))

(define (g v)
  (let ((x (ref v 0))
        (y (ref v 1)))
    (up (f v) y)))

(define h (compose f g))
```

Repeat exercise 9.1 using the computer. Explore both implementations of multiple-argument functions.

References

[1] Harold Abelson and Gerald Jay Sussman with Julie Sussman, *Structure and Interpretation of Computer Programs*, 2nd edition, MIT Press and McGraw-Hill, 1996.

[2] Ralph H. Abraham and Jerrold E. Marsden, *Foundations of Mechanics*, 2nd edition, Addison-Wesley, 1978.

[3] Ralph H. Abraham, Jerrold E. Marsden, and Tudor Raţiu, *Manifolds, Tensor Analysis, and Applications*, 2nd edition, Springer Verlag, 1993.

[4] V. I. Arnold, "Small Denominators and Problems of Stability of Motion in Classical and Celestial Mechanics," *Russian Math. Surveys*, **18**, 6 (1963).

[5] V. I. Arnold, *Mathematical Methods of Classical Mechanics*, Springer Verlag, 1980.

[6] V. I. Arnold, V. V. Kozlov, and A. I. Neishtadt, "Mathematical Aspects of Classical and Celestial Mechanics," *Dynamical Systems III*, Springer Verlag, 1988.

[7] June Barrow-Green, "Poincaré and the Three Body Problem," *History of Mathematics*, vol. 11, American Mathematical Society, London Mathematical Society, 1997.

[8] Max Born, *Vorlesungen über Atommechanik*, Springer, 1925–30.

[9] Constantin Carathéodory, *Calculus of variations and partial differential equations of the first order*, (translated by Robert B. Dean and Julius J. Brandstatter), Holden-Day, 1965–67.

[10] Constantin Carathéodory, *Geometrische Optik*, in *Ergebnisse der Mathematik und ihrer Grenzgebiete*, Bd. 4, Springer, 1937.

[11] Élie Cartan, *Leçons sur les invariants intégraux*, Hermann, 1922; reprinted 1971.

[12] Boris V. Chirikov, "A Universal Instability of Many-Dimensional Oscillator Systems," *Physics Reports* **52**, 5, pp. 263–379 (1979).

[13] Alonzo Church, *The Calculi of Lambda-Conversion*, Princeton University Press, 1941.

[14] Richard Courant and David Hilbert, *Methods of Mathematical Physics*, 2 vols., Wiley-Interscience, 1957.

[15] Jean Dieudonné, *Treatise on Analysis*, Academic Press, 1969.

[16] Albert Einstein, *Relativity, the Special and General Theory*, Crown Publishers, 1961.

[17] Hans Freudenthal, *Didactical Phenomenology of Mathematical Structures*, Kluwer, 1983.

[18] Giovanni Gallavotti, *The Elements of Mechanics*, Springer Verlag, 1983.

[19] F. R. Gantmakher, *Lektsii po analiticheskoĭ mekhanike* (Lectures on analytical mechanics), Fizmatgiz, 1960; English translation by G. Yankovsky, Mir Publishing, 1970.

[20] Herbert Goldstein, *Classical Mechanics*, 2nd edition, Addison-Wesley, 1980.

[21] Michel Hénon, "Numerical Exploration of Hamiltonian Systems," *Chaotic Behavior of Deterministic Systems*, North-Holland Publishing Company, 1983.

[22] Michel Hénon and Carl Heiles, "The Applicability of the Third Integral of Motion: Some Numerical Experiments," *Astronomical Journal*, **69**, pp. 73–79 (1964).

[23] Robert Hermann, *Differential Geometry and the Calculus of Variations*, Academic Press, 1968.

[24] IEEE Std 1178-1990, *IEEE Standard for the Scheme Programming Language*, Institute of Electrical and Electronic Engineers, Inc., 1991.

[25] E. L. Ince, *Ordinary Differential Equations*, Longmans, Green and Co., 1926; Dover Publications, 1956.

[26] Jorge V. José and Eugene J. Saletan, *Classical Dynamics: A Contemporary Approach*, Cambridge University Press, 1998.

[27] Res Jost, "Poisson Brackets: An Unpedagogical Lecture," in *Reviews of Modern Physics*, **36**, p. 572 (1964).

[28] P. E. B. Jourdain, *The Principle of Least Action*, Open Court Publishing Company, 1913.

[29] Cornelius Lanczos, *The Variational Principles of Mechanics*, 4th edition, University of Toronto Press, 1970; Dover Publications, 1982.

[30] L. D. Landau and E. M. Lifshitz, *Mechanics*, 3rd edition, *Course of Theoretical Physics*, vol. 1, Pergamon Press, 1976.

[31] Edward Lorenz, "Deterministic Nonperiodic Flow," *Journal of Atmospheric Science* **20**, p. 130 (1963).

[32] Jerrold E. Marsden and Tudor S. Raţiu, *Introduction to Mechanics and Symmetry*, Springer Verlag, 1994.

[33] Philip Morse and Hermann Feshbach, *Methods of Theoretical Physics*, 2 vols., McGraw-Hill, 1953.

[34] Lothar Nordheim, *The Principles of Mechanics*, in *Handbuch der Physik*, vol. 2, Springer, 1927.

[35] Henri Poincaré, *Les Méthodes nouvelles de la Mécanique céleste*, Paris, 1892; Dover Publications, 1957; English translation by the National Aeronautics and Space Administration, technical report NASA TT F-452.

[36] H. C. Plummer, *An Introductory Treatise on Dynamical Astronomy*, Cambridge University Press, 1918; Dover Publications, 1960.

[37] Florian Scheck, *Mechanics, From Newton's Laws to Deterministic Chaos*, 2nd edition, Springer-Verlag, 1994.

[38] P. Kenneth Seidelmann, editor, *Explanatory Supplement to the Astronomical Almanac*, University Science Books, 1992.

[39] Jean-Marie Souriau, *Structure des Systèmes Dynamiques*, Dunod Université, Paris, 1970; English translation: Birkhäuser Boston, 1998.

[40] Michael Spivak, *Calculus on Manifolds*, W. A. Benjamin, 1965.

[41] Stanly Steinberg, "Lie Series, Lie Transformations, and Their Applications," in *Lie Methods in Optics*, J. Sánchez Mondragón and K. B. Wolf, editors, Springer Verlag, 1986, pp. 45–103.

[42] Shlomo Sternberg, *Celestial Mechanics*, W. A. Benjamin, 1969.

[43] E. C. G. Sudarshan and N. Mukunda, *Classical Dynamics: A Modern Perspective*, John Wiley & Sons, 1974.

[44] J. B. Taylor, unpublished, 1968.

[45] Walter Thirring, *A Course in Mathematical Physics 1: Classical Dynamical Systems*, translated by Evans M. Harell, Springer-Verlag, 1978.

[46] E. T. Whittaker, *A Treatise on Analytical Dynamics*, Cambridge University Press, 1937.

[47] J. Wisdom, "The Origin of the Kirkwood Gaps: A Mapping for Asteroidal Motion Near the 3/1 Commensurability," *Astron. J.*, **87**, p. 557 (1982).

[48] J. Wisdom, "A Mapping for the Hénon-Heiles System," unpublished notes, (1987).

[49] J. Wisdom and M. Holman, "Symplectic Maps for the N-Body Problem," *Astron. J.*, **102**, p. 1528 (1991).

List of Exercises

Index

Any inaccuracies in this index may be explained by the fact that it has been prepared with the help of a computer.

Donald E. Knuth, *Fundamental Algorithms* (Volume 1 of *The Art of Computer Programming*)

Page numbers for Scheme procedure definitions are in italics.
Page numbers followed by n indicate footnotes.

Printed in the United States
by Baker & Taylor Publisher Services